AMINES
Synthesis, Properties and Applications

The understanding of amine chemistry is of paramount importance to numerous chemical industries, as well as to academic research. This book provides an authoritative account of the properties and applications of amines with respect to the characteristics of bonded substituents and the nature of their chemical and physical environments. The synthesis of alkyl, aryl and heterocyclic amines and inorganic amines with a review of their typical reactions is comprehensively treated, whilst practical synthetic and analytical methods for laboratory preparation and detection are provided. The importance of amine chemistry from the nineteenth century to the modern day, with a brief history of the development of ammonia synthesis, is included. This book is an invaluable reference source for undergraduates, postgraduates and chemical researchers working in industry.

STEPHEN A. LAWRENCE is a professionally qualified chemist with over 20 years' experience in the pharmaceutical and fine-chemicals industries. He is currently Director of the chemical consultancy firm Mimas Ltd.

AMINES

Synthesis, Properties and Applications

STEPHEN A. LAWRENCE

CAMBRIDGE
UNIVERSITY PRESS

PUBLISHED BY THE PRESS SYNDICATE OF THE UNIVERSITY OF CAMBRIDGE
The Pitt Building, Trumpington Street, Cambridge, United Kingdom

CAMBRIDGE UNIVERSITY PRESS
The Edinburgh Building, Cambridge, CB2 2RU, UK
40 West 20th Street, New York, NY 10011–4211, USA
477 Williamstown Road, Port Melbourne, VIC 3207, Australia
Ruiz de Alarcón 13, 28014 Madrid, Spain
Dock House, The Waterfront, Cape Town 8001, South Africa

http://www.cambridge.org

First published 2004

Printed in the United Kingdom at the University Press, Cambridge

Typeface Times 11/14 pt.   *System* LaTeX 2_ε [TB]

A catalogue record for this book is available from the British Library

Library of Congress Cataloguing in Publication data
Lawrence, Stephen A., 1960–
Amines: synthesis, properties, and applications / Stephen A. Lawrence.
p.   cm.
Includes bibliographical references and index.
ISBN 0 521 78284 8
1. Amines.   I. Title.
QD305.A8.L38   2004
547′.042 – dc22   2003065415

ISBN 0 521 78284 8 hardback

Contents

Preface

Much of the chemistry of the amines was discovered in the nineteenth century by pioneering chemists such as Hofmann, Leuckart, Gabriel and Knoll. The introduction of the Cyanamide Process and Electric Arc Process at the begining of the twentieth century made amines commercially available for the first time at low cost and no longer an academic curiosity or restricted for use only in high-value products. However, the implementation of the Haber–Bosch Process at Leuna in Eastern Germany in 1917 marks the beginning of the modern age of amine chemistry.

Having worked with amines for many years, both on a small scale in the laboratory producing gram quantities of novel amines and also on a larger scale on full-sized manufacturing plants producing several hundred tonnes per year of amine-based pharmaceutical intermediates, it has always been a problem for me that I could never find a single, up-to-date, resource book specifically dedicated to the synthesis, properties and reactions of amines. The most recent examples that I was able to find that cover the whole area of amine chemistry were David Ginsburg's *Concerning Amines*, which dates from 1968, and the third edition of Neville Sidgwick's *The Organic Chemistry of Nitrogen* (first published in 1910 but revised in 1966 by I. T. Miller and H. D. Springham), although most general organic chemistry books contain a chapter or two on amines. Eventually I decided that the best way to rectify this situation would be for me to write such a book, and so this was my reason for approaching Cambridge University Press, who agreed to my proposal. It is my hope that the readers of this book will find that it contains the answers to many of their unanswered questions about amines and also provides suitable references for further study.

This book contains eight chapters and three appendices. Chapter 1 reviews the history and properties of ammonia and nitrogen. Chapters 2 and 3 are concerned with the properties, syntheses and reactions of alkyl and aryl amines respectively. Heterocyclic amines are found in Chapter 4; inorganic amines including macrocyclic ligands are the subject of Chapter 5, and Chapter 6 contains some laboratory synthetic routes to amines and also details of analytical procedures. Amine oxides, protected amines and amino acids are reviewed in Chapter 7, and Chapter 8 covers the applications and commercial uses of amines. At the end of the text there are three appendices, which shown the structures and isosurface electronic charges of some amines (Appendix 1), a table of the physical properties of selected amines (Appendix 2) and the named reactions and named syntheses of amines (Appendix 3).

The reader's attention is drawn to the safety advice given in, for example, Section 6.1. Some of the chemicals mentioned in this book are dangerous, and although the author can vouch that the experiments described do work, if properly conducted, neither he nor the publisher will be liable for accidents that may take place in the course of experimentation!

I would like to thank Frau Urte Thiele and Dr Klaus Baehr of Chemtec Leuna for the photographs of the plant at Leuna, and also Inegard Rafn and Kai Evensen of Norsk Hydro for the photographs of the plant and laboratories at Notodden in Norway.

I would like also to thank Tim Fishlock, Simon Capelin, Michelle Carey, Robert Whitelock, Jayne Aldhouse and Emily Yossarian at Cambridge University Press for their encouragement and patience while I produced the typescript.

Finally I would like to thank my wife Beverley for her support during the writing process whilst this book came to fruition.

1

An introduction to the amines

1.1 Introduction

The amines are a group of chemical compounds that have the common feature of possessing nitrogen atoms that are sp^3 hybridized with three single bonds to other elements. Although there are many different types of amine, they all have similar features because somewhere in their molecular structures they possess a nitrogen atom. For many people one characteristic is that they possess an identifiable amine smell, which is often described as being rather like rotting fish.

Amines are named according to the IUPAC system from the corresponding parent alcohol, i.e. $C_2H_5-NH_2$ is ethylamine from C_2H_5-OH = ethyl alcohol. Substituents on carbon chains are identified by a number, and the prefix *N*- is used for substituents on the nitrogen atom. Other amines may be named in a similar manner but many aromatic and heterocyclic amines retain their trivial names such as indole, aniline and quinoline, all of which are universally accepted.

An idea of the importance of amines can be obtained from a crude analysis of the worldwide production of ammonia gas, which is around 170 million tonnes per year. Of this figure about 70% is used in the manufacture of fertilizers, 10% is used in the manufacture of nylon and 7% is used in the manufacture of explosives (via nitric acid). The remaining 13% is used for refrigeration applications, directly as a water solution for pH control and also for the production of organic and inorganic chemicals. The percentage of ammonia used in the production of amines (excluding caprolactam and hexamethylene diamine) is about 3% to 4% of the total worldwide output and accounts for around 5–7 million tonnes of ammonia per year. However, a sizeable fraction of the ammonia used in fertilizers may eventually end up in the form of amines during plant metabolism.

The simplest amine is ammonia, NH_3, in which the central nitrogen atom is bonded to three hydrogen atoms which are all chemically equivalent. The replacement of hydrogen atoms in ammonia with other elements produces inorganic

1

amines such as trichloroamine, NCl_3. The replacement of hydrogen atoms with other functional groups such as alkyl and aryl groups gives rise to organic amines which are categorized as follows. Replacement of one hydrogen atom by an alkyl group, R, or an aryl group, Ar, produces a primary alkyl amine $R-NH_2$ or a primary aryl amine $Ar-NH_2$, respectively. Replacement of a second hydrogen with another alkyl or aryl group, R' or Ar', will produce a secondary alkyl amine, $R-NH-R'$, or a secondary aryl amine, $Ar-NH-Ar'$. Where the nitrogen atom is substituted with an alkyl group and an aryl group, the product is called a secondary alkyl aryl amine $R-NH-Ar$. Replacement of all hydrogen atoms with alkyl or aryl groups produces tertiary alkyl and aryl amines of the type $R-NR'-R''$ or $Ar-NAr'-Ar''$, or alkyl aryl variations like $R-NAr'-Ar$. Finally, tertiary amines can add on another molecule of an amine such as an amine hydrobromide to form a quaternary amine (e.g. $[R-N(R'R'')-R''']^+$) with a bromide counter ion. Examples of these amines are shown below.

Primary amine   Secondary amine   Tertiary amine   Quaternary amine

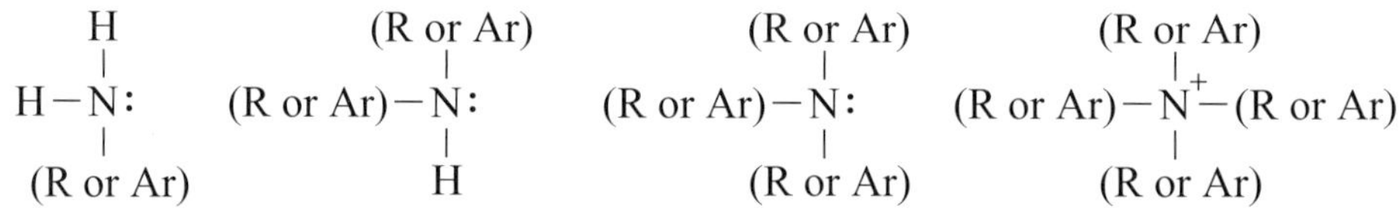

$$
\begin{array}{cccc}
\text{H} & \text{(R or Ar)} & \text{(R or Ar)} & \text{(R or Ar)} \\
| & | & | & | \\
\text{H}-\overset{..}{\text{N}} & \text{(R or Ar)}-\overset{..}{\text{N}} & \text{(R or Ar)}-\overset{..}{\text{N}} & \text{(R or Ar)}-\overset{+}{\text{N}}-\text{(R or Ar)} \\
| & | & | & | \\
\text{(R or Ar)} & \text{H} & \text{(R or Ar)} & \text{(R or Ar)}
\end{array}
$$

The properties of amines are largely controlled by the electronic characteristics of the electron pair on the central nitrogen atom (or atoms for bi- and polyfunctional amines), which is able to act as a Lewis Base and use the lone pair of electrons for donation. The ability of the nitrogen atom to donate its lone pair of electrons in chemical reactions is modified by the presence of the functional groups bonded to the nitrogen atom that can increase or decrease this ability. Generally electron density in s orbitals is held more strongly by the atomic nucleus and is less available for donation. Therefore amines with sp^3 hybridization are generally better at donating electrons to electrophiles than amines or nitrogen-containing molecules with sp^2 hybridization, which have a greater proportion of s character. This helps to explain why acetonitrile is a weaker base than ethylamine. However, because the chemistry and properties of amines are so strongly determined by the electron lone pair on the nitrogen atom it is important to look into the properties of the element nitrogen before trying to understand the behaviour of amines.

1.2 Nitrogen

The Earth's atmosphere contains about 75.5% of nitrogen by weight; this is equivalent to around 4000 billion tonnes of nitrogen gas. Nitrogen is a colourless, odourless, diatomic gas and is actually the single most abundant element accessible to Man in an uncombined state. In its combined state nitrogen is the thirtythird most abundant element next to gadolinium. Without nitrogen, life as we know it would not exist on Earth: nitrogen is an essential part of proteins, which are about 15% nitrogen (by weight). Mankind had known for centuries that air was composed of two fractions, a part which would burn and a part which would not. The discovery of nitrogen as a chemical element in 1772 is credited to Rutherford [1] (although it was isolated at the same time by Scheele and Cavendish [2]). Rutherford's experiment was to burn carbon in a limited supply of air. After combustion Rutherford removed the carbon dioxide with potassium hydroxide and named the remaining gas 'residual air'. The name nitrogen was proposed in 1790 by Jean-Antoine Chaptal when it was discovered that this residual air was a constituent of nitric acid and nitrates.

Lavoisier preferred the name azote (from the Greek meaning no-life) because of its asphyxiating properties; this name is still used in France and is the root of chemical names such as azo, diazo, azide, etc. In German nitrogen is called Stickstoff from the verb 'to stricken' ('to suffocate').

In order to understand the chemistry of the amines it is important to understand how nitrogen's position at the top of Group V in the first row of the Periodic Table (see below) can explain its unique properties which also characterize the behaviour of the amines.

Decreasing metallic character →
Li Be B C N O F
             P
             As
             Sb   ↑
             Bi   Decreasing metallic character

Nitrogen has the smallest atomic number of the elements in Group V, and this usually results in differences in chemical behaviour from the other elements in the group (this is known as 'the first row anomaly'). Nitrogen also lies near the middle of the first period of the p block, which means that it should display a pronounced lack of metallic character. From its position in the Periodic Table it would most probably be predicted that nitrogen would have little tendency to form ions but would have a large variety of oxidation states. However, nitrogen has only four available orbitals in its valence shell ($1s^2\ 2s^2\ 2p^3$) and there are no d orbitals available. This limits to four the maximum number of covalent bonds that a nitrogen atom can make. Also because one valence orbital is occupied by a lone electron pair in neutral molecules, nitrogen can form only three bonds. From its small size (interatomic radius 1.10 Å)

Table 1.1. *The properties of nitrogen*

Boiling point	77.36 K
Freezing point	63.26 K
Critical pressure	33.54 bar (3398.44 Pa)
Transition $\alpha \rightarrow \beta$	35.62 K
Enthalpy of transition	228.9 kJ mol^{-1}
Enthalpy of vapourization	5577 J mol^{-1}
Enthalpy of fusion	719.6 J mol^{-1}
Dissociation energy	944.7 kJ mol^{-1}
Atomic mass	14.0067
Electronegativity	3.04 (Pauling)
Interatomic radius (N_2)	1.10 Å
Isotropic abundance	99.634% ^{14}N, 0.366% ^{15}N

it can be predicted that the element nitrogen would have a tendency to form multiple bonds. For example, the formation of the nitrogen molecule N_2 can be represented as

$$2N\ (1s^2\ 2s^2\ 2p^3) \rightarrow N_2 \left(1\sigma_g^2\ 1\sigma_u^2\ 2\sigma_g^2\ 2\sigma_u^2\ 1\pi_u^4\ 3\sigma_g^2\right)$$

Of the ten molecular orbitals, the four that make up $1\sigma_u$ and $2\sigma_u$ are anti-bonding. The bond order in the nitrogen molecule (N_2) is therefore $(10-4)/2 = 3$.

However, the triple-bonded N_2 molecule has a high stability. For example, the bond energies of double-bonded N, C and O are all similar but there is a big difference in the bond energies between double- and triple-bonded nitrogen and carbon.

$$C{=}C = 272 \text{ kJ mol}^{-1} \qquad C{\equiv}C = 887 \text{ kJ mol}^{-1}$$
$$N{=}N = 260 \text{ kJ mol}^{-1} \qquad N{\equiv}N = 1206 \text{ kJ mol}^{-1}$$
$$O{=}O = 264 \text{ kJ mol}^{-1}$$

As the bond energies of double-bonded carbon–carbon and nitrogen–nitrogen are similar, it might be predicted that the bond energy of triple-bonded nitrogen–nitrogen would similarly be around 615 kJ mol^{-1} higher. However, the bond strength increases by 946 kJ/mol^{-1}, so the nitrogen–nitrogen triple bond is 331 kJ mol^{-1} more stable than might be predicted. The high stability of the nitrogen molecule is the reason why nitrogen gas is such a major constituent of the Earth's atmosphere and why so many compounds of nitrogen are unstable and often explosive. Nitrogen gas reacts only with some transition metal complexes, with carbides and with nitrogen-fixing bacteria. The physical properties of nitrogen are given in Table 1.1.

$$NR_3 + RBr \longrightarrow [NR_4]^+Br^-$$

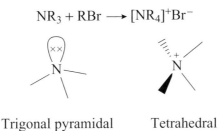

Trigonal pyramidal   Tetrahedral

Figure 1.1. The trigonal pyramidal and tetrahedral geometries of substituted nitrogen compounds.

Because the electronic configuration of nitrogen in its ground state (^4S) is $1s^2$ $2s^2 2p^3$, nitrogen is one of the most electronegative elements (exceeded only by oxygen and fluorine). The nitrogen atom may complete its octet of electrons by:

(i) formation of the N^{3-} nitride ion, e.g. in lithium nitride;
(ii) formation of three electron-pair bonds, e.g. NH_3 or N_2;
(iii) formation of two electron-pair bonds plus electron gain, e.g. $[NH_2]^-$;
(iv) formation of four electron-pair bonds plus electron loss, e.g. $[NH_4]^+$;
(v) formation of the N_5^+ ion (e.g. $N_5(Sb_2F_4)$) or the N_5^- ion.

It is the ease with which nitrogen forms four electron-pair bonds to generate quaternary ions with the loss of an electron that defines the chemistry of nitrogen and also the chemistry of the amines.

The conversion of three electron-pair bonds around a central nitrogen atom to four bonds with electron loss results in a change of geometry, as shown in Figure 1.1.

In instances when the R groups are very large, steric crowding may produce a trigonal planar NR_3 molecule rather than a trigonal pyramidal form. It might be expected that all four bonds in tetrahedral nitrogen would be equivalent (three electrons from 2p and one electron from the $2s^2$) and experimentally this has been proved [3] by measurement of all four bonds in the ammonium ion. These bonds have been shown to have equal valence angles of 109.5°. In ammonia the bond angle is 107° and not 90°, which might be predicted from the contribution from the three electrons in the 2p orbital. This is indicative of orbital hybridization taking place. The importance of ammonia will now be examined in greater detail.

1.3 The history of ammonia and its importance to mankind

In 2002 the total annual usage of fertilizers worldwide was around 135 million tonnes. This figure represents around 90 million tonnes of fixed nitrogen, which is expected to rise to a requirement of 100 million tonnes of fixed nitrogen per year in 2010. The most widely used fertilizer is urea, of which 50 million tonnes alone were used as fertilizer in 2002. The balance of the nitrogen-containing fertilizers consisted of nitrates (such as calcium nitrate and ammonium nitrate), ammonium salts (such as diammonium phosphate, calcium ammonium nitrate, ammonium potassium chloride and ammonium phosphate chloride) and others such as aqua ammonia (aqueous ammonia solution) and dicyandiamide. Until the 1950s, ammonium nitrate was the single most widely used fertilizer but more recently urea (with 46.7% nitrogen content) has become the single most important fertilizer. In the USA the use of aqua ammonia seems to be growing, but its usage is still tiny compared with the widespread use of urea.

Prior to the development of industrial nitrogen-fixation processes, the only fixed nitrogen available for assimilation by plants and animals came either from oxidation of atmospheric nitrogen during lightning storms or from the biological fixation of nitrogen by microorganisms present in the soil and also in certain plants and algae. The microorganisms that can convert nitrogen to ammonia produce an enzyme complex called nitrogenase, which catalyses the reaction.

$$N_2 + 6H^+ + 6e^- \rightarrow 2NH_3$$

Nitrogenase consists of two high-molecular-weight components called molybdoferredoxin and azoferredoxin. They are able to fix atmospheric nitrogen because they contain small metallic iron/sulphur and iron/molybdenum centres, which fix and reduce nitrogen. It is thought that the Fe/Mo centres bind to nitrogen and the Fe/S centres are responsible for its reduction. Reactive sites containing Fe_4S_4 units have been identified in azoferredoxin.

In free soil, bacteria can assimilate about 50 kg of nitrogen per hectare per year. However, when the soil is planted with legumes, clover or alfalfa this figure rises to 250 kg of nitrogen per year. The principles of crop rotation have been known since the Middle Ages in Europe, even though the importance of nitrogen to plant metabolism remained a mystery.

Once fixed, nitrogen can enter the biological nitrogen cycle; it is oxidized by other bacteria, e.g. nitrosomas and nitrobacters, and can be incorporated into amino acids and then into proteins. Animals are able to obtain fixed nitrogen either by eating plants (herbivores) or by eating other animals (carnivores).

Every nitrogen atom in industrially produced chemical compounds today comes either directly or indirectly from ammonia. In addition to the production of fertilizers, once ammonia has been converted to nitric acid it is essential for the production of plastics and fibres such as polyamides, urea–formaldehyde resins, polyurethanes and polyacrylonitrile. However, to get a true appreciation of the importance of ammonia to mankind we have to look back to 1798 when the philosopher T. R. Malthus proposed the principle of population [4], stating:

In the absence of any external constraints, the world's population always increases in geometric progression, while the production of food increases in arithmetic progression.

Put another way, Malthus predicted that at some point in the future the planet would not be able to produce enough food to feed mankind. Malthus was essentially correct in this principle but advances in science and agricultural technology have increased the worldwide production of food and so postponed the Malthusian final outcome.

In 1840 Justus von Leibig studied plant nutrition and established the importance of fixed nitrogen for the growth of plants [5]. Justus von Liebig is nowadays considered to be the father of the concept of fertilizers. In the mid nineteenth century the major natural resources of fixed nitrogen were sodium nitrate from Chilean saltpetre, and guano (bird excrement) from islands in the Pacific Ocean (the term saltpetre alone is misleading as Chilean saltpetre is sodium nitrate whereas Indian saltpetre is potassium nitrate). Synthetic fixed nitrogen could be obtained in limited quantities from the destructive distillation of coal as ammonium sulphate. Competing with fertilizers for the world supply of fixed nitrogen were the dyestuffs industry and, after the discovery of dynamite in 1866, the emerging explosives industry.

In 1898 Sir William Crookes warned in his presidential address to the British Association for the Advancement of Science that the supply of fixed nitrogen for agriculture was rapidly reaching a point where it was insufficient to support the world's rapidly increasing population (i.e. the world's population was not far from reaching the Mathusian limit) [6]. Sir William's address must be understood in view of the fact that the reserves of Chilean saltpetre and guano were known to be limited and it was thought that they could not sustain the needs of agriculture and industry for the rest of the twentieth century.

The fixation of atmospheric nitrogen was seen as the solution to this problem, but at the time there was no satisfactory fixation process that could be carried out on a large scale at a low cost. For example, the atmospheric discharge of high-voltage electricity had been studied by Cavendish at the end of the eighteenth century [2] and had been discovered to produce small quantities of nitric and nitrous acids. However, the thermodynamic equilibrium concentration of nitric oxide is very low at normal temperatures. Although this was not a practical route it did at least provide an example to scientists working at the time that the fixation of atmospheric nitrogen might be technically feasible. In principle in order to 'fix' atmospheric nitrogen it is necessary first to break the nitrogen–nitrogen triple bond, and at the end of the nineteenth century there were three methods by which this was thought to be possible.

(i) By combining nitrogen with oxygen to form nitric oxides (as studied by Cavendish).

$$2N_2 + 3O_2 \rightarrow 2NO + 2NO_2$$

(ii) By combining nitrogen with hydrogen to form ammonia.

$$N_2 + 3H_2 \rightarrow 2NH_3$$

(iii) By the use of active chemical compounds capable of splitting the N≡N bond under energetic conditions.

(In the twentyfirst century the use of biotechnology and enzymes has emerged as a fourth possibility, but a low-cost method for producing fixed nitrogen by this process is still one of the remaining goals of science.)

At the beginning of the twentieth century the two most advanced methods for the fixation of atmospheric nitrogen were the Cyanamide Process developed by Adolf Frank and Nikodem Caro in 1895 [7] and the Electric Arc Process, first successfully developed commercially in 1904 by Kristian Birkeland and Samuel Eyde [8]. In 1902 the Electric Arc Process was the first ammonia-synthesis process to be run on a commercial scale at Niagara in the USA. The process used an electric arc, which heated air to 3000 °C and produced dilute nitric acid. The commercial venture was started by two Americans, Bradley and Lovejoy, who raised around USD 1 million in share capital. The concentration of the air after passing through the arc was around only 2% nitric oxide and the plant was closed two years later as it was not commercially viable and the share capital was exhausted. Although the energy consumption of the Arc Process was enormous the process was commercialized for a second time by Norsk Hydro in Notodden, Norway, in 1904. By 1908 this plant was producing 7000 tonnes of fixed nitrogen per year. Eyde and Birkeland succeeded where Bradley and Lovejoy had failed primarily because of three factors: first, because of the availability of cheap hydroelectric power in Notodden; second because of relatively cheaper labour costs in Norway compared with the USA (Norway was not a rich country in the early 1900s and it was common for several families working at Notodden to share single rooms together with their children in hostels near the plant); and third, because of the Birkeland furnace, which was the most energy-efficient furnace of its time.

The plant at Notodden operated throughout the First World War and, together with a sister plant at Rjuken, reached a production of over 30 000 tonnes per year with a power consumption of 60 000 kW per tonne of fixed nitrogen. Fortunately in Norway cheap hydroelectric power was available to operate the arc but if fossil fuels had been used the energy requirement would have been around 600 GJ per

Figure 1.2. Birkeland at work in his laboratory on the Arc Process. (Photo © Hydro Media, reproduced with permission of Norsk Hydro a.s.)

tonne of fixed nitrogen. The Arc Process was abandoned in the 1920s as it was by then obsolete and had been overtaken by newer technology. However, the 36 furnaces at Notodden and at the sister plant in Rjukan were in use right up until 1940. At the time of writing (in 2003), Norsk Hydro have established a museum at the Notodden site in Norway. This museum contains many historical documents and some of the original laboratory equipment used by Birkeland. A photograph of Birkeland's laboratory is shown in Figure 1.2, and Figure 1.3 shows a photograph of the Notodden plant.

Schoenherr, working at BASF in Germany, also demonstrated an electric arc furnace in 1905, and in 1912 BASF and the Norwegians decided jointly to build a new plant in Norway with increased capacity. However, this project was shelved as the First World War intervened and also, shortly after agreeing to enter this joint venture, BASF had demonstrated another more energetically feasible ammonia synthesis process called the Haber–Bosch Process.

In the Cyanamide Process coke and limestone were reacted in a furnace to form calcium carbide, which then reacts with nitrogen to form calcium cyanamide. The hydrolysis of calcium cyanamide with steam, or more slowly with water, produces

Figure 1.3. The plant at Notodden. (Photo © Hydro Media, reproduced with permission Norsk Hydro a.s.)

ammonia. This allows calcium cyanamide to be used directly as a fertilizer or as a source of ammonia gas.

$$Ca + 2C \rightarrow CaC_2$$
$$CaC_2 + N_2 \rightarrow Ca(CN)_2$$

Although this process uses 190 GJ per tonne of fixed ammonia this was about a quarter of the amount used by the Arc Process and so it was useful for operation in areas without cheap hydroelectricity. The Cyanamide Process was first operated commercially in 1905 in Westeregeln in Germany and by 1934 about 11% of the world's fixed nitrogen was obtained from this process. The Cyanamide Process fell into decline after the Second World War and is no longer operated on a commercial scale.

At the beginning of the twentieth century the synthesis of ammonia by the direct combination of hydrogen and nitrogen was thought to be impractical. Although Bertholet had established in 1784 that ammonia consisted of hydrogen and nitrogen and had also established the elemental ratio between them [9], there seemed to be no feasible way of combining them until Fritz Haber began to investigate the ammonia equilibrium in 1904. Haber found that at atmospheric pressure and at an applied temperature of 1000 °C, hydrogen and nitrogen existed in equilibrium with 0.012% ammonia. At a meeting of the Deutsche Bunsen Gesselschaft in 1907 Nernst declared that the direct synthesis of ammonia from its elements was never going to be feasible because the yield of ammonia in the equilibrium mixture was too low [10].

Although Haber collaborated with Nernst in 1905 they disagreed over the interpretation of the laboratory data and Haber decided to pursue his own research. Haber might have been encouraged by the work of Henri Le Chatelier, who managed to synthesize ammonia in 1900 in the laboratory from nitrogen and hydrogen with a catalyst – but unfortunately managed to blow up his equipment in the process and was persuaded not to restart the research project [11]!

Haber and his research group were brave enough to challenge Nernst, who was one of the most eminent scientists of his day, and managed to make small quantities of ammonia at 200 atm applied pressure at 550 °C by using osmium- and later uranium-based catalysts. Haber patented this process [12] and in 1908 he approached the company Badische Aniline und Soda Fabrik (BASF) for support for his work. On 2 July 1909 the company sent two engineers (Karl Bosch and Alwin Mittesch) to see a demonstration of his work. The experiment they witnessed used 98 g of an osmium catalyst at 175 bar and yielded 80 g of ammonia per hour. This experiment was reported at a scientific meeting in Karlsruhe in March 1910 and was described as being so revolutionary that it was like 'obtaining bread from air' [13]. (A full-scale replica of the equipment used by Haber in his laboratory to synthesize ammonia may be seen in the Jewish Museum in Berlin.)

The industrial development of the ammonia process was started at BASF almost straightaway and Bosch was made responsible for controlling the project. BASF agreed to pay Haber 1 pfennig for every kilogram of ammonia he produced from his patented process. It should be noted that in 1910 pressures of 50 bar and temperatures of 250 °C were regarded as the practical limit of the chemical engineering technology at the time. Ammonia synthesis required 500 °C and 100 bar according to Haber's experimental findings. Bosch, who was a chemical engineer, designed double-lined reactor tubes made of soft iron inside a lined steel jacket into which tiny holes were drilled to allow any diffused hydrogen to escape. The decarbonization and embrittlement of steel by hydrogen is pronounced at these conditions.

The development of an effective catalyst, for which Mittesch was responsible, proved to be a long process and over 6500 experiments were carried out before the iron/alumina/magnesia catalyst was selected. The combination of iron with Swedish magnetite, which contains alumina and potash, proved to be the best combination for ammonia synthesis.

Industrial development of the ammonia process was started immediately. A lead plant was built at Oppau, which it was envisaged would be capable of producing 30 tonnes of ammonia per day. The reactors were 30 cm in diameter and each contained 300 kg of catalyst. Initially 3–5 tonnes of ammonia was produced per day, but by 1915 this had increased to 20 tonnes per day and by December 1917 up to 230 tonnes per day was produced. A second BASF ammonia plant was built in 1917 at Leuna in Eastern Germany that was capable of producing 36 000 tonnes of ammonia per day. The plant in Leuna remained in existence until the mid 1990s and was essentially using the Haber–Bosch Process that had been discovered some 80 years earlier. Leuna was chosen as a convenient location for an ammonia plant as in 1917 it was far enough away from the French artillery to prevent any military bombardment, and also the presence of the Saale River and Eastern German coal fields provided convenient sources of raw materials. I visited the Leuna plant in 1997 while it was being demolished so that the land could be used for redevelopment as a chemical business park. My lasting memory is of seeing several huge iron corkscrews from inside the reactors some 6 m or more across and over 100 m long that were being cut up for recycling. There had been little or no investment at the site since before 1939 and most of the equipment used for chemical manufacturing was too antiquated and worn out to repair so the authorities in Germany (the Treuhand) took the decision in 1994 to start the demolition of everything on the site (except for some buildings of historic interest) and to rebuild the site as a chemical business park (see Figures 1.4 and 1.5).

Haber was awarded the Nobel Prize in 1919 for his studies on ammonia synthesis, and Bosch also received the Nobel Prize for his work on high-pressure chemistry together with Bergius in Haber's own laboratory in Karlsruhe (interestingly Birkeland did not win a Nobel Prize for his work on the Arc Process). Haber was later expelled from Nazi Germany in the early 1930s because of his Jewish ancestry and he worked at the University of Cambridge in England until 1934 when he accepted a position in a research institute in Palestine. However, he died in Switzerland in 1934 while *en route* to the Middle East.

After 1918 and the end of the First World War, ammonia plants were built in England, France and Italy under BASF licences. However, new Cyanamide plants also continued to be built as it took about five years for the Haber–Bosch chemistry to be accepted worldwide.

Haber's original process used hydrogen from water gas formed from the reaction of steam with white hot coke, but this was soon replaced by the catalytic reaction between water gas and steam. This is known as the Water Gas Shift (WGS) Reaction and is catalysed by iron oxide.

$$\text{Water gas formation} \quad H_2O + C \rightarrow H_2 + CO$$
$$\text{The WGS Reaction} \quad CO + H_2O \rightarrow CO_2 + H_2$$

Figure 1.4. Part of the ammonia-synthesis plant at Leuna during the demolition that began in 1994. The holding tank in the foreground would have contained an aqueous solution of ammonia (25%).

Figure 1.5. Another view of the ammonia-synthesis plant at Leuna being demolished (1994).

Table 1.2. *Energy requirements of nitrogen fixation*

Process	Energy per tonne of fixed nitrogen
Arc	7174 kJ
Cyanamide	1794 kJ
Haber–Bosch	633 kJ
Haber–Bosch + hydrocarbon feed	317 kJ
Theoretical minimum for NH_3	253 kJ

The carbon dioxide was removed by scrubbing and could be recycled in fertilizer production downstream. The advantage of the WGS Reaction is that the carbon monoxide content of water gas is used to generate more hydrogen from water and so enrich the hydrogen stream prior to entry into the reactor.

During the First World War Germany was effectively cut off from supplies of Chilean saltpetre and a shortage of fixed ammonia led to food shortages. The situation was made worse by the fact that ammonia was needed to produce munitions for the war effort as well. According to some historians the First World War effectively ended when Germany ran out of food for its citizens [14]. In 1913 the world consumption of fixed nitrogen was around 700 000 tonnes, and by 1929 it had trebled to around 2.1 million tonnes, of which the Haber–Bosch Process accounted for just less than half of the total, as shown in the data below.

Fixed nitrogen sources in 1929
43%   Haber–Bosch Process
24%   Chilean saltpetre
20%   Coke-oven gas
12%   Cyanamide Process
1%   Arc Process

Although there have been significant process improvements in the original Haber–Bosch Process for ammonia synthesis since its inception, such as the use of natural gas or naphtha as feedstock materials (reaction with steam at high temperatures $C_nH_{2m} + nH_2O \rightarrow (n+m)H_2 + nCO$), and the energy and plant efficiencies introduced by Kellogg [15], the process operated worldwide today is still fundamentally no different from the process demonstrated in 1915 at Oppau. A comparison of the energy requirements for the different processes for fixing nitrogen is given in Table 1.2.

During the ammonia-manufacturing process the hydrogen and nitrogen gas mixture is in contact with the catalyst for only around 30 s and chemical equilibrium is not attained. Instead a product mixture of about 11% ammonia is drawn off from the reactor at 500 °C; this represents the most economic way of operating a reactor using rapid gas flow. Two types of reactor are commonly used. The older tubular type had metal heat exchangers running along the reactor parallel to the reactor walls with the voids being filled with catalyst. More-modern reactors use layers of heat-plate-like heat exchangers with the catalyst sandwiched in between in layers. The rate determining step in the Haber–Bosch Process is the dissociative adsorption of molecular nitrogen onto the α-iron catalyst surface to form a nitride. The catalyst is usually a mixture of iron oxide and other promotor oxides such as Al_2O_3, MgO and K_2O. These act as stabilizers preventing the iron particles sintering and also as electronic promotors. The α-iron is made by reduction with hydrogen at around $400\ ^\circ C$ and the presence of K_2O actually doubles the α-iron activity by donating an electron to the iron and so strengthening the π back bonding. Schematically this may be represented as

$$:Fe \leftarrow N\equiv N: \rightleftharpoons Fe \rightleftarrows N=N::$$

Once the nitrogen is adsorbed onto the iron surface it is easily hydrogenated to ammonia via an imide and amide and is then dissociated.

Because the adsorption of nitrogen onto the iron surface is exothermic the Haber–Bosch Process usually operates without external heating and the heat generated in the reactor is used to preheat the hydrogen and nitrogen gas mixture before it enters the reactor.

$$N_2 \rightleftharpoons 2N_{adsorbed} \quad \Delta H = -210\ \text{kJ mol}^{-1}$$
$$3H_2 \rightleftharpoons 6H_{adsorbed} \quad \Delta H = -100\ \text{kJ mol}^{-1}$$

It is likely that for the foreseeable future the Haber–Bosch Process will be the main world source of fixed nitrogen with natural gas as the preferred feedstock. The current level of worldwide ammonia production is around 170 million tonnes per year (equivalent to 138 million tonnes of fixed nitrogen) and has risen in line with world population, except for a period in the 1980s when there were economic problems in the former Soviet Bloc (very approximately every person on the planet consumes about 25 kg of fixed nitrogen per year either as food or in the form of chemical products). In 2002 there were about 500 ammonia-synthesis plants worldwide each with an average capacity of 380 000 tonnes per year.

The current worldwide capacity is just under 200 million tonnes of fixed nitrogen per year. At the turn of the twentyfirst century there were just over 5 billion people on the planet, and this number is expected to increase to 6 billion by 2005. The Agriculture Department of the United Nations Organization (www.fao.org/ag) predicts that by 2030 the world population will be over 8 billion with over 60% of people living in towns and cities. Therefore before 2030, at present rates of consumption, there will be insufficient fixed-nitrogen production capacity to continue to supply the world's population with its total requirement. However, as the majority of this population growth will be in the developing world, rather than in Western Europe and North America where the average consumption of fixed nitrogen is lower, any future crisis will probably be moved forward a few years into the future. Interestingly, despite the current high rate of population increase the extent of agricultural land has remained fairly static over the past 10 years at 49 billion hectares worldwide. Therefore it is clear that this land will have to be used more effectively in the future to feed the Earth's growing population.

A comparison of wheat production in France against fertilizer usage over 50 years from 1950 to 2000 carried out by the United Nations gave the following findings.

	Fertilizer usage (kg ha^{-1})	Wheat production (t ha^{-1})
1950	230	1.8
2000	450	7.0

Crop yields in the developing world are not as high as this and in many cases no fertilizer other than animal dung is used, or if synthetic fertilizers are applied incorrectly they do not result in optimum crop yields. In sub-Saharan Africa the fertilizer usage is as low as 10 kg per hectare and crop yields are far lower than in the developed world.

In 2000 the United Nations Agriculture Department found that the rice yields per hectare in the USA, India and Thailand were 7037 kg, 3008 kg and 2329 kg, respectively. These results were partly explained on the basis that poorer farmers in India and Thailand were unable to afford the high levels of fertilizers and agrochemicals used in the USA.

It is clear then that with the increasing urbanization of the world population this increase in food production will have to come from the developing world, as the developed world is already almost at full capacity. An efficient distribution system will also be required to feed the population because at present overproduction in the developing world usually leads to a fall in prices, with farmers being worse off and the excess food being left to spoil. At the time of writing in 2003 there is a rapid improvement in the quality of life in many developing countries; this is leading to an increased demand for the consumer products and chemicals that are currently readily available in the developed world. One example of this is the development of a generic pharmaceuticals industry in India over the past 50 years; this has enabled the people of India to become self-sufficient in basic medicines and less reliant on imported goods. A second example is the growth of the market for cosmetics and personal-care products in China. This market is small by levels in the developed world, but it represents an existing market of USD500–600 million per year and is rapidly growing. Obviously in the future increasing local demand for fixed nitrogen in the developing world will limit the amount of fixed-nitrogen products available for export in the form of those cheap chemical products on which the chemical and pharmaceutical industries of the developed world are starting to become increasing reliant.

The answer to this problem in the longer term is to develop a new, more-effective method of nitrogen fixation. However, in 2008 the Haber–Bosch Process will be 100 years old and despite the efforts of academic and industrial research the basic process of the elemental combination of nitrogen and hydrogen remains unchanged. The development of new, more-effective catalysts for the Haber–Bosch Process remains a goal of research and is one in which I was involved without success in the mid 1980s during my postdoctoral research.

Public opinion in Europe seems to be set against the development of GMO (genetically modified) crops, which may be able to tolerate lower levels of nitrogen in soil than at present and be better at fixing atmospheric nitrogen. The symbiotic bacteria that are present in legumes could possibly be transferred to other crops to reduce their dependence on soil nitrogen. The advantage of these types of crop is that they could free up fixed nitrogen capacity for other uses, but for the time being the 'green' political lobby in Europe is very much against this type of genetic manipulation.

At the beginning of the twentieth century it was thought that science would provide most of the answers to mankind's problems. In the longer term this may prove to be the case, but if it is to happen then investment must be made into research in the area of ammonia synthesis. The Haber–Bosch Process has served mankind well for over 90 years and will probably continue to do so well into the future, but in order to supply the ever-increasing demand for fixed nitrogen new technology will eventually be required. In my opinion the carbon crisis and any future fixed-nitrogen crisis are linked together. At present, barring the discovery of new coal and oil fields in Antarctica, the world's reserves of fossil fuels and methane are known to be limited and will start to run out some time in the next 150–200 years. The development of a clean renewable source of cheap energy therefore has to be a priority. As the major cost factor in ammonia production is the energy cost, the discovery of a low-cost energy source will make ammonia synthesis more cost effective. At present most of the major ammonia-manufacturing plants in the world are already operating at near optimum efficiency so any minor capacity or process improvements are going to have only a minimal effect on the overall world production unless new plants are built. In the USA, for example, plant efficiencies of large-scale producers are in the region of 95%. Admittedly plant efficiencies are much lower for some smaller, village, plants in China and in the former Soviet Union where efficiencies of 70% or lower are common, but in general these plants

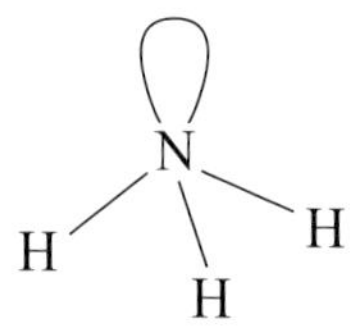

Figure 1.6. The pyramidal geometry of ammonia.

produce less than 30 000 tonnes per year each and even if their process efficiencies were dramatically improved they would not make a major contribution to the overall worldwide production of fixed nitrogen. Even in the event of a radical development in ammonia process technology it will probably take at least 10 years for this to become significant in worldwide ammonia production.

The development of new catalysts may improve the efficiency of the Haber–Bosch Process particularly if they enable the synthesis of ammonia to be performed at milder conditions than at present. However, in the near future with the increasing world population and requirement for more and more fixed nitrogen the only way to ensure that there is sufficient food and consumer products for everyone is to build more ammonia-synthesis plants and to introduce effective transport systems and trade agreements in the developing world. These measures will encourage farmers to make more effective use of the fertilizers to ensure that they achieve the maximum sustainable yield of their crop, which they can then sell in the worldwide marketplace.

1.4 The chemistry of ammonia

Ammonia, NH_3, is a colourless gas that is lighter than air and possesses a unique odour. The ancient Egyptians knew about ammonium salts as there were sublimed deposits of ammonium chloride from the burning of camel dung inside the Temple of Ammon in Eygpt (which is how the name 'ammonia' came about). Old maps of Egypt from the time of Alexander the Great show the town Ammonium in western Egypt. This still exists although its modern name is Siwa. Alexander the Great visited Ammonium in 331 BCE and the ancient Egyptians used ammonium chloride for textile dyeing. Water pollution by ammonium salts is mentioned in the Bible [16]. Ammonia was first isolated as a pure gas by Joseph Priestley in 1774 [17].

The geometry of the ammonia molecule is that of a nitrogen atom sitting at the apex of a triangular pyramid with three hydrogen atoms forming an equilateral triangle (Figure 1.6).

The $H-N-H$ bond angle is $107°$ and although the three N–H bonds are covalent they have a significant polar contribution because of the higher electronegativity of nitrogen compared with hydrogen. For example, the N–H bond is only 18% ionic

Table 1.3. *The physical properties of ammonia*

Molecular weight	17.03
Gas density	0.7714 (at 0 °C and 1 bar)
Liquid density	0.6386 (at 0 °C and 1 bar)
Critical pressure	11.28 MPa
Critical temperature	132.4 °C
Triple point	−77.71 °C
Boiling point	−33.43 °C (at 1 bar)
Enthalpy of formation	−46.22 kJ mol^{-1} (at 25 °C)
Dielectric constant	23 (liquid)
Lower limit of human perception	53 p.p.m.
pH of 1 M solution	11.6
Solubility in water at 0 °C	47%
15 °C	38%
20 °C	34%
30 °C	31%
50 °C	28%

whereas the O–H bond is 39% ionic. This is given by the formula

$$\text{amount of ionic character of a bond A–B} = 1 - e^{-1/4(X_A - X_B)},$$

where X_A and X_B are the electronegativities of the two constituent elements.

Some of the physical properties of ammonia are given in Table 1.3. The hydrogen bonds in crystalline and liquid ammonia are weaker than in water ice for two reasons. Firstly, the smaller ionic character of the N–H bond gives ammonia weaker hydrogen-bond-forming power only and, secondly, the lone-pair electrons must serve for all of the bonds formed by the molecule with other N–H groups. In crystalline ammonia each nitrogen atom has six neighbours at a distance of 3.38 Å. This distance represents a weak N−H$\cdots$H bond (cf. O−H$\cdots$O in water ice = 2.76 Å).

The relatively stronger bonds in ammonium azide, NH_4N_3, show bond distances of 2.94–2.99 Å. From measurement of the heat of sublimation of NH_4N_3 (27.2 kJ mol^{-1}) calculation can be made of the energy of the N−H$\cdots$H bond, which has been found to be 5.44 kJ mol^{-1}, with 10.9 kJ mol^{-1} for the van der Waals energy.

Liquid ammonia is very similar to water as it can self-ionize and can act as a solvent. Also, both chemicals possess the same electronic configuration and similar bond angles and dipoles. However, liquid ammonia is a better solvent than water for organic chemicals as a consequence of its lower dielectric constant. The dielectric constant of ammonia is about 15 times greater than for most other condensed gases.

Liquid ammonia is relatively easy to handle both in the laboratory and also on larger scale as its relatively high heat of evaporation (1.37 kJ g^{-1}) and its boiling point (−33.35 °C) enable it to be used in a wide range of organic and metallo-organic reactions. For example, ammonia has a lower reactivity than water with electropositive metals and forms blue solutions with sodium and potassium, for example, that can be used for demethylation reactions. The solutions are effectively metal ions with solvated electrons and have similar properties to Na/NH_3 and K/NH_3.

$$2NH_3 \rightleftharpoons [NH_4]^+ + [NH_2]^- \quad K_{-50} = [NH_4]^+/[NH_2]^- = 10^{-30}$$
$$2H_2O \rightleftharpoons [H_3O]^+ + [OH]^- \quad K_{25} = [H_3O]^+/[OH]^- = 10^{-14}$$

The use of Group I and II metals in ammonia is the basis for the Birch Reduction of aromatic compounds to give dihydro derivatives. Methoxybenzoic acids can be converted to 2-alkyl cyclohexanones under similar conditions. Liquid ammonia will also react with diborane to form an addition complex which can reduce carbonyls and will react selectively with aldehydes in the presence of ketones.

Liquid ammonia can be used as a solvent for palladium-catalysed hydrogenolysis, and this method is often used for sulphur-containing peptides and sensitive molecules.

Ammonia can also form stable hydrates such as $NH_3.H_2O$ and $2NH_3.H_2O$ with water, in which the water and ammonia molecules are linked by hydrogen bonds. However, these hydrates are stable only at low temperatures and melt at 194 K. In water, ammonia is hydrated and the use of the term ammonium hydroxide is technically incorrect as there is no evidence for the existence of undissociated NH_4OH.

$$NH_3 + H_2O \rightleftharpoons NH_4^+ + OH^-$$
$$K_{25} = [NH_4^+][OH^-]/[NH_3] = 1.81 \times 10^{-5} \ (pK_b = 4.75)$$

NMR (nuclear magnetic resonance) studies have shown that there is exchange of protons between the hydrogen atoms of ammonia and water, and also that the four hydrogens on the ammonia ion are equivalent [18].

There are many stable water-soluble salts of the tetrahedral ammonium ion NH_4. Most salts of strong acids are fully ionized and their solutions are slightly acidic. Ammonium salts generally resemble those of potassium or rubidium in solubility as the three ions are comparable in size: $NH_4^+ = 1.48$ Å, $K^+ = 1.33$ Å and $Rb^+ = 1.148$ Å. Ammonium salts dissociate when heated or, if the anion is oxidizing, they can decompose at lower temperatures to yield nitrogen or nitrogen oxides.

$$(NH_4)_2Cr_2O_7 \rightarrow N_2 + 4H_2O + Cr_2O_3$$
$$(NH_4)NO_3 \rightarrow N_2O + 2H_2O$$

Ammonia is oxidized by air or oxygen to mixtures of nitrogen oxides, depending upon the reaction conditions, and this is the basis for the Ostwald Synthesis of nitric acid [19], where ammonia is converted to nitric acid on a large scale. Usually this process takes place on the same site where the ammonia is manufactured. Under extreme conditions nitrogen and water are generated from the reaction between ammonia and oxygen but this is very unusual. If the reaction between oxygen and ammonia takes place on an alkaline surface, such as in the presence of sodalime, then nitrate and nitrite ions can be produced.

$$\text{(i)} \quad 4NH_3 + 5O_2 \rightarrow 4NO + 6H_2O \quad \Delta H = -906.11 \text{ kJ mol}^{-1}$$
$$\text{(ii)} \quad 4NH_3 + 4O_2 \rightarrow 2N_2O + 6H_2O \quad \Delta H = -1105 \text{ kJ mol}^{-1}$$

Oxidation reaction (i) proceeds extremely quickly, with the time taken for 50% of a given quantity of ammonia to be used up ($t_{\frac{1}{2}}$) being 10^{-11} s.

Ammonia reacts with aqueous sodium oxychloride to form hydrazine in the Raschig Process. The overall reaction is

$$NH_3 + NaOCl \rightarrow NaOH + NH_2Cl$$
$$NH_2Cl + NH_3 + NaOH \rightarrow N_2H_4 + NaCl + H_2O$$

However, once formed the hydrazine needs to be separated from the reaction mixture immediately before it reacts with chloramine to form ammonium chloride.

$$NH_2Cl + N_2H_4 \rightarrow 2NH_4Cl + N_2$$

In commercial manufacture the addition of a sequestration agent is also important to remove any heavy-metal contaminants that might catalyse the reaction of hydrazine with chloramines. Gelatine is often used for this purpose. Anhydrous hydrazine may be obtained by distillation or by isolation as the sulphate by precipitation as $N_2H_6SO_4$ and secondary reaction with ammonia.

Hydrazine is essentially a substitution product of ammonia where hydrogen has been replaced by $-NH_2$. It is a weaker base than ammonia.

$$N_2H_4 + H_2O \rightarrow [N_2H_5]^+ + [OH]^- \quad K_{25} = 8.5 \times 10^{-7}$$
$$[N_2H_5]^+ + H_2O \rightarrow [N_2H_6]^{2+} + [OH]^- \quad K_{25} = 8.9 \times 10^{-16}$$

For organic chemists, however, the most important reaction of ammonia is ammonolysis (this literally means 'splitting with ammonia') to form amides from acid chlorides, acid anhydrides or esters.

$$\begin{array}{c} R-C=O + 2NH_3 \rightarrow R-C=O + NH_4Cl \\ \quad | \qquad\qquad\qquad\quad | \\ \quad Cl \qquad\qquad\qquad\; NH_2 \end{array}$$

The reaction mechanism is understood to be as shown in Figure 1.7, where Z may be $-Cl$, $-R$ or $-OCOR$.

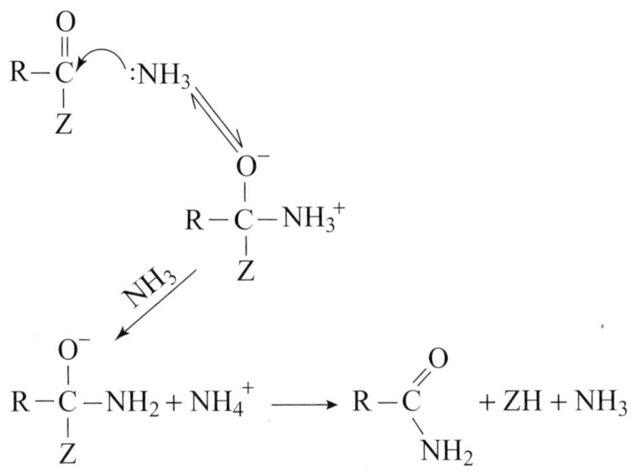

$$R-\overset{\overset{\textstyle O}{\|}}{\underset{\underset{\textstyle Z}{|}}{C}} \quad :NH_3$$

$$R-\overset{\overset{\textstyle O^-}{|}}{\underset{\underset{\textstyle Z}{|}}{C}}-NH_3^+$$

$$R-\overset{\overset{\textstyle O^-}{|}}{\underset{\underset{\textstyle Z}{|}}{C}}-NH_2 + NH_4^+ \longrightarrow R-\overset{\overset{\textstyle O}{\|}}{\underset{\underset{\textstyle NH_2}{\diagdown}}{C}} + ZH + NH_3$$

Figure 1.7. The reaction of ammonia with an acid chloride.

Ammonium salts of carboxylic acids also decompose on heating to form amides. However, this method of synthesis is not suitable for the manufacture of heat-sensitive amides from the acid-chloride derivatives of amino acids as they may decompose under the reaction conditions.

When hydrazine (NH_2-NH_2) is used in this reaction, hydrazides are produced ($R-CO-NH-NH_2$); and with hydroxylamine (NH_2-OH), hydroxamic acids are produced ($R-CO-NH-OH$). When phthalic acid (with two adjacent carboxylic acid groups) is reacted with ammonia, phthalimide is produced.

The ammonolysis of alcohols or alkyl halides produces amines and is discussed later in Chapter 2, which deals with amine synthesis.

$$R-OH + NH_3 \rightarrow R-NH_2 + H_2O$$

Phenols also undergo this reaction but require heating under pressure and the use of a catalyst such as zinc chloride.

Ammonia will also react with ethylene oxide and other epoxides to form hydroxylamines.

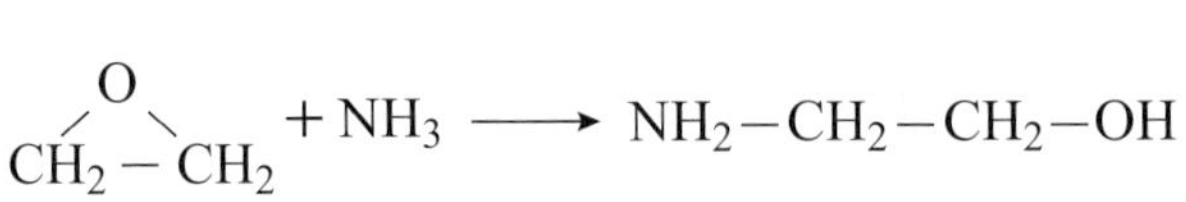

$$\underset{CH_2-CH_2}{\overset{O}{\diagup\diagdown}} + NH_3 \longrightarrow NH_2-CH_2-CH_2-OH$$

With lactones there are two possible reaction products plus a variety of doubly and triply bridged products as shown in Figure 1.8. The major reaction product in each case depends upon the reaction conditions and the specific lactone used.

Liquid ammonia can be used as a solvent for ozonolysis reactions and is not converted to nitrogen oxides. For example, the ozonolysis of indene and indole gives quinazoline and isoquinoline respectively.

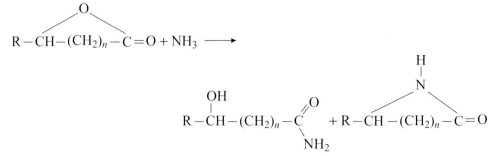

Figure 1.8. The reaction of a lactone with ammonia.

Replacing the hydrogen atoms on ammonia with other substituents produces different types of amine, which may be either organic or inorganic. The synthesis, properties and reactions of these amines will be investigated in the following chapters of this book with a view to understanding how changing the substituents affects the characteristics of the central nitrogen atom for different amines.

References

[1] D. Rutherford, *Disseratio inauguralis de aere fixo dicto aut mephitico*, 12 September 1772.
[2] M. E. Weeks, *J. Chem. Ed.* (1968), p. 191 (and references therein).
[3] D. A. House, *Comp. Coord. Chem.*, **2** (1987), 23.
[4] T. R. Malthus, *Essay on the Principle of Population* (London: J. Johnson, 1798).
[5] J. von Leibig, *Ann.*, **36** (1850), 88.
[6] W. Crookes, *Report of the 68th meeting of the British Association for the Advancement of Science, Bristol* (London: J. Murray, 1898).
[7] J. Hess, *Chem. Ind.*, **45** (1922), 538.
[8] *Profile Magazine (90 Years of Progress – Norsk Hydro's 90th Anniversary)* (Oslo: Hydro Media, 1995), p. 12.
[9] C. A. Vancini, *The Synthesis of Ammonia* (London: McMillan, 1971).
[10] W. Nernst, *Z. Elektrochem., Angew. Phys. Chem.*, **13** (1907), 521.
[11] F. A. Ernst, *The Fixation of Atomospheric Nitrogen* (New York: Van Nostrand, 1928).
[12] A. Mittasch, *Geschichte der Ammoniak Synthese* (Berlin: Verlag Chemie, 1951); also DE Patent 238450 (1912).
[13] F. Haber, *Z. Elektrochem.*, **16** (1910), 244.
[14] M. Gilbert, *The First World War* (London: Harper Collins, 1995).
[15] I. B. Dybkjaerin in *Ammonia Catalysis and Manufacture*, ed. A. Nielsen (Berlin: Springer, 1995), p. 202.
[16] Exodus 7.21.
[17] J. Priestley, *Experiments and Observations on Different Kinds of Air* (Birmingham: Thomas Pearson, 1775; revised in 1790).
[18] K. Mislow, A. Ruak and L. C. Allen, *Angew. Chemie*, **82** (1970), 453.
[19] L. Andrussov, *Angew. Chemie*, **39** (1926), 332.

2

Aliphatic, fatty and cyclic amines

2.1 The physical properties and geometries of amines

From an inspection of the molecular geometries of trimethylamine and 2-methylpropane it might be expected that their physical properties would be similar. Both are trigonal pyramidal molecules with similar molecular weights. However, their physical properties, which are summarized in Table 2.1, show that although these two chemicals have similar melting and boiling points they have very different characteristics.

The major difference in the structures of these two molecules (see below) is that trimethylamine has a lone pair of electrons on the nitrogen atom, and this controls the molecular geometry. The geometry of 2-methylpropane is largely controlled by steric factors as the central carbon atom does not have a lone pair and all available electrons are involved in bonding.

Trimethylamine   2-Methylpropane

The lone pair of electrons on the nitrogen atom of trimethylamine controls the trigonal pyramidal structure but the trimethyl groups also exert an effect on the ability of the nitrogen atom to donate its lone pair of electrons to form a tetrahedral ammonium ion. 2-Methylpropane does not possess a lone pair of electrons on its central carbon atom. It is not a strong base and also is unable to form hydrogen bonds with water. 2-Methylpropane is barely soluble in water but is soluble in organic solvents.

The dipole moments of both chemicals show a large difference of 3.54 D (0.44 D for 2-methylpropane and 3.95 D for trimethylamine), where D is the Debye.

Table 2.1. *A comparison of the properties of trimethylamine and 2-methylpropane*

	Trimethylamine	2-Methylpropane
Molecular weight	59	58
Physical form at ambient conditions	Gas	Gas
Boiling point	2.87 °C	−0.5 °C
Melting point	−117 °C	−138 °C
Solubility in water	40% wt in H_2O	0.15% vol. in H_2O
Dipole moment	3.95 D	0.44 D
Density	0.67	0.6
Flashpoint	−6 °C	−138 °C

Interestingly the 2-methylpropane radical cation has a dipole of only 2.43 D, with 0.18 of a lone electron localized on the central carbon atom, which is still 1.52 D less than the dipole of trimethylamine. Trichloromethane is able to form hydrogen bonds with water and has a dipole moment of 1.02 D.

The properties of amines are largely defined by the presence of the non-bonding electron pair on the sp^3-hybridized central nitrogen atom (or nitrogen atoms for bi- and polyfunctional amines). In order to understand the reactivity and biological activity of amines, a good place to start is to examine the lone pair on the ammonia molecule and then see how its properties change with the introduction of substituents.

Gaseous ammonia has a pyramidal structure with the radius of the H–N bond = 0.015 nm and the bond angle of 106.6°. Within C_{3v} symmetry the $2p_z$ and 2s orbitals of the nitrogen atom and the (+) combinations of the 1s orbital of the hydrogen all give orbitals of a_1 symmetry. The $2p_x$, $2p_y$ and (−)1s (H) all give orbitals of e symmetry in ammonia. The atomic orbitals (AO) are shown below.

AO	2s	$2p_z$	$2p_x$	$2p_y$	(+)1s	(−)1s	(−)1s
Symmetry	a_1	a_1	e	e	a_1	e	e

The $1a_1$ orbital is equivalent to the 1s orbital of nitrogen and is not involved in chemical bonding. The frontier orbital $3a_1$ is 90% made up of the nitrogen $2p_z$ orbital and is only about 10% from other orbital components. Its molecular orbital (MO) coefficients are shown below. The atomic orbitals that combine to form the molecular orbitals in the ground state of ammonia are shown in Figure 2.1.

		H1	H2	H3	N	N	N	N	N
MO	ε (eV)	$1s_1$	$1s_1$	$1s_1$	1s	2s	$2p_x$	$2p_y$	$2p_z$
$3a_1$	−11.22	0.07	0.07	0.07	0.08	−0.3	0.0	0.0	0.94

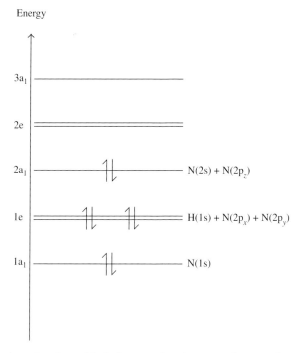

Figure 2.1. A molecular-orbital diagram for the ground state of ammonia. The molecular orbitals are shown on the right, with equivalent atomic orbitals on the left. The pairs of single-headed arrows represent electron pairs.

The 1e orbital contributes to N–H bonding and to H⋯H interactions. Interestingly if the ammonia molecule distorts to planar geometry, the $3a_1$ and $2a_1$ orbitals destabilize because of an increasing contribution from the nitrogen 2s orbital. The 1e orbital is stabilized in planar geometry but to a smaller extent and is insufficient to counteract the destabilization of the a_1 orbitals. A marked tendency to pyramidal geometry is observed in alkyl substituted amines. For example, in trimethylamine the valence angle at the N atom increases by $1.4°$ to $108°$ compared with ammonia.

At ambient temperature the ammonia molecule inverts via a planar transition state with a frequency of about 10^{10} Hz and with an energy barrier to inversion of 24.28 kJ mol^{-1} (see Figure 2.2(a)). For alkylamines where the energy barrier to inversion is typically between 16 and 40 kJ mol^{-1} nitrogen inversion is slightly slower but is still between 10^3–10^5 Hz. This makes optical separation of enantiomers impossible.

A consequence of this rapid inversion is that although it is possible to draw chiral enantiomers for trisubstituted tetrahedral nitrogen it is impossible to isolate them for most amines except at very low temperatures. An exception to this general rule is Tröger's Base (see Figure 2.2(b)), where the nitrogen atoms are fixed into

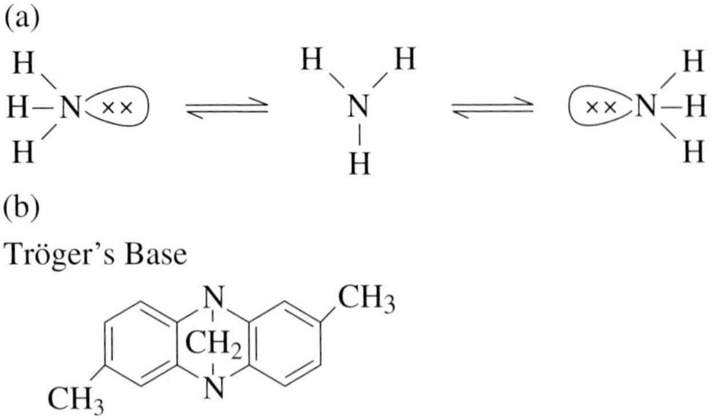

(a)

(b)

Tröger's Base

Figure 2.2. (a) Inversion through the nitrogen centre in ammonia. (b) Steric hindrance in Tröger's Base prevents inversion.

position by the cyclic structure and cannot invert. Both (+) and (−) enantiomers are known.

The highest occupied molecular orbital (HOMO) of trimethylamine is formed not only by the $2p_z$ and 2s orbitals of the nitrogen but also by combination of the $2p_z$ orbital from the methyl carbons. The effect of this orbital mixing can be seen below in the decrease in ionization energy (IE) across the series from ammonia to tributylamine.

	Ammonia	Trimethylamine	Triethylamine	Tri-*n*-propylamine	Tri-*n*-butylamine
IE (eV)	10.92	8.53	8.08	7.92	7.90

The decrease in ionization energy can be explained by the inductive effects of the methyl groups and also by the π overlap of corresponding orbitals.

For example, in the gas phase, piperidine exists in two configurations with the lone pair in either the axial position or in the equatorial position. The axial configuration is favoured and is found in 60% to 70% of molecules studied (see below). If the terminal hydrogen on the nitrogen atom is replaced by a methyl group, the resulting *N*-methyl piperidine is found to be 99% in the axial configuration. Therefore the orientations of the nitrogen lone pair and of the adjacent N−C bond have a large effect on the frontier molecular orbital and the spatial geometry.

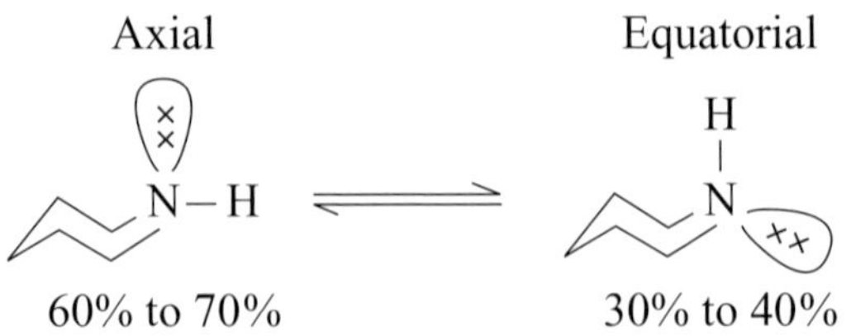

Axial   Equatorial

60% to 70%   30% to 40%

The ionization energy of piperidine is 8.63 eV, whereas for *N*-methyl piperidine it is 8.23 eV. This difference can be explained by the overlap of the nitrogen lone pair and the adjacent C—H bond on the neighbouring carbon. This $n(N)$—$\sigma(C—CH_3)$ is possible only when the nitrogen lone pair is in the axial position and gives the geometry increased stability.

However, there are other types of interaction that can be made by the nitrogen lone pair with neighbouring bonds. The substitution of a phenyl group into ammonia has the effect of drastically decreasing its basic properties; for example, aniline is a weaker base than ammonia by a factor of a million.

The low basicity of aniline cannot be explained by the negative inductive effect of the phenyl group alone and is accounted for by delocalization of the nitrogen lone pair by $n(N)$ and $\pi(Ph)$ overlap. The interaction is facilitated by the fact that the aniline molecule is not planar. The angle between the amino group and C—N is 37°, which helps to facilitate overlap and donation of the nitrogen lone pair into the π-orbital.

Chemically the phenyl group in aniline is far more reactive to attacking electrophiles (e.g. bromine) than benzene. The π-electron densities in aniline are shown below (cf. benzene where every carbon atom has a π-electron density of 1).

The inversion barrier for aniline is 6.70 kJ mol^{-1} compared with 18.42 kJ mol^{-1} in dimethylamine, which does not have any π-orbitals available. Dimethylamine is virtually planar when in the gas phase, with a bond angle Me—N—Me of 116° compared with 113° for H—N—H in aniline.

Inversion of amines with sp^3 configuration occurs via an sp^2-hybridized intermediate with the nitrogen lone pair occupying a p orbital. Similarly oximes, which have sp^2 hybridization in their ground states, invert via an sp-hybridized intermediate (see Figure 2.3).

Because of the constant inversion of amines at room temperature, it might be thought that there would be no chiral amines. However, there are three special cases in which amines may exhibit chirality. These are:

 (i) when an amine has a side chain with a chiral centre;
(ii) when a quaternary amine has four different substituents (i.e. without a lone pair it cannot invert unless one or more of the substituents is labile); and

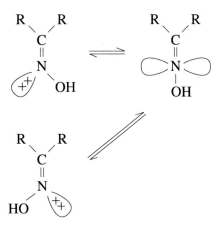

Figure 2.3. Inversion about a nitrogen centre in an oxime.

(iii) when an amine is constrained, as in the case of fused heterocycles (e.g. Tröger's Base) or highly substituted small-ring heterocycles like aziridine with *N*-chloro or *N*-methoxy substituents where the energy barrier to inversion is large enough for the two enantiomers to be isolated. 1-Chloro-2,2-diphenylaziridine is an example, where inversion between enantiomers takes several days at ambient conditions and the individual (+) and (−) forms can be completely resolved at −60 °C.

The amino group attached to phenyl or other aromatic or heteroaromatic groups can also be involved in intermolecular interactions. These have been extensively studied in the β-lactams, which have a tertiary nitrogen atom bridging two fused rings. Generally donation of the nitrogen lone pair decreases the charge on the nitrogen atom and it becomes less pyramidal. In a normal amide N, C and O are planar as the nitrogen lone pair is assumed to be delocalized. In a strained ring system such as that found in β-lactams, the greater the ring strain the more readily the ring opens at the target sites and so it might be reasonably expected that drugs with a high ring strain might have better bioactivity than those with less ring strain. The ring-strain energies of β-lactams are of the order of 109–122 kJ mol^{-1}. In an attempt to quantify the ring strain, infra-red $v_{C=O}$ values have been measured for a range of β-lactams (see below).

Penems	1770–1790 cm^{-1}
Monocyclic unfused lactams	1730–1760 cm^{-1}
Non-planar, 3-penems	1780–1790 cm^{-1}
Planar 2-penems	1750–1780 cm^{-1}

Generally, a higher $v_{C=O}$ value is associated with a shorter C=O bond length and more-pyramidal nitrogen geometry, which indicates a more-strained ring. More-planar nitrogen geometry allows resonance stabilization of the four-membered ring

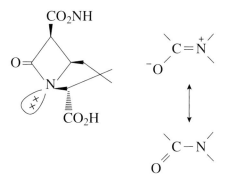

Figure 2.4. The β-lactam ring structure and resonance stabilization.

with a zwitterionic form. This stabilization is equivalent to around 75 kJ mol^{-1} (see Figure 2.4).

Many natural products such as alkaloids contain bicyclic amines, which have bioactivities and find application in medicines. Bicyclic amines are very easily ionizable and form stable cations with long half lives. Interaction of adjacent σ-orbitals helps to delocalize the positive charge and this stabilizes the cation. If C=C or C=O π-orbitals are available for overlap with the nitrogen lone pair the extent of this stabilization can be between 0.5 and 1.0 eV.

The relative strength of amine bases (**B**) is usually expressed as either the pK_b, or as the pK_a of the conjugate acid.

$$\mathbf{BH^+} \rightleftharpoons \mathbf{B}\text{:} + \mathbf{H^+}$$
$$pK_a = [\mathbf{B}\text{:}][\mathbf{H^+}]/[\mathbf{BH^+}]$$

The pK_a is related to pK_b by the equation

$$pK_a + pK_b = pK_w$$

where K_w is the autopyrolysis constant of water $= 10^{-14}$, or p$K_w = 14$ (at 25 °C).

Therefore strong bases have low pK_b values and correspondingly high values for the pK_a of their conjugate acids. The basicities of amines are governed largely by two parameters that are electronic and steric in nature. The electronic effects are primarily inductive and resonance. The presence of electron-releasing (+I) groups attached to the nitrogen atom of amines enhances the electron density around the nitrogen and increases its basicity. Similarly electron-withdrawing groups (−I) have the opposite effect.

Resonance effects (also called mesomeric effects) are the effects of substituents to delocalize any formal electronic charge on the nitrogen atom by molecular re-arrangement. Electron-donating groups have +R effects (base strengthening) and electron-withdrawing groups have −R effects (base weakening). An example of

Table 2.2. *Inductive and resonating groups*

Effect	Examples
+I	RO^-, alkyl groups
−I	CF_3, Cl, Br, I, CO_2H, aryl groups
+R	F, Cl, Br, I, OH, OR, alkyl groups
−R	NO_2, CN, CO_2H, aryl groups

these two effects can be seen by examining the basicities of *meta*-methylaniline and *para*-methylaniline. The methyl group is +I (electron donating) in both molecules but the +R effect of a methyl group in a *para* position is much greater than for a methyl group in a *meta* position, so *para*-methylaniline is a stronger base than *meta*-methylaniline. Some examples of inductive and resonating groups are given in Table 2.2.

Some texts refer to resonance effects as mesomeric effects and use the notation +M or −M but the overall effect is the same.

There are two ways in which steric parameters can affect the base strength of amines. These are defined as primary and secondary effects.

Steric impedance to protonation when bulky substituents are bonded to the amine nitrogen and result in greater strain in the protonated cation are called the primary steric effects. The net effect of bulky substituents is to create weaker bases. Hindrance to solvation is a secondary steric effect; it is also a base-weakening effect and occurs when substituents reduce the available space around the nitrogen centre for solvent molecules. The relative weakness of tertiary amines compared with secondary amines has been attributed to this effect.

In the liquid phase in the presence of solvents the order of basicity is $NH_3 < RNH_2 = R_2NH > R_3N$, where R = alkyl. The increased +I effect going from primary to secondary amines is offset by the solvation effects. This is why there is often no increased basic character in many amines by substitution of a second alkyl group. However, in the gas phase, with no solvent effects, the order of basicity is found to be $NH_3 < RNH_2 < R_2NH < R_3N$.

The inversion of the basicity series observed in solution is an effect of the ease of solvation. In aqueous solution an ammonium ion has four potential sites for solvation, whereas an ammonium salt formed from a tertiary amine R_3NH^+ has only one potential site available for solvation.

In addition to being electron donors, amines can also be deprotonated by very strong bases such as organolithium salts and some Grignard reagents. For example, butyl lithium (BuLi) will deprotonate diethylamine to form lithium diethylamide with the $[N(Et)_2]^-$ ion. Amide bases of this type are sometimes called non-nuclophilic bases as the bulky alkyl substituents hinder their approach to carbon reaction sites but they can react at less-crowded hydrogen positions. Two useful reagents for the formation of lithium amide salts are 1,3-diazobicyclo[3.4.0]nonane (DBN) and 1,5-diazobicylo[5,4,0]undecane (DBU), both of which have very sterically hindered bridgehead nitrogen atoms.

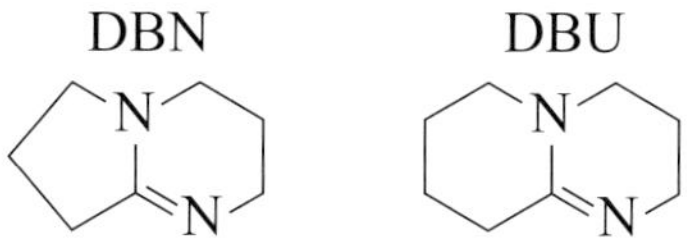

In general it is the lone pair on the nitrogen atom of amines and the ability of this lone pair to be shared that forms the basis of the chemistry of the amines. Excluding reactions that take place in the side chains of amines, there are only two classes of reaction in which amines take part. These are reactions in which amines act as bases and reactions in which amines act as nucleophiles, as shown below. In both instances the amine nitrogen atom loses its lone electron pair and gains a positive charge.

Under specific reaction conditions, therefore, amines are able to act as bases and nucleophiles.

An amine acting as a base (e.g. adding a proton)

$$(CH_3)_3N + H^+ \rightarrow [(CH_3)_3N{-}H]^+$$

An amine acting as a nucleophile (e.g. with acyl and alkyl halides)

$$(CH_3)_3N + R{-}X \rightarrow [(CH_3)_3N{-}R]^+ + X^-$$

The synthesis and chemical properties of amines will now be examined in greater detail.

2.2 Aliphatic amines

Amines with up to six carbons per alkyl chain are known as lower alkyl amines and those with over eight carbons per alkyl chain are called fatty alkyl amines. Methyl and ethyl amines were first prepared by the hydrolysis of the corresponding alkyl isocyanate by Wurtz in 1849[1]. However, Hofmann was the first chemist to use the terms primary, secondary and tertiary amines [2]. In general most naturally occurring amines have complex structures; however, some simpler members are also known, such as tetramethyldiamine (putrescine), pentamethylenediamine (cadaverine) and di(4-amino-*n*-butyl)amine (spermidine). The major applications for amines are in the areas of agrochemicals, dyestuffs, pharmaceuticals, surfactants, plastics, chemical auxiliaries (e.g. for rubber, textiles and paper manufacture), anti-corrosion agents and as process chemicals.

The lower alkyl amines are gases at room temperature (methylamine, dimethylamine, trimethylamine and ethylamine); higher amines up to about 12 carbons

Table 2.3. *The physical properties of some alkylamines*

Amine	Melting point (°C)	K_b ($\times 10^{-4}$)
Methylamine	−92	4.5
Dimethylamine	−96	5.4
Trimethylamine	−117	0.6
Ethylamine	−80	5.1
Diethylamine	−39	10.0
n-Propylamine	−83	4.1
t-Butylamine	−67	4.0
Pentylamine	−55	4.2
Cyclohexylamine	17.7	5.0
Octylamine	0	4.4
Decylamine	17	4.5
Ethylenediamine	8	0.7
Tetramethylethylenediamine	27	0.85
Hexamethylethylenediamine	39	5.0

per chain are liquid and longer carbons chains are solids at ambient temperature. Some melting points and basicities of amines are given in Table 2.3 and also in Appendix 2.

Most of the lower alkyl amines are soluble in water up to tri-*n*-butylamine, which is only partially miscible with water above 18 °C. All alkyl amines have high vapour pressures and can form explosive mixtures with air. Also odour thresholds are around 0.1 p.p.m. and the characteristic amine smell is often described as being like rotten fish. Hence great care is needed with handling and especially with transportation where international air, road and sea shipping regulations must be observed.

2.2.1 Synthetic methods for the preparation of aliphatic amines

(a) From alcohols (amino-dehydroxylation)

The reaction between ammonia, alcohol (or alcohols) and a catalyst is the most common method of manufacture for lower alkyl amines. The product ratio of primary : secondary : tertiary can be controlled by the reaction conditions although most commercial processes produce a mixture of all three, which needs downstream fractionation.

$$xs\ NH_3 + ROH \rightarrow (NH_2R + NHR_2 + NR_3 + H_2O)$$

Undesirable or unmarketable mono- and diamines can be recycled in the process to tertiary amines, although it is more favourable to have the correct product mixture in the first instance if this can be achieved by catalyst selectivity and appropriate reactor heat, pressure and residence times. In the past, dehydrogenation catalysts such as Al_2O_3, W_2O_5 and Cr_2O_5 or mixed oxides were used and these are satisfactory for the production of lower-chain alkyl amines. However, when using high-molecular-weight alcohols, other catalysts based on nickel, cobalt, iron and copper are used.

Typically the reaction between ammonia and alcohol takes place at 0.5–20 MPa applied pressure at temperatures of 100–250 °C in a fixed-bed reactor under continuous conditions. A two to eight times stoichiometric excess of ammonia is used, and hydrogen is also added to the mixture to maintain the activity of the metal surface of the catalyst.

This reaction can also be used with a primary or secondary amine in place of ammonia although copper catalysts are generally preferred to avoid transalkylation [3]. An example of this would be the reaction between dimethylamine and ethanol to produce dimethylethylamine.

$$(CH_3)_2NH + C_2H_5OH \rightarrow (CH_3)_2NC_2H_5 + H_2O$$

The ammonolysis of alcohols is complex because in many reactions there are several equilibria and an overall balance has to be achieved between the reagents and the various mono-, di- and trisubstituted reaction products.

It was initially thought that this reaction proceeded via a nitrile intermediate but, more recently, azomethane has been identified during the reaction, as a transient intermediate, and it is now thought that this is hydrogenated to form the product amine or amines [4].

The conversion of R–OH to R–NH$_2$ can be accomplished with other reagents apart from ammonia such as hydrazoic acid, diisopropyl azodicarboxylate and a stoichiometric excess of triphenyl phosphine in tetrahydrofuran/water. These are in effect a type of Mitsunobu Reaction, which will be covered in more detail in Chapter 3 on aryl amines. With chiral aliphatic alcohols, the Mitsunobu Reaction proceeds with inversion at the chiral centre. Conversion of alcohols to amines takes place with reaction of the alcohol with phthalimide (Pht–NH) in the presence of diethyl azodicarbonate (DEAD) and triphenyl phosphine followed by hydrazinolysis of the product to form the amine.

$$CH_3-CH_2-\underset{\underset{OH}{\blacktriangle}}{CH}-CH=CH_2 + Pht-NH + DEAD/P(Ph)_3 \rightarrow$$

$$CH_3-CH_2-\underset{\underset{Pht}{\overset{\|}{N}}}{\overset{\overset{H}{|}}{C}}-CH=CH_2 + N_2H_4 \rightarrow CH_3-CH_2-\underset{\underset{NH_2}{\|}}{\overset{\overset{H}{|}}{C}}-CH=CH_2$$

Alcohols can also be converted to amines via alkyloxyphosphonium perchlorates, which can monoalkylate primary and secondary amines to form secondary and tertiary amines respectively.

$$R^1-OH + (CCl_4/P(NMe_2)_3) + NH_4ClO_4 \rightarrow [R^1OP(NMe_2)_3]^+ \, CClO_4^-$$
$$+ (R^2R^3NH/DMF) \rightarrow R^1R^2R^3N + OP(NMe_2)_3 + HClO_4$$

(b) From carbonyl groups

Generally, aldehydes and ketones are used in preference to alcohols as raw materials for amine synthesis. The reaction is in two stages, as shown below.

$$RHC=O + NH_3 \rightarrow RHC=NH + H_2O$$
$$RHC=NH + H_2 \rightarrow R-CH_2-NH_2$$

In the laboratory this hydrogenation can be conducted using formic acid (the Leuckart–Wallach Procedure) but in industry, for large-scale manufacture, hydrogen is used with a reaction catalyst and the water is removed by distillation before the second reaction step.

The main difference between amination of a carbonyl and amination of an alcohol is that in the first case hydrogen is consumed stoichiometrically. Also the higher enthalpy of this process (60.4 kJ mol^{-1} for acetone vs 7.1 kJ mol^{-1} for isopropyl alcohol) means that a different reactor design is required. Generally for the amination of carbonyls a reactor temperature of 100–$160\,°C$ is used at atmospheric pressure or slightly above.

(c) From nitriles

This method of amine synthesis is used if the process economics favours the use of a nitrile feed over a corresponding alcohol. Nitriles are converted into amines via catalytic hydrogenation.

$$R-CN + H_2 \rightleftharpoons R-CH=NH \rightleftharpoons R-CH_2-NH_2$$
$$\Updownarrow \; R-CH_2NH_2$$
$$R-CH=N-CH_2-R \rightleftharpoons R-CH_2-NH-CH_2-R$$
$$\Updownarrow \; R-CH_2NH_2$$
$$R-CH_2-N-CH_2-R$$
$$|$$
$$CH_2-R$$

Group VIII catalysts are commonly used for this reaction and allow mild conditions of 20–$100\,°C$ and 1–5 atm applied pressure [5]. Nickel and cobalt catalysts require $150\,°C$ and up to 25 atm applied pressure. Side reactions can be suppressed by the addition of sodium hydroxide or ammonia. Alternatively, if secondary and tertiary amines are required the reaction times can be extended. This method of amine synthesis from nitriles is commonly used for the synthesis of di- and triethyl-amines from acetonitrile. Secondary amines can be prepared from the reaction of sodium or calcium cyanamide salts with alkyl halides to give dicyanamides, which can be hydrolysed and then decarboxylated to give amines. Cyclic secondary amines are produced when this reaction is carried out with alkyl dihalides with the halides at opposite ends of an *n*-alkane chain.

(d) The Ritter Reaction

This process utilizes the reaction of olefins with hydrocyanic acid [6]. However, because of the severe difficulties in handling hydrogen cyanide gas this process is relatively expensive and is only used to produce sterically hindered amines such as t-butyl amine.

$$(i)\ (CH_3)_2{-}C{=}CH_2 + HCN + H_2O \rightarrow (CH_3)_3{-}C{-}NH{-}CHO$$

$$(ii)\ (CH_3)_3{-}C{-}NH{-}CHO + H_2O \rightarrow (CH_3)_3{-}C{-}NH_2 + HCO_2H$$

(e) From alkyl halides

The Hofmann Amination Reaction was discovered in 1849 by Hofmann in his London laboratory [2]. Hofmann's Amination Reaction and the discovery that amines can be prepared from the direct reaction of alkyl halides and ammonia pre-dates the better-known Hofmann Degradation Reaction by 33 years. However, this amination process is not used industrially except for laboratory synthesis because of corrosion problems associated with the production of hydrogen chloride at high temperatures and pressures.

$$R{-}Cl + NH_3 \rightarrow R{-}NH_2 + HCl$$

Another problem with this synthesis is that the primary amines produced from the reaction are stronger bases than ammonia and preferentially attack the alkyl halide substrate. The reaction is useful for the production of tertiary amines and even more so for the production of quaternary ammonium salts (called the Menshutkin Reaction [7]). Primary amines are best prepared from alkyl halides either by the Gabriel Reaction [8] or via the formation of azide intermediates. One notable exception to the above is the preparation of cyclic amines, such as pyrrolidine and aziridine, which can be conveniently prepared from 4-chloro-1-aminobutane and 2-chloroethylamine respectively. In both cases the addition of a non-nucleophile base is necessary to catalyse the reaction.

Four-membered cyclic amines (azetidines), however, are best prepared by a slightly different route.

$$Ph-NH_2 + TsO-CH_2-CH_2-CH_2-OTs \rightarrow Ph-N \begin{matrix} \overset{H_2}{C} \\ \diagup \quad \diagdown \\ \qquad\qquad CH_2 \\ \diagdown \quad \diagup \\ \underset{H_2}{C} \end{matrix}$$

(Here, Ts = tosyl.)

Active halides such as allylic and benzylic halides can also be converted to primary alkyl amines by the Delépine Reaction [9], where alkyl halides are reacted with hexamethylenetetramine and the resulting salts are cleaved with a solution of ethanolic hydrogen chloride.

(f) From nitroalkanes

This method is limited in application because of the poor availability and stability of nitroalkanes [10]. In general, the reduction of nitro groups may be performed with tin and hydrochloric acid although other reducing agents have been reported.

$$R-NO_2 \rightarrow R-NH_2$$

(g) From the Fischer–Tropsch Reaction

The reaction of carbon monoxide and hydrogen plus ammonia over an activated iron catalyst has been reported as producing methylamine [11]. If ammonia is replaced with an amine in the gaseous phase, secondary and tertiary amines can also be produced, although this reaction has yet to find a large-scale industrial application.

$$CO + 2H_2 + NH_3 \rightarrow NH_2-CH_3 + H_2O$$

(h) From transamination

Lithium amide salts Li[RNH] can react with primary amines to form secondary amines.

$$RNH_2 + Li[R^1NH] \rightarrow RR^1NH + LiNH_2$$

A variation on this reaction is to reflux a primary amine in xylene in the presence of Raney nickel to yield a secondary amine and ammonia.

$$2R-NH_2 \rightarrow R_2NH + NH_3$$

(i) Amination of alkanes

Alkanes can be aminated at tertiary positions by the reagent mixture trichloroamine/aluminium chloride at 0 °C. Although trichloroamine is difficult to handle this reaction is useful for the synthesis of tertiary alkyl amines. The mechanism reaction is S_N1 with the H^- ion as the leaving group.

$$NCl_3 + AlCl_3 \rightarrow [Cl_2N-AlCl_3]^-[Cl]^+$$
$$R_3CH + Cl^+ \rightarrow R_3C^+ \rightarrow R_3CNCl_2 \rightarrow R_3C-NH_2$$

(j) Amination of alkenes

In 1956, Brown and Subba-Rao reported that dialkylboranes react with carbon–carbon double bonds to give anti-Markovnikov addition products [12]. These reactive borane intermediates can then further react with aminating agents such as potassium phthalimide or lithium amides to form amines.

$$R^1HC=CH_2 + H-BR^2R^3 \rightarrow R^1-CH_2-CH_2-BR^2R^3$$
$$+ \text{ aminating agent} \rightarrow R^1-CH_2-CH_2-NH_2$$

For example, the reaction between styrene and dimethylborane followed by an aminating agent yields 2-amino-1-methylethylbenzene (3-phenyl-1-isopropylamine).

(k) From amides

Amines can be synthesized from amides by two main methods. In the first method the carbonyl group $R-CO-NH_2$ can be reduced either catalytically or by sodium metal in ethanol to yield a methylene group to give $R-CH_2-NH_2$.

$$R-CO-NH_2 + (Na/C_2H_5OH) \rightarrow R-CH_2-NH_2 + H_2O$$

This reaction may also be performed by first forming a cyanide intermediate from the amide by using a dehydrating agent such as phosphorous pentoxide or thionyl chloride. Heating at elevated temperatures also results in amides being dehydrated, although this reaction is not very precise and must be carried out in the presence of excess ammonia to prevent the parent carboxylic acid being produced as the major reaction product. Excess ammonia recycles the carboxylic acid back to the amide.

$$2R-CONH_2 \rightarrow R-CN + R-CO_2H + NH_3$$
$$(R-CO_2H + NH_3 \rightarrow RCO_2NH_4 \rightarrow R-CONH_2)$$

The Hofmann Degradation Reaction is the conversion of an amide into a primary amine with the loss of one carbon atom [13]. The reaction is usually carried out with bromine in aqueous potassium hydroxide and proceeds via an intramolecular intermediate with a bridging alkyl (or aryl) group that migrates from the nitrogen to the adjacent carbon.

$$R-CO-NH_2 + Br \rightarrow R-CO-NHBr (+KOH) \rightarrow [R-CO-NBr]^-K^+$$
$$\rightarrow KBr + R-N=C=O (+H_2O) \rightarrow R-NH_2 + CO_2$$

The reaction intermediate is thought to be cyclic and of one of the following forms.

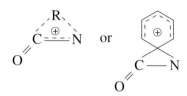

When the Hofmann Reaction is carried out with aryl group substituents, the presence of *para* electron-withdrawing groups slows down the reaction and, conversely, the presence of *para* electron-donating groups accelerates the reaction.

The Hofmann Reaction works well for amides with up to about seven carbons in the alkyl chain. After seven carbons the reaction products are a mixture of cyanides and amines (both with the loss of a carbon). In order for the reaction to work with long-chain alkyl groups it is necessary to change the procedure slightly and to add bromine to a solution of the amide and sodium methoxide in methanol. This produces an *N*-alkylurethane, which on hydrolysis yields the amine.

$$R-CO-NH_2 + 2CH_3ONa + Br_2 \rightarrow RNHCO_2CH_3 + 2NaBr + CH_3OH$$
$$R-NH-CO_2CH_3 + 2NaOH \rightarrow R-NH_2 + Na_2CO_3 + CH_3OH$$

(l) From the Curtius Reaction (Rearrangement)

The Curtius Rearrangement occurs with acyl azides [14]. It is similar to the Hofmann Degradation in that an R− group migrates from an acyl carbon to a nitrogen atom at the same time as an elimination occurs. However, in the Curtius Reaction the leaving group is nitrogen. Acyl azides can be prepared by the action of sodium azide on acyl chlorides. Heating the azide brings about the rearrangement to form an isocyanate, which can be hydrolysed to form the required amine.

$$R-CO-Cl + NaN_3 \rightarrow R-CO-N_3 + NaCl$$
$$R-CO-N_3 \rightarrow R-N=C=O + N_2$$
$$R-N=C=O + H_2O \rightarrow R-NH_2 + CO_2$$

(m) From the Gabriel Reaction

Potassium phthalimide can be used as an aminating agent to prepare primary amines by the Gabriel Reaction [15]. This is actually a variation of the Hofmann synthesis from alkyl halides described in section (c) above, and was discovered by Gabriel who was one of Hofmann's students in 1887. The Gabriel Reaction uses potassium phthalimide as an aminating agent. When alkyl halides are used in this reaction alkyl amines are produced, but when chloro alkyl nitriles are used (with the chloro and nitrile groups at either end of the alkyl chain) amino acids are obtained. Potassium phthalimide can be obtained from the reaction of potassium hydroxide and phthalimide and its subsequent reaction with an alkyl halide followed by hydrazine yields the corresponding alkyl amine.

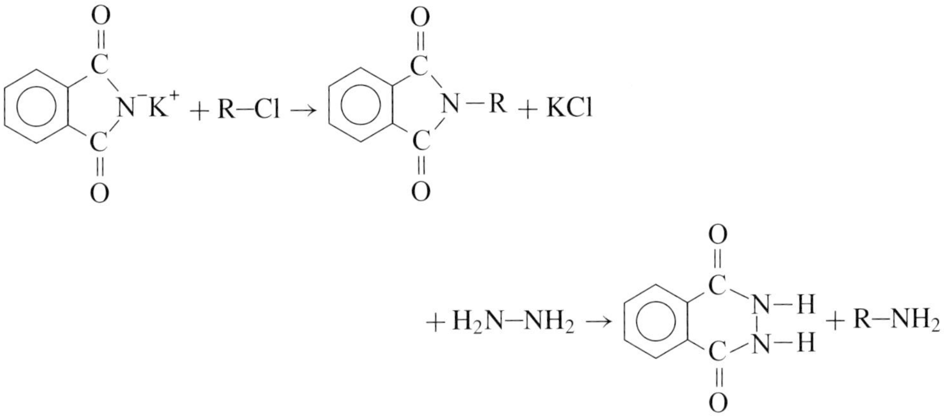

(n) From isocyanates

Alkyl and allyl isocyanates give good yields of amines when heated with dilute hydrochloric acid.

$$CH_2{=}CH{-}CH_2{-}N{=}C{=}S + HCl \rightarrow CH_2{=}CH{-}CH_2{-}NH_2 + COS$$

(o) From allylic and styrenic bonds

Metal complexes can be used to aminate allylic bonds although two reaction products may be obtained depending upon the reagents and reaction conditions employed. With an aminating agent of the form EWG–N=X, where EWG is an electron-withdrawing group and X = O, S or Se, and in the presence of a Lewis Acid, the reaction product is as shown below [16].

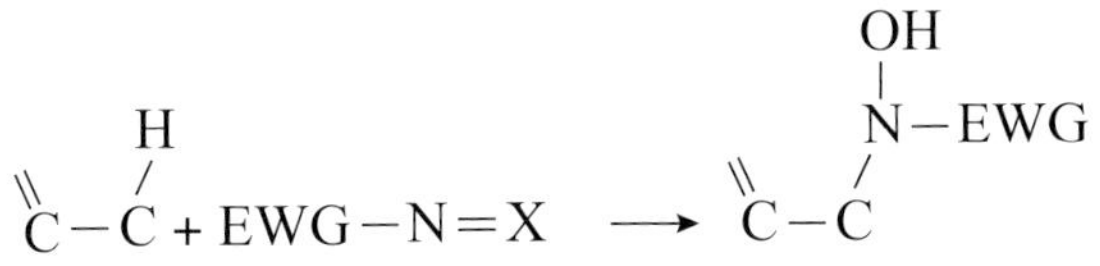

However, if this reaction is performed in the presence of meso-tetra-2,6-dichlorophenylporphyrin manganese perchlorate, Mn(TDCPP)[ClO$_4$], with the same aminating agent then the reaction proceeds via a cyclic intermediate to give a different reaction product [17].

$$\overset{\backslash\backslash}{C}-\overset{/}{\underset{}{C}}\overset{H}{\overset{/}{}} + EWG-N=X \longrightarrow \overset{\backslash\backslash}{C}-\overset{/}{\underset{}{C}}\overset{\overset{H}{\overset{|}{N}-EWG}}{\overset{/}{}}$$

Sharpless and co-workers developed a new aminating agent in 1978 based on a molybdenum complex with phenyl hydroxylamine [18] that was able to reduce non-functionalized alkenes.

The molybdenum-based aminating agent developed by Sharpless is able to aminate alkenes to form secondary alkyl phenylamines with rearrangement of the carbon–carbon double bond.

$$\overset{\backslash}{\underset{/}{C}}=CH-CH_2-CH_2-CH_3 \longrightarrow \overset{\backslash\backslash}{\underset{/}{C}}-\overset{\overset{H-N-C_6H_5}{\overset{|}{CH}}}{}-CH_2-CH_2-CH_3$$

The formation of cyclic transition states is very common with amination reactions and has also been found in amination reactions with an organozirconium complex discovered by Luker *et al.* in 1995 [19] formed between dicyclopentadienyl zirconium dichloride and diphenyldiazomethane.

$$(Cp)_2ZrCl_2 + (C_6H_5)_2CN_2 \rightarrow$$

Here, C_p represents the cyclopentadienyl group. This zirconium complex is able to aminate terminal alkenes to form primary amines.

$$-\overset{|}{\underset{|}{C}}-CH=CH_2 \longrightarrow -\overset{|}{\underset{|}{C}}-CH_2-CH_2-NH_2$$

2.2.2 *The reactions of alkyl amines*

(a) *Salt formation*

Because of the alkyl substituents, aliphatic amines are stronger bases than ammonia. With acids they form salts that are very soluble in water but usually insoluble in organic solvents. This property of amines is exploited industrially in the manufacture of semi-synthetic penicillins in the solvent-extraction process.

$$NR_3 + HCOOH \rightarrow [NHR_3]^+ [HCOO]^-$$

When in the presence of a hydrogen halide, amines act as proton acceptors and form quaternary salts. Tertiary amines are widely used as proton acceptors in organic synthesis and help to catalyse reactions such as Friedel–Crafts alkylation and acetylation by mopping up excess halogens formed as reaction by-products. The most widely used tertiary amines for this purpose are diisopropylethylamine (Hünig's Base), dimethyl aniline and 2,6-dimethyl pyridine. The quaternary salts formed by the amines after reaction with a halide are water soluble and can easily be removed from organic solution. The total worldwide requirement for auxiliary bases of this sort is probably less than 1000 tonnes per year.

(b) *Conversion to carboxamides*

Aliphatic amines react readily with carboxylic acids, esters and acid chlorides to produce carboxamides of the general formula R^1–CO–NR^2R^3, where R^1, R^2 and R^3 may be similar or different alkyl groups or hydrogen. Reaction yields are typically in excess of 90% and this reaction is widely used for the synthesis of herbicides. This reaction does not always proceed at room temperature and may require heating if a stable carboxylic acid salt is produced with a quaternary amine. Alternatively a catalyst can be used if the required amide is heat sensitive.

$$R–CO–X + H_2N–R' \rightarrow R–CO–NH–R' + H–X$$

(Where X = OH, OR or Cl.)

Ammonia reacts with carboxylic acids to form ammonium salts, which can be pyrolized to yield amides.

$$R–CO_2H + NH_3 \rightarrow [R–CO_2]^-[NH_4]^+ \rightarrow R–CO–NH_2 + H_2O$$

When this reaction is carried out with γ- and δ-amino acids, a lactam is obtained. However, as the lactam ring is sensitive to heat the reaction is best catalysed enzymatically. Lactams can also be produced by the action of ammonia or a primary amine on a lactone.

(c) Conversion to sulphonamides

The reaction between amines and benzenesulphonyl chloride is utilized for distinguishing between primary, secondary and tertiary amines (the Hinsberg Test – see Chapter 6). Primary amines produce alkali-soluble benzenesulphonamides, secondary amines produce alkali-insoluble benzenesulphonamides, and tertiary amines give no reaction at all.

$$R–NH_2 + C_6H_5–SO_2–Cl \rightarrow C_6H_5–SO_2–NH–R + HCl$$

$$R_2–NH + C_6H_5–SO_2–Cl \rightarrow C_6H_5–SO_2–NR_2 + HCl$$

$$R_3–N + C_6H_5–SO_2–Cl \rightarrow \text{no reaction}$$

Sulphonamide drugs are an important class of pharmaceuticals, the life-saving properties of which were discovered almost by accident by Gerhard Domagk of the company I. G. Farben. He used a newly developed drug called Prontosil to save his daughter from dying from a streptococcal infection [20]. The biochemical activity of Prontosil was explained by Ernest Fourneau of the Pasteur Institute in 1936. He found that Prontosil was metabolized in humans to produce sulphanilimide, which was the active agent against streptococcal infections. By replacing the terminal hydrogen atom in sulphanilimide by a heterocyclic ring the toxic effects of sulphanilimide were reduced and a range of sulphanilimide drugs were produced. One of these, Sulphapyridine, which was developed in 1938, was the first ever drug to be effective against pneumonia. In recent years the sulpha drugs have been replaced by penicillins as the drug of first choice to treat bacterial infections, but the use of sulpha drugs in the Second World War saved tens of thousands of lives.

(d) With carbonyl compounds

Depending on the condensations and compounds used there are two possible outcomes to the reactions between the amines and carbonyl compounds: either (i) Schiff Bases or (ii) enamines. The reaction between a carbonyl group and an amine to produce a Schiff Base is called the Leuckart Reaction [21], and was discovered in 1885 by Leuckart.

$$R–NH_2 + R'–CH_2–CO–R'' \rightarrow R–N{=}\overset{\overset{\displaystyle R''}{|}}{C}–CH_2R' + H_2O$$

Schiff Bases can be reduced to amines by formaldehyde by the Mannich Reaction, which was discovered in 1912 [22].

$$R–N{=}\overset{\overset{\displaystyle R''}{|}}{C}–CH_2R' + H_2C{=}O \rightarrow R–NH–\overset{\overset{\displaystyle R''}{|}}{C}H–CH_2R'$$

Schiff Bases are characterized by the N=C double bond and can rearrange as follows to produce enamines with a carbon–carbon double bond. Enamines can be hydrogenated to produce more highly alkylated amines. This is an important method of synthesis for secondary and tertiary amines.

$$R-N=C \overset{CH_2R'}{\underset{R''}{}} \rightleftharpoons R-NH-C \overset{\overset{CHR'}{\|}}{\underset{R''}{}}$$

(Here R = alkyl group and R' and R'' are alkyl groups or hydrogen atoms.)

When secondary amines react with carbonyl groups, enamines are produced because formation of an N=C double bond and disubstituted nitrogen is unfavourable compared with formation of a C=C double bond at the α-carbon.

(e) Reactions with carbon dioxide or carbon disulphide

Primary amines can react with carbon dioxide or carbon disulphide to form carbamic acids and dithiocarbamic acids. Both of these are unstable but can be isolated in a stable form as their salts or esters. Dithiocarbamates obtained from amines are important in the rubber industry as vulcanization accelerators.

$$R-NH_2 + CX_2 \rightarrow R-NH-C \overset{\overset{X}{\|}}{\underset{X-H}{}}$$

(Here X = O or S.)

(f) Reaction with epoxides

Primary amines react with epoxides to give a mixture of mono- and di-oxylated derivatives. Secondary amines give only mono-oxylated products and tertiary amines yield only quaternary salts. This reaction with epoxides is industrially important for the synthesis of ethanol amines and propanolamines.

$$R'R''-NH + (CH_2CH_2)O \rightarrow R'R''-N-CH_2-CH_2-OH$$

When trimethylamine is heated in aqueous solution with ethylene oxide, choline is produced.

$$CH_2-CH_2 + (CH_3)_3N + H_2O \rightarrow [(CH_3)_3N-CH_2-CH_2-OH]^+[OH]^-$$

If choline is refluxed with aqueous barium hydroxide solution it eliminates water to form neurine, $[(CH_3)_3N-CH=CH_2]^+[OH]^-$, a biochemical found in the brain.

(g) Alkylation

The reaction of amines with alkyl halides and dialkyl sulphates to give quaternary amines is utilized in drug synthesis, the preparation of anti-corrosion agents, biocides and phase-transfer catalysts.

$$R^1R^2R^3N + R^4Cl \rightarrow [R^1R^2R^3R^4N]^+ \, Cl^-$$

(h) Phosgenation

Alkyl amines undergo a two-step reaction with phosgene leading to the formation of alkylisocyanates via carbonyl chloride intermediates.

$$R-NH_2 + COCl_2 \rightarrow R-NH-CO-Cl + HCl$$
$$R-NH-CO-Cl \rightarrow R-N=C=O + HCl$$

The phosgenation of alkyl amines is important in the production of herbicides with urea, carbamate or thiocarbamate structures and also in the synthesis of polyurethanes.

(i) Reaction with acrylonitrile

The addition of primary or secondary amines to acrylonitrile to form aminopropionitriles is utilized industrially for the manufacture of higher diamines and polyamines. The nitriles produced from the reaction with acrylonitriles are easily hydrogenated to form amines.

$$R-NH_2 + CH_2=CH-CN \rightarrow R-NH-CH_2-CH_2-CN$$
$$R-NH-CH_2-CH_2-CN + H_2 \rightarrow R-NH-CH_2-CH_2-CH_2-NH_2$$

(j) Oxidation

Free akylamines are generally susceptible to oxidation, giving a range of products depending upon the conditions and oxidant used. For example, tertiary amines are oxidized by hydrogen peroxide to form amine oxides, whereas primary and secondary amines give hydroxylamines and aldoximes.

$$R_3N + H_2O_2 \rightarrow R_3N-O + H_2O$$
$$R-CH_2-NH_2 + H_2O_2 \rightarrow R-CH_2-NH_2O$$
$$\downarrow$$
$$R-CH_2-NH-OH + R-CH=N-OH$$

Oxidation with nitrous acid can be used to distinguish between primary, secondary and tertiary amines, because primary amines yield alcohols and secondary amines yield *N*-nitrosyl compounds ($R_2-N-N=O$). Tertiary amines and amines without α-hydrogens on the alkyl chain do not react with nitrous acid.

(k) Dealkylation

Tertiary amines can be dealkylated by heating the quaternary ammonium salts. Methyl groups are eliminated as methanol. Higher alkyl groups are eliminated as alkenes (along with water). This reaction is useful in the structural analysis of unknown amines.

(l) Reactions specific to primary, secondary and tertiary amines

Primary, secondary and tertiary amines behave differently in the isocyanide and Schotten–Baumann reactions, and in the formation of diazonium salts.

	Isocyanide reaction	Schotten–Baumann	Diazonium salts
Primary amine	Yes	Yes	Yes
Secondary amine	No	Yes	No
Tertiary amines	No	No	No

For the isocyanide reaction, a few drops of chloroform should be added to about 0.2 g of the amine to be tested in 5 ml of ethanolic sodium hydroxide, and the characteristic pungent smell of isocyanides should be detected. Addition of hydrochloric acid will stop the reaction and also remove the terrible smell. For the Schotten–Baumann Reaction, about 1 g of the amine to be tested should be mixed in a test tube with about 20 ml of 10% sodium hydroxide solution. To this solution about 1.5 g of toluene-*p*-sulphonyl chloride in acetone solution should be added. The test tube should then be stoppered and shaken for about 10 min. After shaking, the solution in the test tube should be acidified with 1 M hydrochloric acid to precipitate the product from the diluted solution. This product can then be filtered off and recrystallized from methylated spirit to give the *para*-toluene-sulphonyl derivative.

Finally, diazotisation may be accomplished by adding about 0.2 g of the amine under test to a solution of 1 ml of concentrated hydrochloric acid in 3 ml of water, cooled to 0 °C. Addition of a few drops of sodium nitrite solution should produce the diazonium salt, which can be tested for by adding the solution to another solution of 2-naphthol in excess 10% sodium hydroxide. If the diazonium salt has been formed it will react with the 2-naphthol to form a brilliant red dye. Secondary and tertiary amines give green *N*-nitroso compounds in this reaction.

(m) The Cope Elimination

On heating, tertiary amine oxides eliminate a dialkylhydroxylamine via a cyclic transition state. The other reaction product is an alkene. This reaction is known as the Cope Elimination [23].

$$R-CH_2-CH_2-\overset{\overset{\displaystyle O}{|}}{N}(CH_3)_2 \rightarrow R-CH=CH_2 + CH_3-\overset{\overset{\displaystyle OH}{|}}{N}-CH_3$$

(n) Reaction with diazo compounds

Amines are not acidic enough to react directly with diazo compounds but will do so in the presence of a catalyst such as boron trifluoride or copper(I) cyanide. Diazomethane can be used to methylate amines and ammonia but the reaction proceeds to give a mixture of different mono-, di-, tri- and tetramethylated products. Secondary amines and primary aromatic amines also give this reaction, but very reactive reagents like diazomethane are rather blunt tools and other alkylating agents such as dimethyl sulphate and diethyl sulphate allow better control of the reaction.

(o) Reaction with lithium

Tertiary amines such as tetramethylethylene diamine (TMEDA) can be used as auxiliary reagents in butylation reactions with butyl lithium (BuLi). TMEDA is a very effective sequestration agent for positively charged metal ions and is an effective way of removing spent lithium from the reaction.

Secondary amines can react with lithium to form lithium salts, which can then be used as aminating agents. This reaction works well with aziridine, because this can be lithiated and then reacted with an alkyl amine to form an *N*-alkyl aziridine salt. Ring opening of this salt with a halogen yields a quaternary amine with a terminal halide on the ethyl chain.

$$2(CH_2)_2N-H + 2Li \rightarrow 2(CH_2)_2N-Li + H_2\uparrow$$
$$(CH_2)_2N-Li + R-Cl \rightarrow (CH_2)_2N-R + LiCl$$
$$(CH_2)_2N-R + Cl_2 \rightarrow [R-NH_2-CH_2-CH_2-Cl]^+Cl^-$$

(Aziridine can also be reacted with tosyl chloride to form *N*-tosyl aziridine, which can react with other amines in the presence of Yb(Otf)$_3$ to form diamines.)

(p) Formation of isocyanides

Primary amines react with chloroform in the presence of ethanolic potassium hydroxide to yield isocyanides. This reaction works reasonably well and proceeds via the dichloromethylene (CCl$_2$) diradical. The isocyanide produced from the reaction can be recovered in diethyl ether and purified by fractional distillation; yields of up to 40% can achieved relatively easily.

$$R-NH_2 + CHCl_3 + 3KOH \rightarrow R-NC + 3KCl + 3H_2O$$

(*Author's note*. In the mid 1980s I made a series of alkyl and aryl isocyanides while working as a postdoctoral student. At the time I was working on a project to find new catalysts for the Fischer–Tropsch Reaction, and it was thought that metal isocyanide complexes might be a good idea. The terrible smell of some isocyanides cannot be adequately described but – in my opinion – they must have one of the most awful odours known to Man! After my first synthesis of a few grams in a fume hood I was told by the departmental head to continue my work on the roof, preferably on a windy day, as the vile smell of a few grams of phenyl isocyanide had managed to permeate the whole department. My advice to anyone attempting this synthesis would be to wear old clothes that can be burned afterwards and also to take several showers as I found that one shower is rarely sufficient. In addition it is a good idea not to attempt this synthesis before an important date or meeting as the smell of isocyanides lingers. It was not until several weeks after this incident that my colleagues would speak to me again.)

When secondary amines are used in this reaction the adduct with dichloromethylene cannot eliminate two molecules of hydrogen chloride as there are insufficient available hydrogens, so instead hydrolysis to *N,N*-disubstituted formamides takes place.

$$NR_2H + CCl_2 \rightarrow Cl_2HC-NR_2 \ (+H_2O) \rightarrow H-CO-NR_2$$

(q) Reaction with amides

This is an example of an exchange reaction and it is usually best carried out with an amine salt in the presence of a boron trifluoride auxiliary, which complexes the ammonia generated by the reaction.

$$R-CONH_2 + [R'NH_3]^+Cl^- \rightarrow R-CO-NHR' + [NH_4]^+[Cl]^-$$

This reaction is often used to convert urea into an *N*-alkyl substituted urea.

$$NH_2-CO-NH_2 + [RNH_3]^+Cl^- \rightarrow NH_2-CO-NHR + NH_4Cl$$

2.2.3 Applications of alkylamines

The total worldwide capacity of aliphatic amines is about 400 000 tonnes per year. The single most commercially important alkyl amine is ethylamine, which accounts for about 35% to 40% of the world's annual requirement for alkyl amines. Its main usage is in the production of triazine-type herbicides such as atrazine, simazine and cyanazine. Diethylamine is used to produce vulcanization accelerators such as zinc diethyldithiocarbamate and tetraethylthiuram disulphide. Triethylamine is used as a proton acceptor and is used in organic synthesis as an auxiliary reagent.

The worldwide demand for polyamines and isopropyl amines is roughly about the same as the demand for ethylamines (35% to 40% of the world's annual requirement). Monoisopropylamine is used in the production of agrochemicals, and diisopropylamine is used in the synthesis of herbicides. Triisopropylamine is of only minor importance and has only a few specialist applications. Diisopropylethylamine (Hünig's Base), like triethylamine, is used as an auxiliary in organic synthesis as a proton acceptor.

Butylamines are less widely used but have applications is the production of herbicides (butylate) and as vulcanization accelerators, e.g. sulphenamides.

Dibutylamine is used for the formulation of flotation agents, cutting oils, corrosion protection, and agrochemical and pharmaceutical products. Monobutylamine is used for the production of plasticizers.

Of the other alkyl amines only three are of significant commercial importance. These are:

 (i) octylamine, which is used for the production of Suloctidil;
 (ii) 2-ethylhexylamines, which are used as anti-corrosion additives for engine oils; and
(iii) diallylamine, which is used in crop protection and as an auxiliary in paper production.

2.2.4 Fatty alkyl amines

Fatty alkyl amines have typical alkyl chain lengths of 8–24 carbon atoms, and many of major commercial importance, such as tallow amine, oleylamine, cocoamine and soya amine, are naturally derived. Fatty amines are soluble in polar and non-polar solvents, but solubility in water is limited to fatty amines with fewer than 10 carbons per unit chain. Fatty alkyl amines can also be produced synthetically from paraffins or from naturally occurring fatty acids such as the cocoamines, soya amines and tallow amines.

The problem for chemists is that many fatty amines are in fact mixtures of different alkyl chain lengths. For example, cocoamine, tallow amine, soya amine and oleylamine contain the following alkyl chains.

Cocoamine $= 7\%\ C_{10} + 50\%\ C_{12} + 18\%\ C_{14} + 6\%$ unsaturated C_{18}
                    $+ 19\%$ others.
Oleylamine $= 5\%\ C_{18} + 76\%$ unsaturated $C_{18} + 19\%$ others.
Tallow amine $= 29\%\ C_{16} + 23\%\ C_{18} + 37\%$ unsaturated $C_{18} + 11\%$ others.
Soya amine $= 16\%\ C_{16} + 15\%\ C_{18} + 50\%$ unsaturated $C_{18} + 13\%$ doubly
                    unsaturated $C_{18} + 6\%$ others.

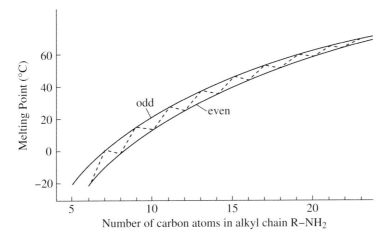

Figure 2.5. The variation of melting point with molecular weight for primary fatty amines.

The ionization constants of the fatty amines do not change significantly with increases in the alkyl chain length, but their melting points are strongly dependent upon molecular weight.

	Molecular weight	Melting point
Octylamine	129	−0.1 °C
Decylamine	157	15.9 °C
Dodecylamine	185	28.2 °C
Cocoamine	204*	12–17 °C
Soya amine	270*	27–30 °C
Dihexadecylamine	465	67 °C
Dicocoamine	410*	40–47 °C

(The asterisk (*) indicates the maximum molecular weight of a typical commercial blend.)

If molecular weight is plotted against melting point for primary fatty amines an oscillation is observed between odd and even carbon chain homologues, although this effect is reduced on moving to higher molecular weights (see Figure 2.5).

All fatty alkyl amines, like short-chain alkyl amines, are stronger bases than ammonia and have the following characteristics.

(i) They form salts with inorganic and organic acids. The acetate salts of fatty alkyl amines are slightly soluble in water but more readily soluble in organic solvents.
(ii) They can be converted into *N*-alkyl carboxamides with carboxylic acids and their derivatives.
(iii) They react with carbon dioxide to give carbonates. These are also formed when fatty alkyl amines are exposed to air for too long.

(iv) They can be alkylated in the presence of a stoichiometric excess of a strong nucleophile such as methyl chloride or dimethyl sulphate. The reaction products are quaternary salts and the reaction can be carried out in aqueous ethanol as the solvent.

 (v) Fatty alkyl amines can be oxylated with ethylene oxide and propylene oxide. The reaction proceeds uncatalysed with ethylene oxide at 200 °C with primary amines to give 2-(alkylimino)diethanols (*N,N*-bis(2-hydroxyethyl) fatty amines. Under mild conditions (350 Pa and 80 °C) secondary and tertiary amines react with ethylene oxide to produce mono or bis(2-hydroxyethyl) quaternary amine salts.

(vi) Fatty amines react with halogenated carboxylic acids or with lactones to yield betaines.

(vii) Fatty amines react with activated double bonds (the Michael Addition [24]). For example, 1,3-propanediamines can be formed by reaction of a primary fatty amine with acrylonitrile followed by hydrogenation of the intermediate 2-(akylamino)-ethanenitrile.

(viii) Fatty amines give the Mannich Reaction with formaldehyde and nucleophiles such as aldehydes, nitroparaffins or phenols.

(ix) Fatty amines react with phosgene to form isocyanates.

 (x) Oxidation of fatty amines with hydrogen peroxide, peracids or ozone produces amine oxides.

2.2.5 The manufacture and production of fatty alkyl amines

Fatty amines can be made from natural fats and oils or from synthetic raw materials. Both types are widely used.

(a) From fatty nitriles

The amines are made by reacting carboxylic acids with ammonia over dehydrating catalysts such as alumina, zinc oxide, or manganese or cobalt salts in liquid-phase reactors, or liquid/vapour-phase reactors at 280–360 °C. Hydrogenation of the fatty nitrile thus produced gives different end products depending upon the reaction conditions.

Saturated primary amines are formed if the reaction is conducted at 80–140 °C and an applied pressure of 10 to 40 bar over a nickel catalyst.

If Raney cobalt or copper chromite catalysts are used, unsaturated amines are produced; ammonia can be added to the reaction as a suppressant to avoid formation of secondary amines. Secondary amines can be formed if ammonia is continually vented from the reactor and temperatures of 160–210 °C and applied pressures of 5–20 MPa are maintained. Under these conditions yields of over 90% can be obtained.

Tertiary amines are produced from nitriles via the imine $RCH=NH$ and the Schiff Base $RCH=N-CH_3$ by using supported nickel catalysts and hydrogen at 230 °C and 0.7 MPa.

N-alkyl, N-methyl alkylamines and N,N-dimethylalkylamines can be synthesized from the Leuckart Reaction and N,N-dimethylalkylamines can also be produced from the reaction of fatty nitriles and dimethylamine.

(b) From alcohols and carbonyl compounds

Fatty amines are produced by the reductive alkylation of ammonia or substituted amines with primary alcohols at 90–190 °C under low pressure. If water is continuously removed from the reactor, secondary and tertiary amines only are obtained. Aldehydes may also be converted to primary amines by reductive amination at 110–120 °C and 1.5 MPa by using rare-earth-promoted cobalt-based catalysts. Aldehydes can also react with methylamine to form imines, which may then be hydrolysed to give N-methyl alkylamines at 115 °C and 3 MPa by using Raney nickel catalysts. Both alcohols and aldehydes can be converted into N,N-dimethyl tertiary amines by using dimethyl amine. Alcohols can be converted into amines at 230 °C and atmospheric pressure by using copper chromite catalysts. Fatty aldehydes require noble metal, copper chelate or copper carboxylate catalysts.

(c) From alkenes

Straight-chain primary amines can be produced from t-bromoalkenes by direct amination. In the Ritter Reaction, α-alkenes can react with HCl gas or with acetonitrile in concentrated sulphuric acid or aqueous HF to give branched alkyl primary amines (β-amines). However, the reaction is not specific and produces a range of positional isomers.

(d) From nitroalkanes

Nitroalkanes can be hydrogenated over palladium/carbon at 190 °C and 3.9 MPa to give secondary alkylamines in up to 93% yield.

(e) From glycerides

Glycerides can be converted into fatty nitriles at 220–300 °C provided that ammonia is vented continuously from the reactor. Hydrogenation occurs with Zn/Al catalysts at high temperature and pressure (350 °C and 10–30 MPa). Coconut glycerides can be converted to cocoamines in this way.

2.2.6 Specific reactions of fatty amines

Although fatty amines undergo the same reactions as short-chain aliphatic amines some of these reactions are particularly useful for the manufacture of fatty-amine derivatives for commercial applications.

(a) Salt formation: fatty amines form salts with acids and also with water. Cationic fatty amine salts are widely used as corrosion inhibitors and in ore flotation.

$$RNH_2 + CH_3COOH \rightarrow [RNH_3]^+[CH_3CO_2]^-$$

(b) Non-ionic surfactants such as ethoxylates and propoxylates are formed by the reactions between fatty amines and ethylene oxide and propylene oxide, respectively. Treatment of the produced surfactant with oxidizing agents such as peroxides and peracids yields amine oxides.

$$RNH_2 + nCH_2O \rightarrow RN-((CH_2O)_n-H)_2$$
$$R_3N + [O] \rightarrow R_3N-O$$

(c) Fatty amines react with chloroacetic acid to form betaines, which are widely used as shampoo bases.

$$RN(CH_3)_2 + Cl-CH_2COOH \rightarrow RN^+(CH_3)_2CH_2COO^- + HCl$$

(d) Substituted ureas, which are widely used as agrochemicals, are formed from the reaction of fatty amines with ammonia via isocyanate intermediates.

$$R-NH_2 + COCl_2 \rightarrow R-N=C=O + 2HCl$$
$$R-NH_2 + R-N=C=O \rightarrow R-NH-CO-NH-R$$

2.2.7 Applications of fatty amines

The largest single use of fatty amines is in fabric softeners. Prior to the 1950s most detergents were based on tallow-derived soap formulations and some of the detergent would remain on the fabric after washing. This gave a soft finish to the laundered items. After the 1950s the use of synthetic and vegetable-based formulations created a demand for fabric softeners. These are typically added into the rinse cycle at concentrations of 5% or less, of which about 0.1% to 0.2% substants into the item being rinsed. Most fabric softeners are based on quaternary amines such as dimethylalkyl ammonium salts with long-chain alkyl groups, typically C_{12} to C_{14}. Fabric softeners can also be used in the form of impregnated paper or tissue that is added after the wash cycle during tumble drying. This formulation has to include a transfer agent to allow the active fabric-softening agent to be transferred during the drying process. Typically the transfer agent is a fatty ester.

Another application of fatty amines is in the formulation of flotation agents.

In general, any particles with negatively charged surfaces can be removed from a system by complexation with amines. Originally this property was used to remove KCl from NaCl in the beneficiation of potash. However, many low-grade rocks and ores can be concentrated by using flotation processes with fatty-amine based detergent formulations. In addition, salts of fatty amines are often added to mineral salts to prevent caking during processing.

Some other applications of fatty amines are as corrosion inhibitors (e.g. *N*-alkyl-1,3-propanediamines), as anti-static additives for plastics (e.g. 2-(alkyliminodiethanol)-bis(2-hydroxyethyl)alkylamines), as pigment dispersants (*N*-tallow-1,3-propanediammonium dioleate) and in drilling fluids (e.g. methyl and benzyl quaternary fatty-amine salts, which react with anionic clay to form thickening agents).

Fatty amines are used as asphalt emulsifiers as their positive charge increases the adherence of the asphalt to the aggregate. Typically between 0.5% and 1.0% is added depending on the speed of setting required (1% gives the fastest set). Fatty-amine derivatives such as 2-(alkylimino) diethanol are used as anti-static agents to polypropylene and polyethylene to prevent charging at low humidities.

Finally, fatty amines are used as bactericides, and in sanitizing solutions and disinfectants. Typically alkyl(benzyl)dimethylammonium chloride and alkyl trimethylammonium chlorides with alkyl chains between C_{12} to C_{14} and also dialkyldimethylammonium chlorides with alkyl chains between C_8 to C_{10} are used.

Worldwide production of fatty amines is estimated to be in excess of 300 000 tonnes per year, with the greatest usage in North America, Western Europe and Japan.

2.3 Cyclic amines

In general, there are four commercially important cyclic amines: these are ethylene imine (aziridine), morpholine (1,4-oxazine), piperazine and piperidine. Ethylene imine is used for the production of resins and for viral inactivation in the preparation of vaccines. Morpholine is used for the production of rubber vulcanization accelerators, corrosion inhibitors, optical brighteners, waxes and polishes, and pharmaceuticals. Piperazine's main application is in the production of anthelmintic pharmaceuticals for the treatment of intestinal worms and parasites, and also for the production of β-lactam antibiotics. Piperidines and their oxo derivatives, piperidones, are widely used for the production of pharmaceuticals with applications in mental health.

The reactions of cyclic amines are typical of secondary amines (they form salts with organic and inorganic acids and can be *N*-alkylated). However, the presence of a carbocyclic ring generally has the effect of exposing the nitrogen lone pair on cyclic amines compared with dialkylamines. This means that the lone electrons are slightly more available for donation. For example, cyclic amines are usually better corrosion inhibitors than corresponding dialkylamines with similar molecular weights. Morpholine is a colourless liquid, piperazine is a white solid and 4-piperidone, which is not very stable in its pure form, is usually encountered as the hydrochloride monohydrate. This is also a solid at ambient conditions or in a BOC-protected form, which can be deprotected before use (see Chapter 7).

Morpholine

Piperidine

4-Piperidone

Aziridine

2.3.1 The manufacture of cyclic amines

Morpholine is commercially produced by two main methods. The first method utilizes the reaction between diethylene glycol and ammonia in the presence of hydrogen and a metal-based hydrogenation catalyst. A wide range of metals, in particular, copper, nickel, cobalt, chromium and the Group VIII metals, catalyse this reaction. The reaction is best carried out between 150 and 400 °C and 3–40 MPa applied pressure. After any excess ammonia has been removed from the crude reaction mixture the morpholine can be recovered by fractional distillation. The main by-product of this process is 2-(2-aminoethoxy) ethanol.

$$(HOCH_2CH_2)_2O + NH_3/H_2/cat \rightarrow O(C_2H_4)_2NH + (H_2NCH_2CH_2)_2O$$

The second method of production involves the dehydration of diethanolamine with strong acids such as concentrated hydrochloric acid, sulphuric acid or oleum. A stoichiometric excess of acid is usually required and the reaction works best over 150 °C.

$$(HOCH_2CH_2)_2NH + H_2S_2O_7 \rightarrow O(C_2H_4)_2NH + 2H_2SO_4$$

(Morpholine can also be made from bis(2-chloroethyl)ether and ammonia but the yield is only 30% based on bis(2-chloroethyl)ether.)

Piperazine can be conveniently prepared from the reaction between ethanolamine and ammonia in a molar ratio of 1 : 3.5 over a Raney nickel catalyst at 195 °C and

13.5 MPa. Diethanolamine can also be used for this reaction but a higher molar ratio of up to 1 : 10 is required.

$$HOCH_2CH_2NH_2 + H_2/cat \rightarrow HN(C_2H_4)_2NH + H_2N-CH_2-CH_2-NH_2$$
$$+ (H_2NCH_2CH_2)_2NH$$

Piperazine can also be produced from the reaction between ethylene dichloride and ammonia, but higher polyamines may also be obtained and the resulting reaction mixture is difficult to separate.

Piperidine, or hexahydropyridine, is obtained commercially from the hydrogenation of pyridine over nickel at 200 °C or in acetic acid over a platinum catalyst at ambient conditions.

$$C_5H_5N + H_2/cat \rightarrow C_5H_{11}N$$

4-Piperidone can be prepared as the hydrochloride monohydrate but it is also commonly encountered in N-alkyl or N-BOC substituted forms, which protect the nitrogen atom during reaction at the carbonyl group.

The simplest method of preparation for 4-piperidone is from 1,5-dihalopentan-3-one and ammonia, but this reaction is not clean as ammonia can also form a Schiff Base with the carbonyl group.

A better method of synthesis is from the Michael Addition with α,β-unsaturated ketones.

$$(CH_3)_2C=CH-CO-CH=C(CH_3)_2 + 2NH_3 \rightarrow$$

If a primary alkyl amine is used in place of ammonia then the corresponding N-alkyl piperidinone is obtained.

This reaction also works with monounsaturated ketones with a single terminal quaternary nitrogen.

Aziridine (ethylene imine) is best prepared by the Gabriel Synthesis starting from 2-aminoethanol, which can be sulphonated with sulphuric acid and then cyclized by heating with aqueous sodium hydroxide. Aziridine is unstable in acid solution and polymerizes explosively. For this reason it is often used *in situ* to prepare resins or used in the form of an alkaline aqueous solution. The chemistry of aziridine is examined in greater detail in Chapter 4, and the applications of aziridine in viral inactivation are discussed in Chapter 8.

2.3.2 Some specific reactions of cyclic amines

Morpholine, piperidine and pyrrolidine can be used to prepare substituted ketones by reaction with unsubstituted ketones initially to form α,β-unsaturated enamines. The double bond in the enamine can be alkylated or acylated and then the enamine can be hydrolysed to recover the original secondary cyclic amine and the substituted ketone.

Morpholine can also be used in the Willgerodt–Kindler Reaction, where aryl alkyl ketones are converted into the ammonium salts of the corresponding acids by refluxing with a secondary amine such as morpholine and sulphur [25]. Generally arylalkyl ketones of the form $Ar-CO-(CH_2)_nCH_3$ yield carboxylic acid salts of the form $Ar-(CH_2)_{n+1}-CO_2^-NH_4^+$. The ammonium salt can rearrange to form the acid amide on heating with the loss of water.

Although cyclic amines can be alkylated by all of the usual methods (alkyl halides, dialkyl sulphates and the Leuckart–Wallach Reaction) a special method for these three cyclic amines works by using primary or secondary alcohols in the presence of hydrogen and a metal-oxide catalyst. Piperazine is generally dialkylated under this reaction regime but mono-*N*-alkylated piperazines can be prepared by protecting one nitrogen with a BOC, CBZ group or just with benzyl chloride. The cyclic amines all react with epoxides to give the corresponding amine alcohols and with isocyanates to form substituted ureas.

4-Piperidones can be converted to 4-perhydroazepines by reaction with diazo-ethylacetate, but the ring nitrogen atom must be protected with BOC.

Finally, cyclic amines can be lithiated and then alkylated or acetylated at the nitrogen centre without loss of ring structure. This reaction works best at low temperatures, and is shown below for aziridine and BOC-protected tetrahydroiso-quinoline.

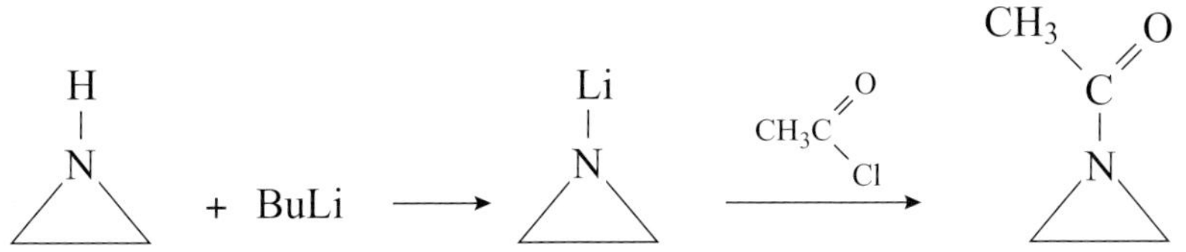

N-acetyl aziridine has been proposed as a viral inactivator but it is generally too reactive to be handled safely except at very low temperatures.

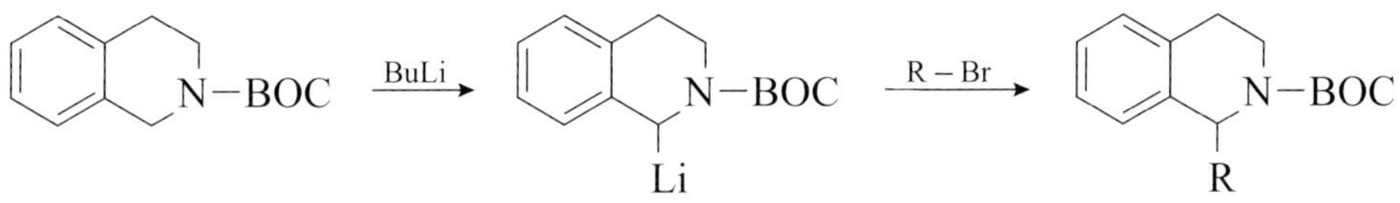

2.3.3 Applications of cyclic amines

The ring structure of piperazine forms the basis for two important classes of drug, anthelmintic agents and phenothiazine anti-psychotic agents. The piperazine-based drugs Diethylcarbamazine and Praziquantel, as well as piperazine itself, can be used for treating intestinal worms in animals although for human applications its derivatives or salts such as the citrate are safer. Diethylcarbamazine and Praziquantel both express antigens that help to bind antibodies to the parasite's surface and also activate the host's immune system. Both drugs also cause paralysis and spastic contractions in the parasitic worms by increasing the permeability of the helminth to calcium ions, which increase in concentration, and also damage its tegument.

Praziquantel

Diethylcarbamazine

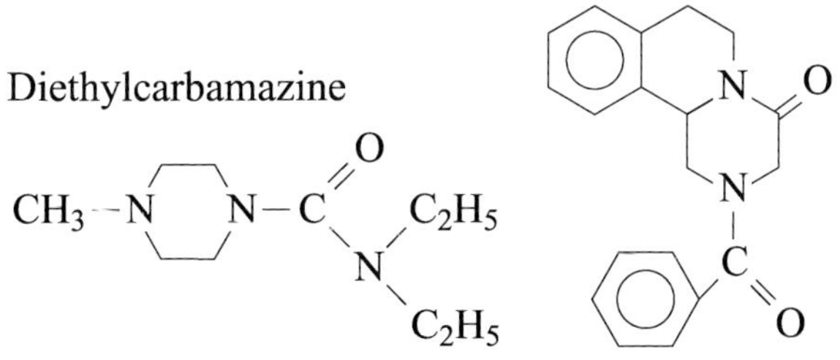

The majority of anti-psychotic agents are used to treat schizophrenia, which is estimated to affect 1% of adults worldwide. Phenothiazine anti-psychotic agents interact with dopamine receptors and reduce their synaptic activity, although the exact mechanism is not clear. The potency of phenothiazine-based drugs corresponds closely to their binding affinity to D_2 receptors. Five common piperazine-type phenothiazine drugs are acetophenazine, fluphenazine and perphenazine, which have the following side chain:

$$R-CH_2-CH_2-CH_2-N \underset{CH_2-CH_2}{\overset{CH_2-CH_2}{\diagup \diagdown}} N-CH_2-CH_2-OH$$

and trifluophenazine and prochlorperazine, which have the following side chain.

$$R-CH_2-CH_2-CH_2-N \underset{CH_2-CH_2}{\overset{CH_2-CH_2}{\diagup \diagdown}} N-CH_3$$

The piperazine ring structure can also be found in the antibiotics cefoperazone and piperacillin, as well as in a range of non-phenotiazine anti-psychotic drugs like Clozapine, Thiothixene and Loxathene.

Many of the applications of morpholine are based on the fact that the molecule has a stong dipole (1.4 D), which makes it an excellent solvent. (When looking at the standard two-dimensional representation of morpholine, it should be borne in mind that the morpholine molecule adopts a chair configuration with the nitrogen and oxygen lone pairs pointing away from each other, hence the high observed dipole.)

Morpholine is chemically stable and is unreactive to both hydrochloric acid and sodium hydroxide up to 160 °C. It is used as a cheap solvent for resins and waxes, as a corrosion inhibitor in steam boilers, as an emulsifier in cosmetics, and as an intermediate in the production of optical brighteners, textile lubricants, rubber vulcanization accelerators, and life-science molecules. Pharmaceuticals incorporating a morpholine ring include Molsidomine, Minaprin and Phenmetrazin. Fungicides incorporating a morpholine ring include Aldimorph, Benzamorf, Carbamorf, Dimethomorph, Fenpropimorph, Flumorph and Tridemorph. Morpholine is a useful reagent in the synthesis of enamines by the Stork Enamine Reaction [26].

Enamines from ammonia and primary amines are unstable and cannot be isolated in the pure form. Enamine formation is catalysed by the presence of trace quantities of acid and also by the presence of drying agents to remove the produced water from the reaction mixture. Enamines are useful in organic synthesis because donation of the nitrogen lone pair to the β-carbon means that the carbon can be alkylated, acetylated and undergo the Michael Addition Reaction. After reaction, the enamine can be hydrolysed to yield the original amine used plus the substituted carbonyl-containing reagent.

4-Piperidone, being bifunctional, offers the chemist the possibility of preparing linear, spiro and fused-chain amines. As examples, the drug Fentanyl has a linear chain, Pethidine and Fluspiriline have spiro chains (where a single carbon atom bridges two cyclic systems), and Tinoridine had a fused chain (where the piperidine group is attached to the rest of the molecule by two adjacent carbons bonded to another cyclic system). However, it is necessary to block the nitrogen atom with a protecting group to allow transformations at the carbonyl group at the 4-position. BOC and CBZ work well, but a benzyl group can also be used. 1-Benzyl-4-piperidone is used in the synthesis of the analgesic drug Fentanyl, and 1-ethoxycarbonyl-4-piperidone is used in the synthesis of the anti-hypertensive drug Endrazoline. Some other examples of drugs prepared from 4-piperidone are Budipine, Mebhydroline, Piperylon and Haloperidol.

4-Alkylated piperidines can be prepared from *N*-substituted piperidones by the Wittig Reaction with phosphorus ylides [27]. For example, 1-benzyl-4-piperidone can react with phosphorus ylide to produce a carbon–carbon double bond in the piperidine 4-position. The subsequent deprotection of the adduct by catalytic hydrogenation to remove the benzyl group also reduces the alkene double bond and yields a 4-alkyl piperidine. This strategy is used in the synthesis of the drug Fentanyl.

The carbonyl group on 4-piperidone can react with amines via the Leuckart Reaction to form Schiff Bases or via the Pictet–Spengler Reaction [28] to form spiro adducts. The Pictet–Spengler synthesis is used to prepare isoquinolines and is discussed further in Chapter 4. Finally, 4-piperidones react with aryl halides to form 4-aryl-4-hydroxy piperidines (in the presence of butyl lithium), which are also useful in drug synthesis. The reactions of 4-piperidones are represented schematically in Figure 2.6.

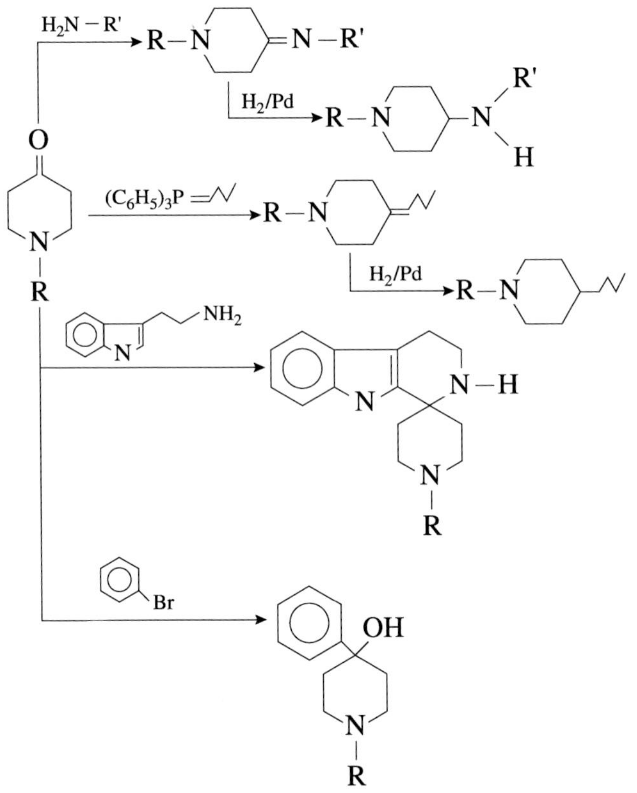

Figure 2.6. The reactions of 4-piperidones.

2.4 Diamines and polyamines

In a letter written to The Royal Society in 1678, Anton van Leeuwenhoek described the slow crystallization of a mysterious substance from semen that he called spermine phosphate [29]. Although the nature of spermine was not elucidated until centuries later, van Leeuwenhoek had discovered the first natural polyamine. In fact it was not until the 1850s that Hofmann synthesized ethylenediamine, which was the first diamine to be identified [30]. In the 1880s Brieger started a series of experiments on putrefying animal organs and found the presence of diamines, such as putrescine and cadaverine, which he was able to isolate as their hydrochlorides [31]. Brieger was later able to isolate these diamines from cell cultures of bacteria and the discovery of the decarboxylation of ornithine to putrescine was published by Dalkin in 1906 [32].

As late as the early twentieth century spermine was thought to be present only in human semen and not in the semen of other mammals despite some rather unclear evidence to the contrary reported by Schreiner in 1878 [33]. In 1921, Mary

Rosenheim isolated spermine from yeast and cod roe and was able to purify it by separation from aqueous solution into butanol and recovered it as the picrate or phosphate salts [34]. The structure of spermine was not finally elucidated until 1926 and 1927 independently by two groups of researchers (Dudley *et al.* [35] and Wrede [36]).

By 1930, five naturally occurring polyamines had been characterized: putrescine, cadaverine, spermine, spermidine and 1,3-diaminopropane.

1,3-Diaminopropane $H_2N-CH_2-CH_2-CH_2-NH_2$

Putrescine $H_2N-CH_2-CH_2-CH_2-CH_2-NH_2$

Cadaverine $H_2N-CH_2-CH_2-CH_2-CH_2-CH_2-NH_2$

Spermidine $H_2N-CH_2-CH_2-CH_2-NH-CH_2-CH_2-CH_2-CH_2-NH_2$

Spermine $H_2N-CH_2-CH_2-CH_2-NH-CH_2-CH_2-CH_2-CH_2-NH-CH_2-CH_2-CH_2-NH_2$

Although spermine is a natural product it can be synthesized easily in the laboratory by the action of iodopropylphthalimide on 1,4-diaminobutane to yield spermine diphthalimide, which can be hydrolysed to the free base and crystallized from solution as spermine phosphate. In fact spermine and spermidine are found in all bacteria and most animal cells. They are growth factors for microorganisms and stabilize the membrane structure of bacteria and also the structures of ribosomes, DNA and viruses. In living cells they are formed from putrescene and *S*-adenoylmethionine. In mammals, spermine is formed from arginine.

Arginine

↓

Ornithine

↓

Putrescine

↓

Spermidine

↓

Spermine

However, in *E. coli* spermine is formed from arginine via an intermediate called agmatine.

Arginine + arginine decarboxylase → $H_2N-\underset{\underset{N-H}{\|}}{C}-NH-CH_2-CH_2-CH_2-CH_2-NH_2$ agmatine

Agmatine + agmatinase → $H_2N-CH_2-CH_2-CH_2-CH_2-NH_2$

putrescine

reactions. TMEDA can be prepared from 1,2-dichloroethane and dimethylamine or from the reaction of ethylene diamine with alkylating agents such as dimethyl sulphate or methyl chloride.

$$H_2N-CH_2-CH_2-NH_2 + 2(CH_3)_2SO_4 \rightarrow (CH_3)_2N-CH_2-CH_2-N(CH_3)_2 + 2H_2SO_4$$

Unlike ethylene diamine, 1,2-diaminopropane cannot be made from a dichloro-alkane (in this case 1,2-dichloropropane) because the reaction with ammonia is difficult to control. Instead it is best prepared from propylene oxide and ammonia in a two-step process.

(i) $CH_3-\underset{\underset{O}{\diagdown\diagup}}{CH-CH_2} + NH_3 \rightarrow CH_3-CH-\underset{\underset{OH}{|}}{CH_2}-NH_2$

(ii) $CH_3-\underset{\underset{OH}{|}}{CH}-CH_2-NH_2 + NH_3 \rightarrow CH_3-\underset{\underset{NH_2}{|}}{CH}-CH_2-NH_2$

The second stage of this reaction requires catalytic activation and high temperatures (190–230 °C). 1,2-Diaminopropane is used in the production of agrochemicals such as Antracol and Basfungin and also in the production of fuel detergents. However, the 1,3-substituted diaminopropane and its derivatives are commercially far more important than 1,2-diaminopropane, as they are used in the manufacture of epoxy resins and the textile finishing agent 1,3-dihydroxymethylhexahydro-pyrimidine.

1,3-Diaminopropane is produced in a two-stage reaction between ammonia and acrylonitrile. The first stage works well at 100–200 bar at 70–100 °C, but the second stage requires a cobalt or nickel catalyst although the reaction conditions are similar. Note that the abbreviation *xs* is used to indicate that an excess of a reagent is used, i.e. above the required stoichiometric quantity.

(i) $CH_2=CH-C\equiv N + xs\ NH_3 \rightarrow CH_3-CH-C\equiv N + HN(CH_2-CH_2-C\equiv N)_2$

(ii) $CH_3-CH-C\equiv N + HN(CH_2-CH_2-C\equiv N)_2 + xs\ NH_3 \rightarrow CH_3-\underset{\underset{NH_2}{|}}{CH}-CH_2-NH_2$

1,3-Diaminopropanes are important as components of epoxy resins. Some commercially important 1,3-diaminopropanes are 1-amino-3-methylaminopropane, 1-amino-3-dimethylaminopropane and 1-amino-3-cyclohexylaminopropane.

Diethylaminopropylamine is widely used in the formulation of anti-corrosion and anti-fouling marine paints for boats. Dimethyldipropylenetriamine is used as an epoxy-resin hardener and tetramethylpropylenediamine is used as a fireproofing agent in textiles and as a plasticizer.

[18] S. L. Liebeskind, K. B. Sharpless, R. D. Wilson and J. A. Ibes, *J. Am. Chem. Soc.*, **100** (1978), 7061.

[19] T. Luker, R. J. Whitby and M. Webster, *J. Organomet. Chem*, **492** (1995), 53.

[20] G. Domagk, *Chemother. Bakt. Infekt.*, **61** (1935), 250.

[21] R. Leuckart, *Ber.*, **18** (1885), 2341.

[22] C. Mannich and W. Krosche, *Arch. Pharm.*, **250** (1912), 647.

[23] A. C. Cope *et al.*, *J. Am. Chem. Soc.*, **62** (1940), 441.

[24] A. Michael, *J. Prakt. Chem.*, **35** (1887), 349.

[25] K. Kindler, *Ann.*, **431** (1923), 193.

[26] G. Stork and H. Landesmann, *J. AM. Chem. Soc.*, **78** (1956), 5128.

[27] U. Schöllkopf, *Angew. Chemie*, **71** (1959), 260.

[28] A. Pictet and T. Spengler, *Ber.*, **44** (1911), 2030.

[29] A. van Leeuwenhoek, Letter to the Royal Society (London, 1678).

[30] A. W. Hofmann, *Phil. Trans. R. Soc. Lond.*, **140** (1850), 93.

[31] S. S. Cohen, *A Guide to Polyamines* (Oxford: Oxford University Press, 1998), p. 16.

[32] H. D. Dalkin, *J. Biol. Chem.*, **1** (1906), 171.

[33] P. Schreiner, *Ann.*, **194** (1878), 68.

[34] M. C. Rosenheim, *J. Physiolog. Soc.*, **51** (1917), vi.

[35] H. W. Dudley, O. Rosenheim and W. W. Starling, *J. Biochem.*, **21** (1927), 97.

[36] F. Wrede, *Hoppe-Seyler's Zeitschr. für Physiol. Chemie*, **161** (1926), 66.

3

Aryl amines

3.1 An introduction to aryl amines

Although there are two types of amine that contain aryl groups,* this chapter is concerned only with those amines where the aryl group is directly bonded to the amino group, e.g. aniline; these are the nuclear amino compounds. Other amines where the amino group is part of an alkyl side chain and remote from the aryl centre behave like aliphatic amines and were discussed in Chapter 2. Nuclear amino compounds have distinct properties and chemistry, as the electron-withdrawing aryl groups attached to the nitrogen atom have the effect of decreasing the basicity of the amine group. For example, attaching a phenyl group to the ammonia molecule dramatically decreases its basic properties, as indicated by the pK_a values below.

	NH_3	CH_3NH_2	$C_6H_5-CH_2-NH_2$	$C_6H_5-NH_2$
pK_a	9.27	10.62	9.34	4.63

Aniline ($C_6H_5-NH_2$, phenylamine) is a weaker base than methylamine (CH_3-NH_2) by a factor of about a million. This effect cannot be explained by the inductive effect of the phenyl group relative to the methyl group and is best accounted for in terms of overlap of the n-(nitrogen) orbitals with π-(phenyl) orbitals. High-resolution electronic spectrocopic studies of the aniline molecule have shown that it is not planar as might have been expected. Instead, the H$-$N$-$H bond makes an angle of 37° with the axis of the C–N bond (i.e. the amine nitrogen is at the apex of a triangular-based pyramid) [1]. The C–N bond is also much shorter than might be expected at 0.1402 nm compared with 0.1474 nm in methylamine. The result of electronic donation via π-N→Ph lowers the basicity of the amine group but increases the reactivity of the benzene nucleus to electrophiles. A MINDO/3 calculation of the π-electron density in the aniline molecule based on a pyramidal nitrogen gave the following results [2].

---

* In this book, the term 'aryl' is used to describe carbocyclic compounds or groups with aromatic character. Rings or groups with heteroatoms may also be aromatic but are not strictly aryl.

Table 3.1. *The effect of ring and nitrogen substitution on the*
pK$_a$ of aniline (unsubstituted aniline has pK$_a$ = 4.63)

	Ortho	Meta	Para
pK$_a$ after ring substitution			
Methyl	4.39	4.69	5.12
Chloro	2.64	3.34	3.98
Methoxy	4.49	4.20	5.29
Amino	4.47	4.88	6.08
pK$_a$ after nitrogen substitution			
N-Methyl	4.85		
N,N-Dimethyl	5.06		
N-Ethyl	5.11		
N,N-Diethyl	6.56		
N-Phenyl	0.9		
N,N-Diphenyl	−5.0		

aniline because the ethyl group is larger than the methyl group and inhibits resonance. The effect of steric resonance can also be used to explain the difference in basicity between 3,5-dimethyl-4-nitroaniline (pK$_a$ = 2.49) and 2,6-dimethyl-4-nitroaniline (pK$_a$ = 0.95). The 3,5-isomer is a stronger base than the 2,6-isomer as in the 2,6-isomer steric effects inhibit planarity of the amino group with the aromatic ring. The effect of ring substituents on the pK$_a$ of aniline is shown in Table 3.1.

The extent of (n) to (π) bonding is greatly reduced in other Group VA elements bonded to aryls. Phosphorus and arsenic do not form strong pπ–pπ bonds as their p orbitals are more spatially diffuse, but they do have empty d orbitals available for back donation and are able to form ylides. Triphenylphosphine is spontaneously oxidized in air and forms an oxide (triphenylphosphine oxide) with a double P=O bond that has partial triple-bond character. In contrast, triphenylamine does not form a stable oxide of the type $(C_6H_5)_3NO$ although triphenylamine oxide has been identified as a transient intermediate in photovoltaic systems. A stable amine oxide can be formed from 4,4′,4″-trimethylphenylamine, but this cannot be produced by direct oxidation and requires the formation of a chloro intermediate.

3.2 Synthesis of (mono) aryl amines

Aniline was first prepared as a pure chemical in 1826 by Unverdorben, who distilled dry indigo dye and obtained a substance he called krystallin [5]. (Helot may have also prepared crude aniline in 1765 [6] but he did not isolate it as a pure chemical.) Runge also identified the same substance in coal tar but was unable to determine its structure. 'Anil' is the Portuguese word for indigo, and the usage of aniline dates from 1841 when Fritzsche [6] repeated Unverdorben's experiment with added potash. The structure of aniline was discovered in 1843 by Hofmann, who managed to reduce nitrobenzene with hydrogen for the first time [7]. Hofmann identified the most important synthetic method for the production of aryl amines by nitration of aryl systems followed by catalytic reduction. This is still the basis for the commercial manufacture of aniline, which is synthesized from benzene via the catalytic reduction of nitrobenzene. However, aryl amines can also be prepared by the ammonolysis of phenols or aromatic halides, and by direct amination of the aromatic ring. All of these synthetic methods will be reviewed in this section.

3.2.1 Nitration and reduction methods of aryl amine synthesis

Although this is the synthetic method route by which most of the world's aniline is produced, this process is not without a few inherent problems that are mainly associated with the nitration step. Nitration is at best a rather blunt tool and if reactions are not controlled they can lead to di-, tri- and poly-substituted products. The reaction is also strongly exothermic and needs a high level of temperature control to avoid runaway exotherms and associated explosions. Nitration leads to the production of waste nitrogen oxide gases, a mixture of NO, N_2O and NO_2, which is commonly called NO_X.

Unfortunately NO_X is detectable in air at around 50 p.p.m., so presents a visible pollution stream from the exhaust outlet of reactors (clean air regulations in many countries specify that the effluent gas streams ejected into the atmosphere from chemical plants must be clear and colourless).

The reaction between benzene and nitric acid is best carried out at low temperatures to avoid di-substitution (benzene freezes at 5.5 °C) and usually requires an initiation catalyst unless fuming concentrated nitric acid is used. The most commonly used nitration system is nitric acid and sulphuric acid, which is the classic mixed-acid system. The overall reaction between the two acids produces the nitronium cation $[NO_2]^+$, which is the active nitrating species.

$$HNO_3 + 2H_2SO_4 \rightleftharpoons [NO_2]^+ + [H_3O]^+ + 2[HSO_4]^-$$

The nitronium ions can actually be isolated as the perchlorate or tetraborate salts, e.g. NO_2ClO_4 or NO_2BF_4, and these salts have a Raman active absorption band at 1400 cm^{-1} that is attributed to the presence of $[NO_2]^+$.

disposal of waste water. However, this method of amine production is still utilized in the dyestuff industry in India and the Far East for small-volume production.

(b) Catalytic reduction

Catalytic reduction is considered to be the most effective method for the large-scale reduction of nitro-aryls to aromatic amines such as aniline. This process is also used for the large-scale manufacture of toluidines, xylidines, phenylenediamines and toluenediamines. The reaction can be carried out in the vapour phase or in solution, and works best with pure hydrogen.

The required hydrogen can be produced from natural gas, water gas or even from the electrolysis of water. The reduction reaction is strongly exothermic with slightly greater production of heat in the liquid-phase reaction. During reduction it is necessary to remove this excess heat from the reactor to avoid thermal detonation of the nitro compounds present at high temperatures. The reaction temperature can also be controlled by starving the reaction of hydrogen, by reducing the amount of catalyst added or by slowing the addition of the nitro compound to be reduced.

$$C_6H_5NO_2 + 3H_2 \rightarrow C_6H_5-NH_2 + 2H_2O \qquad \Delta H_{liquid} = 544 \text{ kJ mol}^{-1}$$
$$\Delta H_{vapour} = 493 \text{ kJ mol}^{-1}$$

Aniline is typically manufactured by a vapour-phase reaction between nitrobenzene and hydrogen at 270–300 °C and 20 psi with short reactor-residence times. This reaction works well with a variety of mainly nickel-based catalysts and also with nickel sulphide on alumina or silica gel, the mixed-oxide system nickel–vanadium pentoxide, and mixtures of copper, cobalt and nickel.

Copper on silica catalysts are generally over 99% effective in converting nitrobenzene to aniline in the vapour phase. Aniline and other aryl amines can also be manufactured by using liquid-phase reactions with Raney nickel or with supported Group VIII metal catalysts such as palladium on alumina.

Lower temperatures of around 100 °C are used with a hydrogen pressure of 150–200 atm in these liquid-phase reactions. Methanol is often used as a solvent and the reaction phase is best described as a slurry rather than as a pure liquid. Hydrogenation of the aromatic ring occurs at these pressures at temperatures in excess of 170–200 °C so lower temperatures are preferred if possible.

The use of lower temperatures also prevents the cleavage or migration of any other functional groups present on the aromatic nucleus. Generally, longer reaction times of several minutes are required by the liquid-phase reaction compared with the vapour-phase reaction, which only takes a few seconds.

Temperature control of the reaction is achieved by the use of cooling coils and in some instances deionized water is added to the reaction and then removed by continuous evaporation. The presence of water seems to activate certain catalysts and helps to produce fewer reaction by-products. Catalytic hydrogenation does not offer the selectivity of chemical reduction with metals with di- and poly-nitrated aromatics, and all nitro groups are reduced during the vapour-phase or liquid-phase reactions. Despite this lack of selectivity, catalytic hydrogenation is the most cost-effective method of producing aryl amines from nitro precursors.

(c) Electrochemical reduction

The electrochemical reduction of nitro-aryls is really a special case of chemical reduction performed in an electrochemical cell. A typical cell is divided by a semipermeable membrane with 30% sulphuric acid on the anode side and 15% to 20% hydrochloric acid on the cathode side. The cathode may be made of lead, tin, nickel or copper. An inorganic compound is added as a reducing agent, and after reaction with the nitro-aryl it can be regenerated at the cathode.

$$Ar-NO_2 + 6H^+ + 6e^- \rightarrow Ar-NH_2 + 2H_2O$$

A typical reaction that can be peformed electrochemically is the reduction of *para*-nitrobenzoic acid to *para*-aminophenol. Although electrochemical reductions have been known for well over 100 years they have yet to become commercially significant. Reasons for this include the stability of the cell membrane, the high power consumption and difficulties in isolating the desired reaction products from the starting material and electrolytes.

(d) Reduction with hydrazine

The reduction of aryl nitro compounds can be performed with hydrazine, which may be used either with or without a catalyst. Hydrazine is used in this reaction as a source of hydrogen and ideally should produce up to two moles of hydrogen (H_2) per mole of hydrazine, but in practice far less is usually obtained. Under alkaline or neutral conditions one mole of hydrogen is theoretically produced and under acidic conditions two moles of hydrogen are possible, but in practice a stoichiometric excess of hydrazine is always required. In the absence of a catalyst hydrazine will reduce nitroaryls to arylamines, and any carbonyl groups present will also be reduced to hydroxy groups. However, in the presence of a catalyst such as Raney nickel or palladium the reaction is more selective and carbonyl groups are stable. The reaction of nitro groups with hydrazine proceeds via three intermediates, all of

3-nitrobenzenesulphonate or perchloric acid. However, this method has been replaced by the more successful nitration and reduction method, which produces less effluent.

3.2.3 Aromatic amines by rearrangement, degradation and other reactions

The Beckmann Rearrangement, the Hofmann, Lossen and Schmidt Degradations, and the Curtius Reaction (Rearrangement) can be used to prepare aromatic amines. Other methods of amine preparation include the Mitsunobu Reaction, the Buchwald Method, direct amination of aryl halides and use of the Lwowski Reagent.

The Hofmann Degradation [11] involves the reaction of an unsubstituted amide with sodium hydroxide and bromine to form an amine via an isocyanate intermediate. This is essentially the same method as that for alkyl amines described in Chapter 2.

$$Ar-CO-NH_2 + NaOBr \rightarrow Ar-N=C=O\ (+H_2O) \rightarrow Ar-NH_2 + CO_2$$

The Curtius Reaction [12] also produces an isocyanate intermediate from the decomposition of an azide $RCON_3$. If the reaction is carried out by pyrolysis in anhydrous conditions the isocyanate can be isolated. If the reaction is carried out in water or ethanol then the isocyanate is hydrolysed and an amine is produced. Acyl azides can be produced from the reaction of acyl hydrazides with nitrous acid. The Curtius Reaction is catalysed by Lewis and protic acids but also works well in the absence of a catalyst and can be used for the synthesis of akyl and aryl amines.

$$Ar-CO-N_3 \rightarrow Ar-N=C=O\ (+H_2O) \rightarrow Ar-NH_2 + CO_2$$

If the Curtius Reaction is performed with phenyl azide ($C_6H_5-N_3$) this gives in the first instance phenyl amine, which on subsequent reaction with any excess phenyl azide forms 2-(phenylamino)-7-hydro-azepine. Ring-expansion reactions are common when aromatic azides decompose.

The Curtius Reaction and Hofmann Degradation are interesting in that they involve the migration of an akyl or aryl group from a carbon to a negatively charged nitrogen. The mechanisms of both the Hofmann Degradation and the Curtius

Reaction rearrangements involve a singly bonded nitrogen (a nitrene) with a single negative charge, but the end result is the same. It is likely that the reaction steps are concerted and that the nitrene is not actually formed. A three-membered intermediate is formed instead, supporting the observation that the reaction is intramolecular rather than intermolecular.

A related reaction is the Lossen Reaction [13], where O-acyl derivatives of hydroxamic acids (which can be formed from acyl halides) give isocyanates on heating; these can be hydrolysed to amines. This reaction from acyl halides to amines can be performed in a single step with the reagent hydroxylamine-O-sulphonic acid

$$Ar-CO-Cl + NH_2OSO_2OH \rightarrow Ar-CO-NH-OSO_2OH$$
$$\rightarrow Ar-NH_2 + CO_2 + H_2SO_4$$

The Schmidt Reaction [14] uses hydrazoic acid (N_3H) to generate amines from carboxylic acids, ketones, alcohols and alkenes. This reaction for aryl carboxylic acids gives variable yields and is best for sterically hindered aryls such as the mesityl group. Sulphuric acid is often used as a catalyst but Lewis Acids work well too.

$$Ar-CO_2H + N_3H \rightarrow Ar-N=C=O (+H_2O) \rightarrow Ar-NH_2 + CO_2$$

With ketones, Ar—CO—R, the reaction product is a secondary amide, Ar—CO—NHR, which has to be reduced to give an amine of the form Ar—CH$_2$—NHR by using ethanolic sodium or lithium aluminium hydride in tetrahydrofuran.

Diazonium salts are generally only metastable but can be isolated as white or colourless solids that decompose on standing. The diazonium nitrate salts are the exception, as they decompose explosively when in the solid state. Fortunately for the synthetic chemist, diazonium salts can be used in solution to prepare derivatives such as cyanides (which may then be reacted to form carbonyls), halides, hydroxyls, aryls, alkyls, hydrazines, cyanates, thiocyanates, ethers, thioethers or even be regenerated as the free amines.

The mechanism of diazotization involves the formation of a nitroso amine and works best in weakly or moderately acidic conditions with sufficient water present to catalyse formation of the NO^+ ion. The nitrous acid required for diazotization is formed *in situ* from the reaction shown below between sodium nitrite and hydrochloric acid.

$$NaNO_2 + HCl \rightarrow HNO_2 + NaCl$$
$$HNO_2 + [H]^+ \rightleftharpoons [H_2O\text{--}NO]^+ \rightleftharpoons H_2O + [NO]^+$$
$$Ar\text{--}NH_2 + [NO]^+ \rightarrow [Ar\text{--}NH_2]^+ [NO] \rightarrow Ar\text{--}NH\text{--}NO + [H]^+$$
$$\rightarrow Ar\text{--}N{=}N\text{--}OH \rightarrow [Ar\text{--}N_2]^+ + [HO]^-$$

The reaction is usually done at $0\,°C$ (in an ice bath) and starch–iodide paper can be used to detect the presence of free nitrous acid.

Secondary aryl amines react with nitrous acid to form *N*-nitroso compounds, and tertiary aryl amines may undergo *para*-nitrosoization.

$$C_6H_5\text{--}NH(CH_3)_2 + HNO_2/H_2O \rightarrow ON\text{--}C_6H_4\text{--}N(CH_3)_2$$

In solution, diazonium salts are actually derived from diazonium hydroxide $[Ar\text{--}N_2]^+[OH]^-$, which is too unstable to be isolated. In the presence of excess hydroxide ions the diazohydroxide ion is formed. This can then form salts with electropositive metals such as sodium and potassium.

$$[Ar\text{--}N_2]^+[OH]^- + xs\ NaOH\ [\rightarrow Ar\text{--}N{=}N\text{--}OH] \rightarrow [Ar\text{--}N{=}N\text{--}O]^-Na^+$$

When first formed in solution the metastable *syn* (*cis*)-form is produced; this changes over to the more stable *anti*-form. The *syn*-form with adjacent nitrogen lone pairs is called the diazoate ion, and the *anti*-form with nitrogen lone pairs diagonally opposite is called the *iso*-diazote ion.

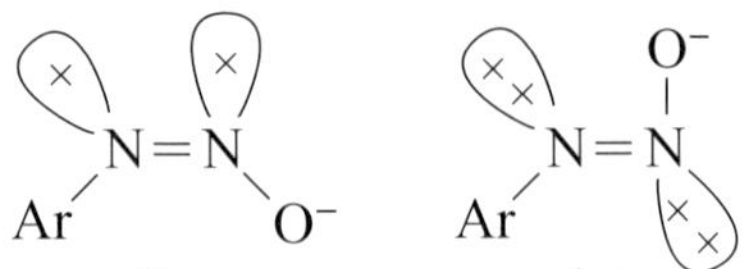

Syn   *Anti*

3.3.2 Formation of biphenyls

Aryl amines such as aniline are able to couple in the presence of powdered copper metal and sodium nitrite or an alkyl nitrite such as amyl nitrite plus an acid such as sulphuric acid to form biphenyls. This is a variation of the Gattermann Reaction [22], which proceeds via the formation of a diazonium salt.

$$C_6H_5-NH_2 + NaNO_2 + H_2SO_4 \rightarrow [C_6H_5-N_2]^+[HSO_4]^- + NaOH + H_2O$$
$$2[C_6H_5-N_2]^+[HSO_4]^- + Cu \rightarrow C_6H_5-C_6H_5 + 2N_2 + H_2SO_4 + CuSO_4$$

If the diazonium salt is added to benzene in alkaline solution, a variation of the Gattermann Reaction called the Gomberg–Bachmann Reaction occurs [23]. This reaction is thought to proceed via a free-radical mechanism to produce biphenyls.

$$[C_6H_5-N_2]^+[HSO_4]^- + [OH]^- \rightarrow C_6H_5-N=N-OH + [HSO_4]^-$$
$$C_6H_5-N=N-OH \rightarrow C_6H_5\bullet + N_2 + \bullet OH$$
$$C_6H_5\bullet + C_6H_6 + \bullet OH \rightarrow C_6H_5-C_6H_5 + H_2O$$

A variation on the Gomberg–Bachmann Reaction is the Pschorr Reaction [24], where α-aryl-*o*-aminocinnamic acids can be formed by intramolecular arylation.

It is possible to produce diazonium salts from primary aryl amines with ring substituents such as halides or nitro groups, but carboxylic and sulphonic substituents may lead to the formation of dipolar ions, e.g. $[R-CO_2]^-[C_6H_4-N_2]^+$.

The Gomberg–Bachmann Reaction between 2,4-difluoroaniline and benzene can be used to produce 2,4-difluorobiphenyl, which is required for the synthesis of the analgesic pharmaceutical Dolobid.

3.3.3 The Sandmeyer and Gattermann Reactions and halide formation

Replacement of a diazonium group with a halogen by using cuprous halides is referred to as the Sandmeyer Reaction [25]. As aryl iodides and fluorides cannot be prepared by direct halogenation, and aryl chlorides and bromides prepared by direct halogenation often contain mixed isomers, the Sandmeyer Reaction offers a useful synthetic pathway to mono-halogenated aryl compounds.

Replacement of the diazo group with chlorine and bromine can be performed by mixing the corresponding diazonium halide salt with cuprous chloride or cuprous bromide at ambient or slightly elevated temperatures. The reaction may take several hours to reach completion with the steady evolution of nitrogen (once the stream of nitrogen stops the reaction is completed).

$$[Ar-N_2]^+X^- \xrightarrow[(CuX)]{} Ar-X + N_2$$

In this scheme, X^- is the negatively charged counter-ion from the diazotization reaction. The cuprous chloride or bromide is oxidized to $[CuCl_2]^-$ or $[CuBr_2]^-$ *in situ*, and is able to halogenate the aryl group after the loss of nitrogen.

To replace a diazonium group with iodine it is not necessary to use a catalyst and the reaction proceeds on mixing potassium iodide with the diazonium salt. The yield is typically around 75% depending upon the reaction conditions.

$$[Ar-N_2]^+X^- + KI \rightarrow Ar-I + N_2 + KX$$

However, to replace a diazonium group with fluorine a slightly different synthetic procedure is required. Firstly, fluoroboric acid must be added to the diazonium salt to precipitate the insoluble diazonium fluoroborate salt. This salt can then be filtered off and vacuum dried. Gentle heating of the fluoroborate salts yields the aryl fluoride. Fluoroborate diazo compounds are stable and can be safely isolated and vacuum dried without the risk of explosion. This procedure is called the Schiemann Reaction [26] and dates from 1927. The experimental yields are typically only 50% to 60%.

$$[Ar-N_2]^+[Cl]^- + HBF_4 \rightarrow [Ar-N_2]^+[BF_4]^- + HCl \rightarrow Ar-F + N_2 + BF_3$$

The thermal decomposition of the fluoroborate salt produces a phenyl cation $[C_5H_5]^+$ and a tetrafluoroborate ion $[BF_4]^-$, which eliminates a fluoride anion to yield boron trifluoride. Better reaction yields can be obtained with hexafluorophosphoric acid; this produces a sparingly soluble hexafluorophosphate diazonium salt, which is easier to isolate than the fluoroborate salt.

$$[Ar-N_2]^+[Cl]^- + HPF_6 \rightarrow [Ar-N_2]^+[PF_6]^- + HCl \rightarrow Ar-F + N_2 + PF_5$$

The replacement of a diazo group with bromine can be carried out by warming a solution of a diazonium bromide salt with copper powder. This is the Gattermann Reaction [22], and it has experimental yields of up to around 50%. The Gatterman Reaction works best when there is an alkyl group *ortho* to, or a nitro group *meta* to, the diazo group on the aromatic ring (i.e. *ortho*-toluidine can be converted to 2-bromotoluene by this route).

3.3.4 Fomation of nitriles (and carboxylic acids)

Aromatic diazo compounds react with cuprous cyanide or copper powder in the presence of aqueous potassium cyanide to form nitriles. This is a really a variation of the Sandmeyer and Gattermann Reactions but it works reasonably well, with a 40% to 50% yield.

$$[C_6H_5-N_2]^+[Cl]^- + (CuCN + 3KCN) \rightarrow C_6H_5-CN + N_2$$

The active species for the formation of the nitrile is $K_3Cu(CN)_4$, which is why the reaction works best in an excess of aqueous potassium cyanide.

Once they have been formed, aryl nitriles can easily be converted into carboxylic acids by hydrolysis with water and sulphuric acid.

$$C_6H_5-CN + 2H_2O \rightarrow C_6H_5-CO_2H + NH_3$$

3.3.5 Replacement of a diazo group with hydrogen

At first sight it might seem rather strange to replace a reactive functional group such as an amine or diazo group with a hydrogen atom to yield a C–H bond that offers no selectivity over other C–H bonds to any attacking reagents, but this method of synthesis does offer a way of producing aromatics with unusual substitution patterns. For example 2,6-dibromoaniline can be converted in a single reaction step to 3-chlorobromobenzene.

The replacement of diazo groups by hydrogen is effected with either hypophosphorous acid or cuprous salts. The reaction mechanism proceeds via a radical intermediate.

$$[Ar-N_2]^+[Cl]^- + H_2PO_3H \rightarrow Ar-N=N-PO_2H_2 + HCl$$
$$\downarrow$$
$$Ar\bullet + N_2 + \bullet PO_2H_2$$
$$\downarrow$$
$$Ar-H + PO_2H$$
$$\downarrow \text{ + } H_2O$$
$$H_3PO_3$$

3.3.6 The Bart Reaction [27]

Aromatic diazonium salts will react with sodium arsenite in the presence of a copper catalyst and a sodium carbonate buffer to produce arsonic acid derivatives such as phenyl arsonic acid. However, the yield is often less than 50%.

$$[C_6H_5N_2]^+[Cl]^- + Na_3AsO_3 + CuSO_4 \rightarrow C_6H_5AsO_3Na_2 + N_2 + NaCl$$
$$\downarrow \text{+ HCl}$$
$$C_6H_5AsO_3H_2$$

3.3.7 The Meerwein Reaction [28]

Activated ethylenic bonds can react with aromatic diazonium salts in the presence of cupric salts such as $CuCl_2$ to yield arylated olefins.

$$Y-CR=CHR' + [Ar-N_2]^+[Cl]^- + CuCl_2 \rightarrow Y-CR=C(R')Ar + N_2$$

Here R and R′ are alkyl groups or H, and may be different or identical, and Y may be C=C, C=O, Ar, CN or H.

A problem with the Meerwein Reaction is that the double bond may be chlorinated by the catalyst or it may react with other aryl groups and so be converted to a carbon–carbon single bond in the reaction.

3.3.8 Other diazonium replacement reactions

The diazonium group can also be replaced by other groups by reaction with a suitable precursor. Some other reactions that have not been covered in the previous sections are as follows.

(i) Replacement with a nitro group by forming a tetrafluoroborate salt that is then decomposed in the presence of copper and sodium nitrite.

$$[Ar-N_2]^+[BF_4]^- + NaNO_2 \xrightarrow[(Cu)]{} Ar-NO_2 + NaBF_4 + N_2$$

(ii) Replacement by a thiocyanate or cyanate group by reaction with copper thiocyanide or potassium cyanate respectively.

(iii) Formation of diaryl thioethers by reaction with ammonium sulphide

$$2[Ar-N_2]^+[Cl]^- + (NH_4)_2S \rightarrow Ar_2S + N_2 + 2NH_4Cl$$

(iv) Reaction with oximes: formaldoxime gives an aldehyde but other oximes give alkylaryl ketones.

$$[Ar-N_2]^+[Cl]^- + R-CH=N-OH \xrightarrow[(NaOH)]{} Ar-C(R)=NOH \xrightarrow[(H^+)]{} Ar-C(O)-R$$

3.3.9 Reactions of diazonium salts in which nitrogen is retained

(a) Formation of azides

When a mixture of elemental bromine and hydrobromic acid is added to a solution of an aromatic diazonium salt, an insoluble precipitate of diazonium perbromide is produced. If this does not immediately dissociate to form a mixture of brominated aromatics plus nitrogen it may be treated with aqueous ammonia (or HN_3 in acetic acid) to produce an azide salt.

$$[C_6H_5-N_2]^+[HSO_4]^- + xs\ Br/HBr \rightarrow C_6H_5-N_2.Br_3$$
$$\downarrow + NH_{3(aq)}$$
$$C_6H_5-N_3$$

(b) Formation of hydrazines

Aromatic diazonium salts can be reduced with sodium sulphite or with tin(II) chloride/hydrochloric acid to form hydrazine derivatives. However, the reaction must be performed carefully as if stronger reducing agents such as zinc/hydrochloric acid are used, or if the reaction is left for too long, the substituted hydrazine will decompose to form an amine.

$$[Ar-N_2]^+[Cl]^- + (SnCl_2/HCl) \rightarrow Ar-NH-NH_2$$
$$\downarrow \text{Zn/HCl}$$
$$Ar-NH_2 + NH_3$$

(c) Coupling reactions

Diazonium salts can react with other aromatic compounds to form intensely coloured products called azo dyes; these are widely used in textile manufacture. Azo dyes can be yellow, orange, blue, red or green. Some azo dyes change colour in acidic and basic conditions and can be used as indicators; one example of this is Methyl Orange.

Methyl Orange

Azo compounds are formed from the reaction of the diazo group $[Ar-N_2]^+$ with another aromatic molecule Ar–H. However, the diazo group is only a very weak electrophile and so the coupling reaction works best when the aromatic molecule under attack contains a powerful electron-releasing group such as a hydroxy, amino or substituted amino group. Attack usually take place *para* to the electron-releasing group on the aromatic ring with the diazonium ion as the attacking reagent. In practice a high level of pH control is required to obtain a satisfactory yield of the desired azo compound. For example, if conditions are not favourable then the diazo ion can react with water to form a phenol, which can then further react with any excess diazonium ions present.

The most favourable conditions for diazo coupling are in mildly alkaline solution for coupling with phenols, and mildly acidic for coupling with amines. Control of the solution pH is achieved by using sodium acetate or sodium bicarbonate.

If the reaction solution is too acidic the diazo ion will become hydrolysed and there will be less of the diazonium ion present in the equilibrium mixture to couple effectively with the other reagent present.

$$[Ar-N_2]^+[OH]^- + NaOH \rightleftharpoons Ar-N=N-OH \, (+ \, xs \, NaOH)$$
$$\rightleftharpoons [Ar-N=N-O]^-Na^+$$

Therefore for the diazo reagent the best condition for coupling is a low concentration of $[OH]^-$ ions. However, if the solution is too acidic the concentrations of the amine and hydroxy ions will be too low and again the reaction is compromised (ammonium ions present on an aromatic ring prevent coupling and the hydroxy groups present on aromatic rings slow down the rate of coupling compared with the phenoxide ion in alkaline solution which activates coupling). The best conditions for coupling reactions for various reagents are shown below.

	Acidic conditions	Basic conditions
Aromatic amine	$[Ar-NH_3]^+$	$Ar-NH_2$ ✓
Phenol	$Ar-OH$	$[Ar-O]^-$ ✓
Diazo	$[Ar-N_2]^+$ ✓	$Ar-N=N-OH$ or $[Ar-N=N-O]^-$

✓ = best for coupling reactions.

Finally, diazonium salts can also couple with compounds having active methylene groups (e.g. diethyl malonate). The reaction first produces an azo intermediate, which then rearranges to the more-stable hydrazone.

$$[Ar-N_2]^+[OH]^- + CH_2(CO_2C_2H_5)_2 \rightarrow Ar-N=N-CH(CO_2C_2H_5)_2 + H_2O$$
$$\updownarrow$$
$$Ar-NH-N=C(CO_2C_2H_5)$$

This coupling reaction is called the Japp–Klingemann Reaction [29] and is sometimes also accompanied by cleavage of a carboxyl group. The reaction works best in basic conditions.

$$CH_3CO-CH(CH_3)-CO_2R + [ArN_2]^+[OH]^- \rightarrow \begin{array}{c} CH_3 \\ \diagdown \\ C=N-NHAr + CH_3CO_2H \\ \diagup \\ CO_2R \end{array}$$

The Japp–Klingemann Reaction can be used for the detection of enols [30]. The applications of azo and azine dyes are reviewed in detail in Chapter 8.

3.3.10 N-acetylation of aryl amines

Primary and secondary aryl amines are able to react with carboxylic acids, acid anhydrides, acid chlorides and esters to form amides. Generally acid chlorides give the best yields, but all combinations give high yields.

The presence of electron donating groups *ortho* and (preferentially) *para* to the amino group promotes the acetylation reaction. The reaction with carboxylic acids also can be catalysed by the addition of phosphorous trichloride or phosphorous oxychloride. The reaction between aniline and acetyl chloride or acetic anhydride produces acetanilide, which is an important intermediate in the production of phar-maceuticals.

$$Ar-NH_2 + CH_3COCl \rightarrow Ar-NH-COCH_3 + HCl$$

$$2Ar-NH_2 + (CH_3CO)_2O \rightarrow 2Ar-NH-COCH_3 + H_2O$$

If the reaction is carried out with an aryl amine and dimethylformamide in sodium methoxide the reaction product is an *N*-arylformamide in a yield of about 50%.

$$Ar-NH_2 + HCON(CH_3)_2 \rightarrow Ar-NH-CHO + HN(CH_3)_2$$

N-methylformilanilide can be prepared in over 90% yield by refluxing *N*-methylaniline and formic acid and toluene.

$$C_6H_5-NH(CH_3) + HCO_2H \rightarrow C_6H_5-N(CH_3)CHO + H_2O$$

It may also be produced in slightly lower yield by refluxing *N,N*-dimethylaniline with hydrated magnesium oxide.

3.3.11 Reaction with phosgene and thiophosgene

Depending upon the reaction conditions, the product from the reaction between a primary aryl amine and phosgene can be either a urea derivative or an isocyanate (or a mixture of both). Generally toluene is used as a reaction solvent for phosgenation.

$$Ar-NH_2 + xs \, COCl_2 \rightarrow Ar-N{=}C{=}O + 2HCl$$

$$xs \, Ar-NH_2 + COCl_2 \rightarrow (ArNH)_2CO + 2HCl$$

Secondary aryl amines form ureas or carbamoyl chlorides because only one position is available on each nitrogen atom for substitution. Actually primary amines form carbamoyl chlorides in the first instance as well, but generally these are unstable and eliminate a molecule of HCl to form isocyanates. Because of the elimination of HCl the reaction works best with the addition of a proton acceptor such as diisopropylethylamine or 2,6-dimethyl pyridine, which can act as a proton acceptor and remove the excess acid from the reaction medium.

$$Ar-NH_2 + Cl-CO-Cl \rightarrow [Ar-NH_2-CO-Cl]^+ \, [Cl]^- \rightarrow Ar-NH-COCl + HCl$$

$$\text{(i)} \;\; Ar-NH-CO-Cl \rightarrow Ar-N{=}C{=}O + HCl \quad \text{or}$$

$$\text{(ii)} \; Ar-NH-CO-Cl \rightarrow ArNH-CO-NHAr + HCl$$

With a secondary aryl amine the *N*-carbamoyl chloride is generally stable and can be isolated. With tertiary aryl amines it is impossible to eliminate HCl as there are no hydrogens available on the nitrogen so instead either an alkyl chloride R–Cl is eliminated or, if this is not possible, a dichloride salt in produced.

$$ArNR_2 + COCl_2 \rightarrow [Ar-NR_2-CO-Cl]^+[Cl]^- \rightarrow Ar-NR-CO-Cl + R-Cl$$
$$\downarrow$$
$$[Ar-NR_2-CO-NR_2-Ar]^{2+} \; 2[Cl]^-$$

With thiophosgene the reaction to produce thiocyanates proceeds similarly and can be performed in toluene. However, if N,N'-diarylthioureas are required it is best to produce these by the following route.

In ethanol or acetone solution aryl amines react with carbon disulphide and potassium hydroxide to form diarylthiourea. With aniline this reaction forms the rubber accelerator N,N-diphenylthiourea (thiocarbanilide) in 70% yield.

$$C_6H_5-NH_2 + CS_2 + 2KOH \rightarrow 2(C_6H_5-NH)_2C=S + K_2S + 2H_2O$$

Diphenylurea is converted into phenyl thiocyanate when treated with hydrochloric acid.

If the reaction between carbon disulphide and aniline is performed in either water or toluene in the presence of ammonia it is possible to produce a dithiocarbamic acid ammonium salt, which can be isolated.

$$C_6H_5-NH_2 + CS_2 + NH_3 \rightarrow [C_6H_5-NH-C(S)-S]^-[NH_4]^+$$

This salt can be treated either with excess phosgene in toluene or with lead nitrate in water to produce phenyl isothiocyanate.

$$[C_6H_5-NH-C(S)-S]^-[NH_4]^+ + H_2O + Pb(NO_3)_2$$
$$\rightarrow C_6H_5-N=C=S + HNO_3 + PbS \quad (75\%)$$

3.3.12 N-alkylation of aryl amines

There are several experimental methods available for the N-alkylation of aryl amines that can offer some selectivity on mono-alkylation versus dialkylation for primary aryl amines.

For example, during the methylation of aniline there are two possible reaction products, N-methylaniline and N,N-dimethylaniline. In the presence of a copper/zinc catalyst, methanol reacts with aniline at 250 °C and 101 kPa to produce N-methylamine in 95%+ yield. However, in the absence of a catalyst methanol and aniline react to form dimethylaniline if there is sufficient methanol present.

In the presence of sulphuric acid at 200 °C N,N-dimethylaniline is the exclusive reaction product between methanol and aniline only if there is a 50% stoichiometric excess of methanol. The reaction kinetics are dependent upon the concentration of aniline in the first step and the concentration of *N*-methylaniline in the second step.

(i) $CH_3-OH + C_6H_5-NH_2 \rightarrow C_6H_5-NH-CH_3 + H_2O$

(ii) $C_6H_5-NH-CH_3 + CH_3-OH \rightarrow C_6H_5-N(CH_3)_2 + H_2O$

At temperatures in excess of 230 °C ring methylation is observed in the above system. *N*-alkylation can also be successfully carried out with alkyl halides, alkyl sulphates and even alkenes. Again the reaction conditions and reaction stoichiometry are critical in defining the product mixture. Lower temperatures and a stoichiometric excess of the amine generally produce a higher yield of the mono-*N*-alkylated product and higher reaction temperatures and a non-stoichiometric excess of the alkylating agent produce an *N*,*N*-dialkylated product. For example, aniline will react with ethylene and a catalyst such as sodium metal below 10 MPa to produce primarily the monosubstituted product, but above 10 MPa the disubstituted product predominates.

Mixtures of mono- and dialkylated products obtained from primary aromatic amines and alkylating agents can be separated by fractional distillation. The mono-alkylated derivative can be easily converted into a non-volatile derivative such as the tosylate from which the dialkylated product can be easily recovered.

Another method of preparation of *N*,*N*-dimethylaniline is from aniline in the presence of formaldehyde and hydrogen over a platinum/carbon catalyst. The reaction gives an overall yield of around 75%.

$$C_6H_5-NH_2 + 2CH_2O + 2H_2 \rightarrow C_6H_5-N(CH_3)_2 + 2H_2O$$

N-methylaniline can also be made from acetanilide by its reaction with sodium and methyl iodide in toluene solution.

(i) $C_6H_5NH-COCH_3 + Na \rightarrow [C_6H_5-NCOCH_3]^-[Na]^+$

(ii) $[C_6H_5-NCOCH_3]^-[Na]^+ + CH_3I \rightarrow C_6H_5-N(CH_3)CO-CH_3 + NaI$

(iii) $C_6H_5-N(CH_3)CO-CH_3 \xrightarrow[(KOH)]{} C_6H_5-NH-CH_3$

3.3.13 The formation of N-nitroso compounds and hydrazines

Secondary aryl amines with an *N*-alkyl group will react with nitrous acid to form *N*-nitrosamines in high yields

$$C_6H_5-NH-CH_3 + HNO_2 \rightarrow C_6H_5-N(CH_3)-N=O + H_2O$$

The presence of *N*-nitrosamines may be detected by the Liebermann Nitroso Reaction [31]. In this test a small amount of phenol and sulphuric acid are added to a test solution that turns green and on the subsequent addition of sodium hydroxide turns deep blue.

N-nitroso compounds may be converted back to amines by treatment with hydrochloric acid and can be reduced to form unsymmetrical hydrazines. The reaction to prepare hydrazines is best carried out in acetic acid in the presence of zinc; however, yields are only around 50%.

$$C_6H_5-N(CH_3)-N=O \rightarrow \quad \begin{array}{c} C_6H_5 \\ \diagdown \\ \quad N-N \diagup \\ CH_3 \diagup \quad \diagdown \\ \end{array} \begin{array}{c} H \\ \diagup \\ \\ H \end{array}$$
$$+ \\ H_2O$$

3.3.14 The Hofmann–Martius Rearrangement

When heated, mono- and dialkyl aryl amines often rearrange to form *ortho-* and *para*-substituted aryl amines. The *para*-substituted product is usually favoured but if this position is already occupied then the *ortho* position may become substituted instead. This migration of alkyl groups is known as the Hofmann–Martius Rearrangement and was discovered in 1871 [32]. The reaction mechanism proceeds via an alkyl carbonium ion that migrates onto the alkyl ring. Trisubstituted amino groups, such as the trimethyl amino group, will rearrange to form a primary amine and will trimethylate the aromatic ring; for example, N,N,N-trimethylphenylammonium chloride will rearrange to form 2,4,6-trimethylaniline hydrochloride.

When N-nitroso derivatives of secondary aryl amines (Ar–N(R)–N=O) are used in this reaction instead of alkyl-substituted amines, the reaction is called the

Fischer–Hepp Rearrangement [33] and it yields a *para*-substituted nitroso compound.

$$C_6H_5-N(R)-N=O + HCl \rightleftharpoons C_6H_5-NHR + NOCl$$
$$\downarrow$$
$$(para)\ O=N-C_6H_4-NHR + HCl$$

Variations of this rearrangement with migration from substituted nitrogen to *para* ring carbon have also been observed for *N*-halides, *N*-nitro and *N*-amino groups. The reaction mechanisms are generally thought to be intramolecular.

3.3.15 The Orton Rearrangement [34]

Although the Orton Rearrangement is really concerned with amides it also known as the chloramine rearrangement as it usually involves *N*-chloroarylamines, such as *N*-chloroacetanilide, which react with aqueous hydrochloric acid to form *ortho*- and *para*-chlorinated derivatives. In the first stage of the reaction the *N*-chloro group is replaced by a hydrogen and a molecule of chlorine gas is produced that can then attack the aromatic ring.

(i) $C_6H_5-N(Cl)-COCH_3 + HCl \rightarrow C_6H_5-NH-COCH_3 + Cl_2$

(ii)

The *para* isomer is usually the major reaction product but some *ortho*-substituted product is also produced in most reactions.

3.3.16 Alkylation of the aromatic ring

The aromatic ring of most aryl amines can be alkylated by using the Friedel–Crafts [35] Reaction with alkenes at over 200 °C and 200 bar applied pressure. The mono-*ortho*- or di-*ortho*-substituted product is usually obtained in preference to the *para*-substituted derivative.

3.3.17 Cyclization reactions

Aryl amines can react with various reagents to form heterocyclic compounds such as quinolines, benzofuroxanes, benzodithiazoles, etc. When aryl amines are condensed with carbonyl compounds in the presence of an acidic catalyst the reaction products are substituted quinolines. This reaction forms the basis of the Skraup Quinoline Synthesis [36] with aniline and glycerol or acrolein. If an aldehyde is used the reaction is known as the Doebner–von Miller Synthesis [37] but two reaction products are possible, either an *N*-alkyl aniline or a 2-alkyl-substituted quinoline.

  Another variation of this synthesis is to react aniline with a 1,3-diketone to make a 2,4-disubstituted quinoline. This is known as the Combes Quinoline Synthesis [38]. The Skraup Reaction with acetone yields 2,2,4-trimethyl-1,2-dihydroquinoline, which is widely used as an anti-oxidant in rubber and is known as TMDHQ.

$$C_6H_5-NH_2 + 2CH_3-CO-CH_3 \rightarrow \qquad [TMDHQ]$$

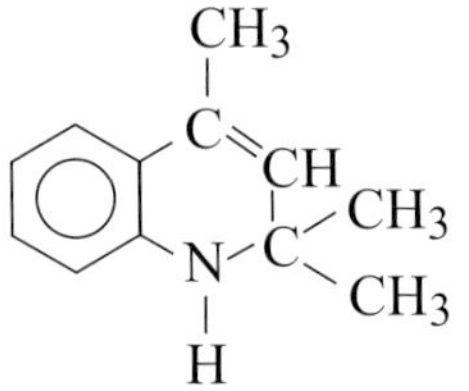

The Combes Synthesis is shown below for 2,4-pentanedione and aniline and yields a 2,4-disubstituted quinoline. With ethylacetoacetate two products are possible from the reaction with aniline. It is possible to perform the reaction in the cold to yield a 4-quinolone (the Conrad–Limpach Synthesis [39]) or hot (boiling) with a sulphuric acid or polyphosphoric acid catalyst to obtain a 2-quinolone. Finally, 2-aminobenzaldehyde can react with ethanal (acetaldehyde) to form an unsubstituted quinoline by the Friedlander Synthesis [40]. The synthesis of quinoline and substituted quinolines and the reaction mechanisms of their formation are examined in greater detail in Chapter 4.

$$C_6H_5-NH_2 + CH_3-CO-CH_2-CO-CH_3 \rightarrow$$

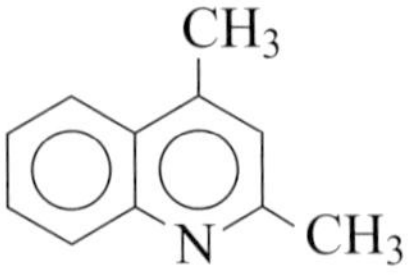

In addition to forming quinolines it is also possible for aromatic amines to form sulphur heterocycles by reacting 2-aminophenols with carbon disulphide (CS_2) to form mercapto-1,3-benzothiazole and 2-methylaniline with sulphur monochloride (S_2Cl_2) to form 1,2,3-benzothiazoles.

2-Methylaniline is a versatile reagent as it also offers the possibility of reaction with formic acid to yield the *N*-formyl derivative, which can cyclize to benzoxazole.

3.3.18 Nitration

Unless a suitable amine-protecting agent is used, nitration of aryl amines results in oxidation of the amino group (aryl amines, like all amines, are very susceptible to oxidation). However, if a protecting group is present, typically by acetylation, then nitration with nitric acid/sulphuric acid can yield *ortho*- and *para*-substituted derivatives. The isomer ratio of *ortho* to *para* is strongly dependent upon the aryl amine used and also upon the reaction conditions, and there are no rules to predict where substitution will occur.

3.3.19 Oxidation

It can be observed that aryl amines develop a brown or black discolouration if left exposed to air for prolonged periods. This is the reason why much of the aniline used in laboratories appears brown – it has become oxidized. The oxidation products of amines depend to a greater extent (but not exclusively) on the oxidizing agent used. With trifluoroperacetic acid, conversion of primary aryl amines takes place to nitro groups, particularly if there are electron-withdrawing substituents on the aryl ring *ortho* and *para* to the amine group.

Oxidation with manganese dioxide or sodium chromate usually yields quinones. With percarboxylic acids and hydrogen peroxide the *N*-hydroxy, *N*-nitroso or nitro group may be produced, depending upon the reaction conditions. Generally higher temperatures and a stoichiometric excess of the oxidizing agent favour the production of the nitro derivative.

3.3.20 Sulphonation

Aryl amines will react with oleum or sulphur trioxide to produce sulphonate derivatives *ortho* and *para* to the amine group. Substitution at the *para* position is preferred, but if this position is already occupied then substitution will occur *ortho*
to the amino group. The reaction with sulphuric acid produces only the sulphate
derivative, which can then be converted to the sulphonate via heating at 220 °C in
vacuum. Aniline can be converted in this way to sulphanilic acid.

$$C_6H_5-NH_2 + H_2SO_4 \rightarrow$$

Naphthyl amines can be hydrolysed to naphthols in excellent yields by water especially if there are *para* sulphonate groups present acting to accelerate the reaction.
This is the reverse of the Bucherer Reaction.

3.4 Aminophenols

Aminophenols are widely used as intermediates for photographic chemicals,
pharmaceuticals and dyestuffs and are an important sub-class of aromatic
amines. 4-Aminophenol is used as a photographic developer, as are amidol (2,4-
diaminophenol) and metol (4-(*N*-methylamino)-phenol). Simple aminophenols
such as *ortho-*, *meta-* and *para*-aminophenol are amphoteric and possess both acidic
and basic properties as a consequence of containing a hydroxy group and an amino
group respectively. At room temperature all three isomers are white crystalline
solids, but interestingly the *ortho-* and *para*-aminophenols are rapidly oxidized in
air whereas *meta*-aminophenol is stable. The solubilities of the three isomers are
also different, as *ortho*-aminophenol can form an intramolecular hydrogen bond.

Generally *ortho*-aminophenol exhibits greater solubility than *meta*-aminophenol, which is turn is more soluble than *para*-aminophenol. This solubility order is reversed for *ortho*-aminophenol and *meta*-aminophenol in acetonitrile and ethanol, but otherwise is valid for all common solvents and water.

All aminophenols can be made by the reduction of the corresponding nitrophenol. This can be performed with dissolving metals or catalytically with hydrogen. In addition, *meta*-aminophenol can be prepared from the fusion of metanillic acid with sodium hydroxide, and *para*-aminophenol can be prepared by refluxing phenylhydroxylamine with sulphuric acid, and also by the reduction of hydroxyazobenzene with sodium dithionite (this reaction also yields aniline as a by-product). Commercially, *para*-aminophenol is prepared by the electrolytic reduction of 4-nitrobenzene in sulphuric acid.

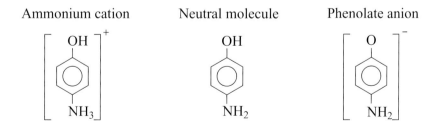

The aminophenols do not possess a doubly charged structure (like amino acids) and exist as anions, cations or neutral molecules depending upon the pH of the solution in which they are dissolved. The acidity of the hydroxyl group is reduced because of the presence of an amino group on the aromatic ring and so aminophenols are less acidic than phenols. The three possible structures for *para*-aminophenol are shown below but apply equally to all three aminophenol isomers.

Ammonium cation   Neutral molecule   Phenolate anion

3.4.1 Reactions of aminophenols

The reactions of the aminophenols may be divided into three types: reactions involving the hydroxy group (such as salt formation); reactions involving the amino group (such as formation of diazonium salts and Schiff Bases); and reactions involving the aromatic ring (such as halogenation or sulphonation reactions). These reactions will now be examined in detail.

(a) Alkylation

It is possible to doubly alkylate the amino group and singly alkylate the hydroxy group of aminophenols with an alkyl halide and Raney nickel or with a dialkyl sulphate such as dimethyl or diethyl sulphate.

Selective alkylation is difficult but it is possible to protect the amino group with a reagent such as acetyl chloride and then selectively to alkylate the hydroxy group to form an anisidine. This reaction is shown below for methylation of the hydroxy group in *para*-aminophenol.

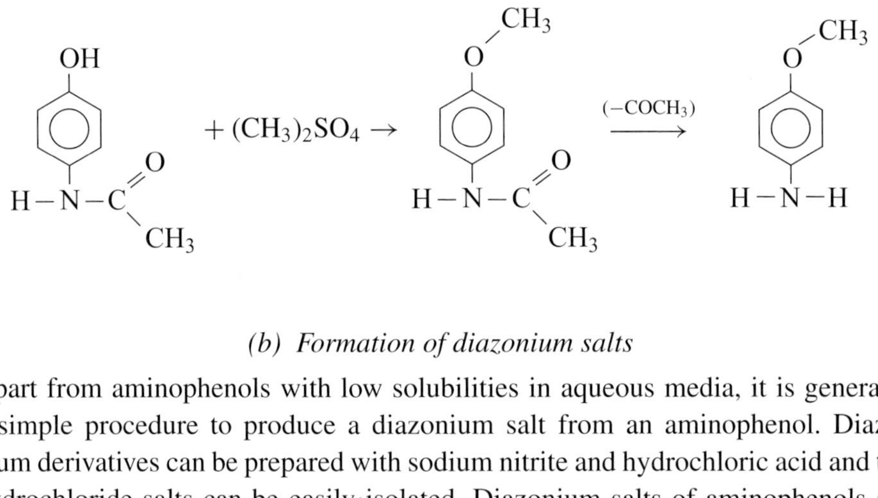

(b) Formation of diazonium salts

Apart from aminophenols with low solubilities in aqueous media, it is generally a simple procedure to produce a diazonium salt from an aminophenol. Diazonium derivatives can be prepared with sodium nitrite and hydrochloric acid and the hydrochloride salts can be easily isolated. Diazonium salts of aminophenols are used in the dyestuff industry.

(c) Acylation

Typical reaction conditions used to acylate aminophenols are acetic anhydride in alkaline media (e.g. aqueous sodium acetate or pyridine) or acetyl chloride in toluene, or dichloromethane plus an auxiliary base such as pyridine. Acylation can also be performed with ketene but this is difficult to handle in the laboratory and better suited to larger applications. The amino group is usually acylated first but, if an excess of the acylation reagent is used, *O,N*-diacylated products are obtained. Ammonium groups are not acylated; only free amino groups are acylated.

Rearrangement reactions are common, and migration of the acyl group from oxygen to nitrogen is often found, as is carbamate formation.

The best-known acylated aminophenol is paracetamol, *N*-acetyl-4-aminophenol, which is a powerful analgesic and widely sold as Calpol, Tylenol, Panadol and Hedex, to name just a few well-known consumer brands.

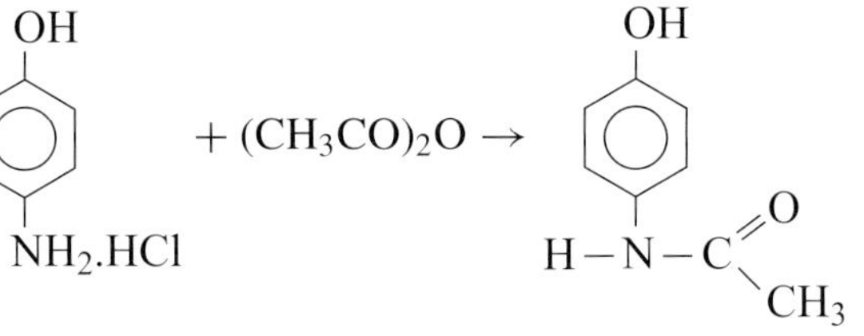

Sodium acetate is usually added to the above reaction, which is performed in aqueous media, to maintain pH control.

(d) Cyclization reactions

The *ortho*-substituted *ortho*-aminophenol is often used for the synthesis of phenoxazines, phenoxazones, benoxazoles and thiobenzoxazoles. The cyclization is usually accomplished by heating in the presence of a catalyst. With acetic acid, *ortho*-aminophenol reacts to form 2-methylbenzoxazole.

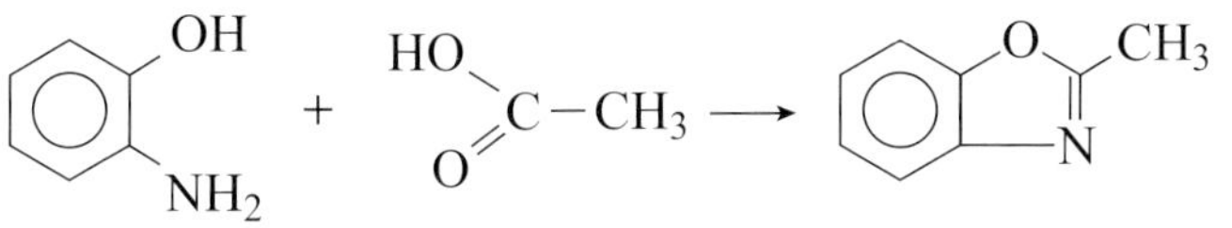

(e) Formation of quinones

Ortho- and *para*-aminophenols are readily oxidized to the corresponding quinones, but *meta*-aminophenol cannot form a quinone and is generally resistant to oxidation. A good oxidizing agent for aminophenols is sodium dichromate in sulphuric acid.

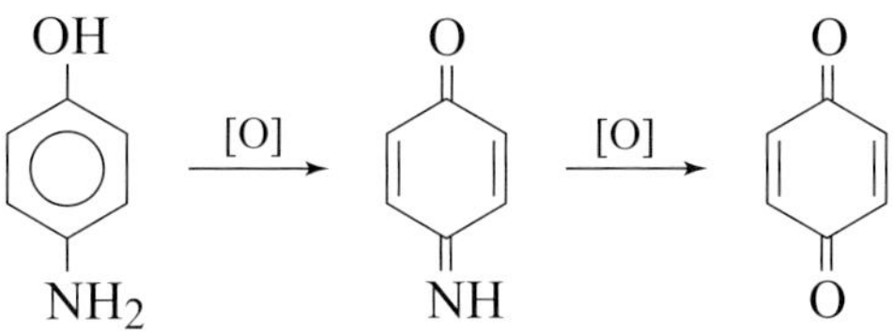

(f) Formation of Schiff Bases

Benzaldehyde, and substituted benzaldehydes, can condense with aminophenols in the presence of a suitable catalyst such as aluminium chloride to give a Schiff Base in up to 90% yield.

(g) Reactions of the aromatic ring

Both the amino and hydroxyl groups are electron donating as a consequence of their both containing electron lone pairs. For this reason any unsubstituted positions on the aromatic ring are susceptible to halogenation and sulphonation. Halogenation can be performed with chlorine or bromine in glacial acetic acid and the reaction will proceed until all unoccupied ring positions are halogenated (i.e. for *ortho-*, *meta-* or *para*-aminophenol the tetrahalogenated product will be formed). Sulphonation of the ring usually occurs is the 2- or 4- positions relative to the hydroxyl group. When oleum is used as a sulphonating agent, disulphonated products can be obtained.

   Finally, it is also possible to introduce a carboxyl group into the aromatic ring and the best-known example of this is the carboxylation of *meta*-aminophenol to yield 4-aminosalicylic acid, which is used as the sodium and calcium salts as an antibacterial agent.

3.5 Aryl diamines

Aryl diamines can generally be prepared by the same synthetic procedures used for the manufacture of aryl monoamines, such as the reduction of dinitro compounds or nitroanilines and occasionally the reduction of aminoazo compounds. The diamines are white or colourless solids and are diacid bases but with a single pK_a value each. The particular reactions of aryl diamines can be studied by reference to the three possible phenylene diamines with *ortho-*, *meta-* and *para*-substitution patterns.

   Ortho-phenylene diamine, or 1,2-diaminobenzene, can be easily prepared by the reduction of *ortho*-nitroaniline with zinc/sodium hydroxide in ethanol as a solvent.

Ortho-phenylene diamine can undergo cyclization reactions to yield 2,3-diaminophenothiazine with ferric chloride, benzimidazole with formic acid, benzotriazole with nitrous acid, quinoxaline with glyoxal, and an insoluble phenazine derivative with phenanthraquinone (which can be used for the quantitative determination of *ortho*-diamines). The reactions of *ortho*-phenylene diamine are shown schematically in Figure 3.2.

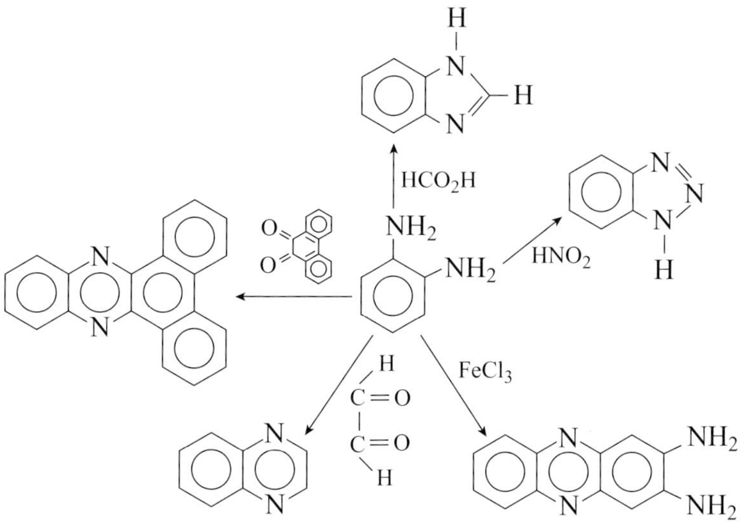

Figure 3.2. The reactions of *ortho*-phenylene diamine.

Meta-phenylene diamine can be prepared by the action of iron powder and hydrochloric acid on *meta*-nitrobenzene. When *meta*-nitrobenzene is treated with nitrous acid it forms Bismarck Brown; this is a brown dyestuff made up of mono-, di- and tri-azo-linked *meta*-phenylene diamine units. By careful treatment with an excess of nitrous acid in hydrochloric acid it is possible to prepare a tetra-azo compound.

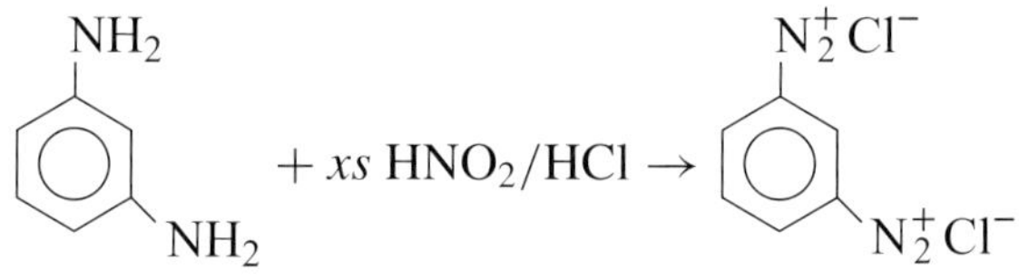

If there is insufficient nitrous acid present, the diazo groups couple with unreacted metaphenylene diamine to form azo linkages. A tetramer is shown below; this will exhibit the typical brown colour, but the molecular weight of the final product from the reaction could be much higher than that of the tetramer shown below.

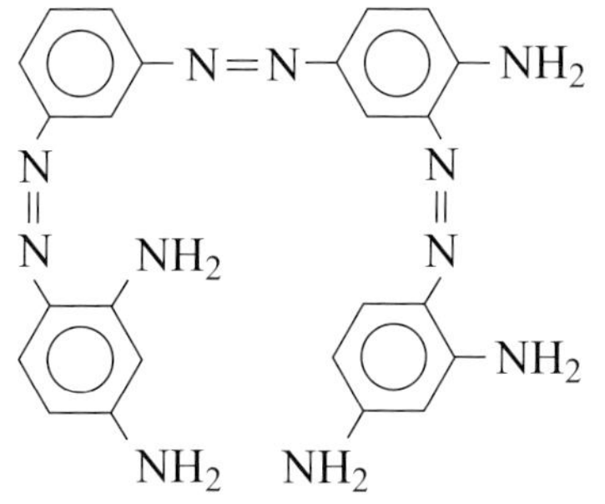

Para-phenylene diamine can be prepared from the reduction of *para*-nitroaniline or from aminoazobenzene. It can be diazotized and used in the formation of azo dyestuffs but, unlike the two other phenylene diamines, it can be oxidized to benzoquinone with potassium dichromate in sulphuric acid.

1,4-Naphthaquinone can be prepared in a similar way from 1,4-diaminonaphthalene.

3.5.1 Toluene diamines

Although there are six possible toluene diamine isomers they are all referred to collectively as TDA. *Meta*-TDA (*m*-TDA) or 2,4-diaminotoluene is an extremely important intermediate and the total worldwide production is around 1 million tonnes per annum. It is the single most important aryl amine in terms of annual production and usage. The principal use for *m*-TDA is in its conversion (with phosgene) to toluenediisocyanate, TDI, which can then undergo a further reaction with a polyol to form polyurethane foams. (Although there are three possible isomers of TDI, the *ortho* and *para* isomers are far less important than the *meta* isomer, so 'TDI' is used to refer to the *meta* isomer unless otherwise stated.)

The phosgenation reaction to produce TDI is usually carried out cold and in an inert solvent to avoid the formation of by-products such as carbonyl-bridged dimers (ureas) and benzimidazolone, which can be formed from the reaction of *ortho*-TDA and phosgene. Phosgene is usually added in excess to the reaction to prevent dimer formation and recycled in the commercial operation.

Diaminotoluenes are produced commercially by the catalytic reduction of dinitro-toluenes with Group VIII metal, palladium/carbon or Raney nickel catalyts, which enable the reduction process to be performed over a wide range of reaction conditions.

The dinitrotoluene is produced by a two-stage nitration reaction starting with toluene. The first-stage nitration to produce mononitrotoluene is performed in mixed nitric and sulphuric acids (typically equal volumes of 30% HNO_3 and 55% H_2SO_4), at 30–70 °C in stirred batch reactions that are very exothermic. The reaction products from this first stage are typically a mixture of around 57% *ortho*-nitrotoluene, 39% *para*-nitrotoluene, and a trace of *meta*-nitrotoluene (usually only around 4% or less). The mixed mononitrotoluene mixture is then nitrated for a second time using a stronger acid solution (again equal volumes) of around 30% nitric acid and 60% to 65% sulphuric acid to give a resulting product mixture of six TDAs. This reaction product mixture from the second nitration contains typically 76% 2,4-dinitrotoluene, 20% 2,6-dinitrotoluene and the remaining 4% is made up of an isomer mixture, which includes residues of mono-nitrated feed, other dinitrated isomers and some tri- and higher poly-nitrated products. Fortunately the boiling points of the isomers are sufficiently distinct to allow for separation by vacuum distillation. Alternatively, the whole reaction mixture can be reduced and the resulting amines can be separated by distillation. This is generally safer, as nitrotoluenes can detonate when heated. Commercially available *m*-TDA is usually over 99% pure and contains less than 0.3% *ortho*-TDA.

Dinitrotoluenes can be reduced catalytically with hydrogen over a palladium/carbon or Raney nickel catalyst. The reaction produces water as a by-product; this has to be removed from the reaction.

Because high-molecular-weight nitroarene tars can poison catalysts by pore plugging, in the commercial reduction of dinitrotoluene the concentration of

Table 3.2. *The physical properties of the TDA isomers*

Isomer	CAS	Melting point (°C)	Boiling point (°C)
2,3-	[2687-25-4]	63.5	225
2,4-	[95-80-7]	99	292
2,5-	[95-70-5]	64	273.4
2,6-	[823-40-5]	106	decomposes
3,4-	[496-72-0]	89.5	265 (sublimes)
3,5-	[108-71-4]	below 0	283.5

dinitrotoluene is kept low in the reactor and reaction conditions to ensure a fast reaction time are used.

The physical properties of the six TDA isomers are shown in Table 3.2. All of the TDA isomers undergo typical aryl amine reactions, which are similar to those of the phenylene diamines discussed in the previous section.

The TDA ring may be alkylated with an alkene together with an organo-aluminium reagent such as an alkyl aluminium halide or an aluminium anilide, or alternatively with an alkene plus catalyst. The catalytic reaction works well for short-chain alkenes and the yield decreases with increasing chain length. Acidic zeolites work well as reaction catalysts and can control the reaction stereochemistry if the correct pore size is selected.

(a) $= CH_2{=}C(CH_3)_2/(t\text{-}Bu)_3Al$

(b) $= CH_2{=}C(CH_3)_2/$acidic zeolite

Mono t-butyltoluenediamine is called t-BTDA and is produced commercially.

The amine side chains on all of the TDA isomers react readily with ethylene oxide at 90–120 °C and 72.5 psi applied pressure in the presence of a basic catalyst such as potassium hydroxide. The reaction also works well with other epoxides such as ethylene oxide.

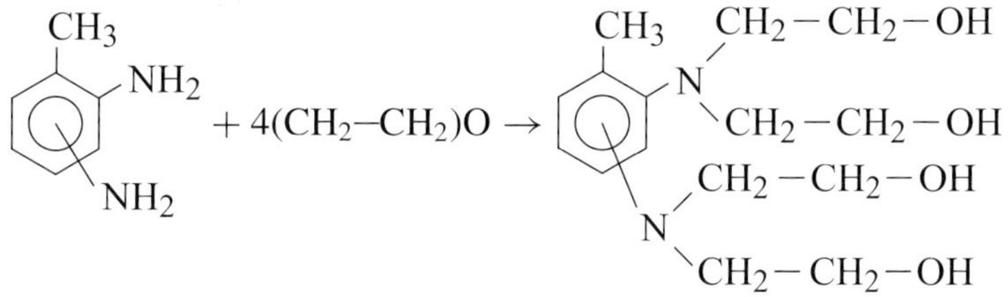

All of the toluene diamines are able to form diazonium salts in a similar manner to the phenylene diamines and, like *ortho*-phenylene diamine, toluene diamine isomers with *ortho* diamine groups can participate in cyclization reactions. With nitrous acid the reaction product is tolyltriazole, as shown below.

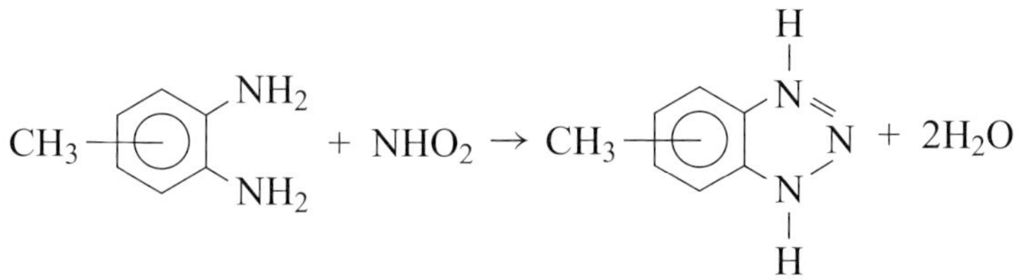

3.6 Diaryl amines

Diaryl amines are secondary amines with a single hydrogen atom attached to the central nitrogen bridge. The simplest diaryl amine is diphenylamine, which was first prepared by Hofmann in 1863 [41] by the destructive distillation of a triphenylmethane dyestuff. Commercially diphenylamine is manufactured today by the reaction between aniline and aniline hydrochloride in the presence of an acid catalyst (typically zinc chloride) at 140 °C under pressure. The reaction yield is around 85%.

$$C_6H_5-NH_2 + C_6H_5-NH_2.HCl \rightarrow C_6H_5-NH-C_6H_5 + NH_4Cl$$

The production of diphenylamine is a major application of aniline and it is widely used as an anti-oxidant in polymers, oils and rubber products such as car tyres.

Diphenylamine ($pK_a = 0.9$) is a much weaker base than aniline ($pK_a = 4.58$) and can be separated from a reaction mixture containing aniline by washing with a solution of a weak acid. The weakly basic nature of diphenylamine is explained by the fact that two phenyl rings are available for resonance with the NH group and there are therefore a larger number of resonating structures possible than for aniline. The fact that the nitrogen atom of diaryl amines donates electron density into the aromatic rings has the consequence that the nitrogen lone pair is less available for sharing with a hydrogen ion and also that the aromatic rings are more susceptible to electrophilic attack.

The N-hydrogen atom in diaryl amines such as diphenylamine can be replaced with alkali metals and aluminium to form diphenylamide ions and will also exchange with deuterium if diphenylamine is dissolved in deuterated ethanol, C_2H_5OD.

Diphenylamines are readily alkylated in the *para* positions with reagents such as styrene and isobutylene in the presence of acidic clay catalysts and can be *ortho* substituted with alkenes by using aluminium amide as a catalyst. Diphenylamine can be dehydrogenated to form carbazole, and dehydrogenation of alkyl-substituted diphenylamines gives acridine. The reaction of diphenylamine with sulphur produces phenothiazine, and it is possible to react diphenylamine with oxidizing agents such as potassium permanganate in acetone to form the diphenylnitrogen radical (this is also called diphenylamidogen). This radical cannot be isolated, but if dipentachlorophenylamine is used instead of diphenylamine then the resulting decachloroaminyl radical can be stabilized because 2,4-substitution creates radical stabilization.

3.6.1 Synthesis of diaryl amines

The standard method of manufacture for diaryl amines is to react two molecules of a primary aryl amine together in the presence of anhydrous hydrochloric acid (typically 0.5 weight%) under high temperature and pressure together with ferrous chloride or ammonium bromide.

(i) $C_6H_5{-}NH_2 + HCl \rightarrow C_6H_5{-}NH_2.HCl$

(ii) $C_6H_5{-}NH_2.HCl + C_6H_5{-}NH_2 \rightarrow (C_6H_5)_2NH + NH_4Cl$

(iii) $C_6H_5{-}NH_2 + NH_4Cl \rightarrow C_6H_5{-}NH_2.HCl + NH_3$

Overall the reaction is as follows.

$$2C_6H_5{-}NH_2 \rightarrow C_6H_5{-}NH{-}C_6H_5 + NH_3$$

The free energy of formation of diphenylamine $\Delta H_{298} = 310.5 \text{ kJ mol}^{-1}$.

The self-condensation of aniline can also be performed in the vapour phase using an alumina or titania catalyst promoted with copper chromite. However, this process has yet to find widespread commercial application or development.

One widely used commercial method is to condense a phenol with an aryl amine. However, the reaction with phenol is slow and works better with naphthols and best of all with hydroquinones.

The reaction between aniline and phenol at 325 °C in the presence of a phosphoric-acid catalyst gives only a 50% yield. The low yield is attributed to the fact that aniline reacts with the keto form of phenol, and this is present only in low concentration. The reaction between aniline and naphthol yields *N*-phenyl-2-naphthylamine, which is widely used in the rubber industry and is referred to as PBNA (phenyl-β-naphthylamine).

$$C_6H_5-NH_2 + C_{10}H_7-OH \rightarrow C_6H_5-NH-C_{10}H_7 + H_2O$$

Diphenylamine may also be prepared by the reaction between chlorobenzene and aniline, but the reaction is rather messy and the product contains a high percentage of high-molecular-weight tars. The reaction works better when the chlorobenzene is activated with an electron-withdrawing substituent. For example, aniline readily reacts with 4-nitrochlorobenzene in the presence of potassium carbonate and copper cyanide at 200 °C to form 4-nitrodiphenylamine, which can then be reduced to 4-aminodiphenylamine. This is a useful intermediate for the production of rubber chemicals.

$$C_6H_5-NH_2 + O_2N-C_6H_4-Cl \rightarrow C_6H_5-NH-C_6H_4-NO_2 + HCl$$
4-nitrochlorobenzene
$$C_6H_5-NH-C_6H_4-NO_2 + 3H_2 \rightarrow C_6H_5-NH-C_6H_4-NH_2 + 2H_2O$$

A useful method for the synthesis of substituted diphenylamines from phenyl-halides is the Ullmann Reaction, which can be performed by refluxing acetanilide, potassium carbonate, bromobenzene and copper-powder catalyst in nitrobenzene solution. The reaction yields are typically only 50% to 60%.

$$C_6H_5-NH-COCH_3 + C_6H_5-Br + K_2CO_3 \rightarrow C_6H_5-NH-C_6H_5 + CO_2$$
$$+ CH_3CO_2K + KBr$$

The Ullmann Reaction [42] can also be performed with *ortho*-chlorobenzoic acid. In this instance the reaction product is 2-[(3-chlorophenyl) amino] benzoic acid, which can be decarboxylated by heating to generate 3-chlorodiphenylamine.

The final method of preparation for diphenylamines discussed here is by the Chapmann Reaction [43] in which a monoarylamine is first converted into benzamide; this then reacts with phosphorous pentachloride to form an *N*-aryl-α-chlorobenzylideneimine. This intermediate is then able to react with sodium phenolate salts to form an arylimino ether, which rearranges to form a benzamide on heating. This rearrangement of the arylimino ether is the eponymously named Chapmann Reaction. Hydrolysis of the benzamide with ethanolic potassium hydroxide then yields the required diarylamine. Benzamides with 2,6-substitution patterns cannot be easily hydrolysed and are to be avoided in this reaction scheme.

3.6.2 Reactions of diphenylamines

(a) With nitrous acid

Diphenylamine reacts with nitrous acid to form *N*-nitrosodiphenylamine. This product can also be formed from the reaction between the diphenylnitrogen radical and nitric oxide (which possesses an unpaired electron). As diphenylamine and other diaryl amines do not possess a primary amino group ($-NH_2$) they do not form diazonium salts.

$$(C_6H_5)_2NH + HNO_2 \rightarrow (C_6H_5)_2-N-N=O + H_2O$$
$$(C_6H_5)_2N\bullet + \bullet NO \rightarrow (C_6H_5)_2-N-N=O$$

If the *N*-nitroso compound is mixed with anhydrous hydrogen chloride at low temperatures, the nitroso group migrates to the *para* position on one of the phenyl rings to form 4-nitrosodiphenylamine. This product can also be formed from the reaction of diphenylamine with nitrosyl chloride.

(b) Akylation

The alkylation of diphenylamine is commercially important for the production of anti-oxidants used in the rubber industry. Diphenylamine can be alkylated at high temperatures with an alkene and aluminium chloride at around 140 °C. This reaction works well for alkenes such as isobutylene, stryrene or α-methyl styrene, but if dialkenes are used the reaction product is often a high-molecular-weight tar.

Diphenylamine alkylates primarily at the 4-position but disubstituted products with 2,4-substitution patterns often also result from this reaction if there is an excess of alkene present, unless a sterically hindered alkene is used that may only be able to substitute at the 4-position.

Acetone will also react with diphenylamine with a Lewis Acid catalyst such as aluminium chloride at 150 °C or with hydrochloric acid at 250 °C. The reaction products are strongly dependent upon the reaction conditions; at low temperatures with a stoichiometric excess of the amine the main reaction product is likely to be 2,2-[4,4′-(dianilino)diphenyl]-propane. At higher temperatures with excess hydrochloric acid present the main reaction product is likely to be 9,9-dimethylacridan.

If *ortho* substitution is required, this can be facilitated by reacting diphenylamine first with powdered aluminium metal and then with an alkene at 300 °C and up to 30 MPa applied pressure. With ethene the reaction product under these conditions is 2,2′-diethyldiphenylamine in 95% yield. However, with higher-molecular-weight alkenes the yield declines dramatically to less than 20%.

(c) Reaction with sodium

If diphenylamine is converted to its sodium salt by heating with sodium metal or sodamide it will react readily with alkyl halides to form 2,4-disubstituted products, or 2,4,6-trisubstituted derivatives with lower-weight alkyls that are not sterically hindered. *N*-alkylation also takes place at the amine nitrogen site.

$$2(C_6H_5)_2NH + 2Na \rightarrow 2[(C_6H_5)_2N]^-Na^+ + H_2$$

Sodium diphenylamine reacts with iodine to form tetraphenylhydrazine.

$$2[(C_6H_5)_2N]^-Na^+ + I_2 \rightarrow (C_6H_5)_2N-N(C_6H_5)_2 + 2NaI$$

If tetraphenylhydrazine is dissolved in an inert, non-ionizing solvent such as benzene, it will partially dissociate to form diphenylnitrogen radicals, the presence of which can be observed as a green colouration. This green colour fades on diluting the solution and can be intensified by heating, and does not obey Beer's Law.

Tetraphenylhydrazine may also be prepared by treating diphenylamine dissolved in acetone with potassium permanganate. Diphenylamine will also form derivatives with gallium, germanium, phosphorus and silicon in addition to the alkali metals and aluminium salts already discussed.

(d) Reaction with carbon disulphide

Diaryl amines do react with carbon disulphide (unlike dialkyl amines, which readily form dithiocarbamates). However, carbon disulphide will react with the metal salts of diaryl amines to form diaryldithiocarbamates. With sodium diphenylamine the reaction product with carbon disulphide is diphenyldithiocarbamic acid sodium salt, which is an exceptionally stable product compared with other diaryldithiocarbamic acids.

$$[(C_6H_5)_2N]^-Na^+ + CS_2 \xrightarrow{\text{NaOH}} (C_6H_5)_2N-C(S)-S^-\ Na^+$$

(e) Acylation

The secondary nitrogen atom in diaryl amines can be readily acylated with acyl halides or acid anhydrides. The reaction with acyl halide works best in an inert solvent with an auxiliary base that acts to remove the build up of halogen acid in the reaction medium.

$$(C_6H_5)_2NH + CH_3COCl \rightarrow (C_6H_5)_2N-CO-CH_3 + HCl$$

(f) Halogenation

The aromatic rings of diaryl amines are very susceptible to electrophilic attack, and diphenylamine reacts with chlorine or bromine to form 2,2′,4,4′-tetrahalide products. If forcing reaction conditions are used it is possible to obtain complete replacement of the ring hydrogen atoms.

$$C_6H_5-NH-C_6H_5 + 4Cl_2 \rightarrow$$

(g) Mannich Reactions

Diaryl amines do give Mannich Reactions [44] but the reaction rate is much slower than with dialkyl amines. With diaryl amines it is best to attempt the Mannich Reaction with substrates containing an active hydrogen site. For example, diphenylamine reacts with formaldehyde to form *N,N,N′N′*-tetraphenylmethylenediamine. However, in the presence of acids and acidic catalysts this product will rearrrange to form (4,4′-dianilinodiphenyl) methane, which then polymerizes to form high-molecular-weight tars.

3.6.3 Diaryl amines as anti-oxidants

Diaryl amines are extensively used as anti-oxidants in rubber and polymers as they are able to react with, and neutralize, the chain-propagating properties of peroxy radicals. Diaryl amines react with peroxy radicals to form the aminyl radical, which can then react with

  (i) other peroxy radicals,
 (ii) other aromatic rings to form polymers, or
(iii) other aryl amines, phenols or thiols to regenerate the diaryl amine.

If the diaryl amine used as an anti-oxidant is already substituted in the *ortho* and *para* positions, this helps to negate the formation of high-molecular-weight polymers. For this reason most commercial anti-oxidants are based on substituted diaryl amines, which cannot undergo coupling reactions.

The reaction of aminyl radicals with peroxy radicals is a desirable outcome as the nitroxy radicals formed from this reaction are potent oxidation inhibitors and can trap alkyl radicals to form hydroxylamine ethers. These ethers can then react with a further peroxy radical to regenerate the original nitroxide.

Although aminyl radicals are stable towards oxygen they can oxidize other amines and so regenerate the diaryl amine. For this reason mixtures of diaryl amines, phenol and aniline are often formulated as anti-oxidant blends in applications such as car tyres.

If diaryl amines are mixed with hydrogen peroxide in the presence of vanadium, nitroxyl ions are formed. In many cases these ions can be isolated. For example, the radical formed from 4,4′-dimethoxydiphenylamine is stable for several years and can be isolated.

Raw rubber polymers are rather sticky semi-solids and require processing with other chemicals to turn then into useful materials. The auxiliary chemicals used by the rubber industry are worth around USD1 billion per year and amines can be estimated to account for around one third to one quarter of this value.

Amines are used in the synthesis of rubber accelerators such as dithiocarbamates (from dialkyl amines and carbon disulphide), guanidines (e.g. *ortho*-tolylbiguanide from aryl amines and cyanogen chloride) and hexamethylenetetramine. One problem associated with some rubber accelerators (which cause the cross linking of elastomers via sulphur linkages) is that they may require the presence of an additional secondary accelerator to improve their performance. Cyclohexylethylamine is often used as a secondary accelerator with dithiocarbamates. However, aryl amines such as the *para*-phenylene diamines are widely used as anti-cracking agents and anti-ozonants (good at reducing ozone-related cracks; see Chapter 8).

Although aniline alone was used for this purpose, diaryl amines blended with aniline and phenols have been found to give the best performance.

Unfortunately the best known anti-ozonant and anti-cracking inhibitor N,N'-di-β-naphthyl-*para*-phenylene (BNPD) is too toxic and is no longer produced.

Some replacement diarylamines with good performance are shown below for the structure aryl′−NH−aryl″.

PAN   aryl′ = α-naphthyl   aryl″ = phenyl
PBN   aryl′ = β-naphthyl   aryl″ = phenyl (seldom used)

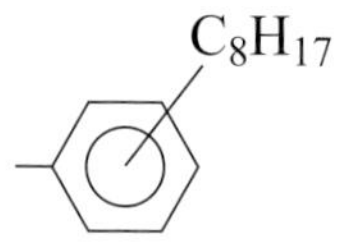

C_8H_{17}

ODPA aryl′ and aryl″ = octylphenyl

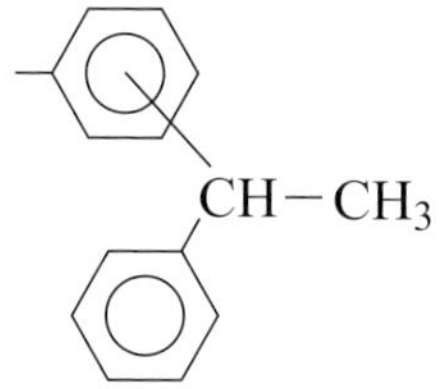

$CH - CH_3$

SDPA aryl′ and aryl″ = 1-phenylethylbenzene

Two other important amines used as anti-ozonants are *N*,*N*-diphenylethylene diamine and 4-aminodiphenylamine. A side effect of amine-based anti-degradation agents in rubber is that they can degrade to form black products that can discolour rubber and so are really only suitable for black products such as car, truck and aircraft tyres or applications where the product colour is not important.

3.7 Triaryl amines

The best-known triaryl amine is triphenylamine, which is an extremely weak base and forms isolatable salts only with large anions such as the tetrafluoroborate and chlorate ions. In concentrated sulphuric acid triphenylamine dissolves to form a blue solution. Although it might be thought that triphenylamine would be an effective auxiliary base, on the basis that the steric crowding around the central nitrogen atom would make it nearly impossible to alkylate or acylate, its very low solubility rules out its application in most situations. Triphenylamine is almost insoluble in water and poorly soluble in organic solvents.

Triphenylamine can be made by the Ullmann Reaction with diphenylamine in nitrobenzene solution in the presence of a copper catalyst. The experimental yield is over 80%.

$$2(C_6H_5)NH + 2C_6H_5I + K_2CO_3 \xrightarrow[(Cu)]{} 2(C_6H_5)_3N + 2KI + H_2O$$

The phenyl rings in triphenylamine behave similarly to those in diphenylamine and they can be halogenated, nitrated, sulphonated and alkylated by the same synthetic procedures. However, triphenylamine does possess interesting electronic properties and can be incorporated into light-emitting diodes and is effective in electron-hole transport mechanisms [45]. Triphenylamine can also form van der Waals complexes with nitriles and other reagents. Many of these complexes have been studied spectroscopically [46].

References

[1] W. E. Sinclair and D. W. Pratt, *J. Chem. Phys.*, **105** (1996), 7942.
[2] L. V. Moram and V. H. Galicia, *Int. J. Quant. Chem.*, **78** (2000), 99.
[3] R. K. Pankratov, I. M. Ucheva and R. K. Chernova, *J. Serb. Chem.*, **66** (2001), 161.
[4] M. L. Verdonk, R. W. Tjerksta, I. S. Ridder, J. A. Kanters, K. Karoon and W. J. M. van der Kamp, *J. Comput. Chem.*, **15** (1994), 1436.
[5] O. Unverdorben, *Ann. Physik*, **8** (1826), 397.
[6] *Beilstein Handbuch der Organischen Chemie* (Berlin: Springer Verlag, 1929), Auflage 4, Band 12.
[7] A. W. Hofmann and C. A. Martius, *Ber.*, **4** (1871), 742.
[8] A. J. Béchamp, *Ann. Chim. Phys.*, **42** (1854), 186.
[9] US Patent 3272865 (1966, assigned to Halcon Corp.).
[10] H. T. Bucherer, *J. Prakt. Chem.*, **69** (1904), 49.
[11] A. W. Hofmann, *Ber.*, **14** (1881), 2725.
[12] S. Patai (ed.), *The Chemistry of the Azido Group* (New York: Interscience, 1971), p. 397.
[13] W. Lossen, *Ann.*, **175** (1874), 271.
[14] R. F. Schmidt, *Ber.*, **57** (1924), 704.
[15] L. G. Donaruma and W. Z. Heldt, *Org. React.*, **11** (1960), 1.
[16] S. A. Lawrence, *Pharmachem.*, **1**, 9 (2002), 12.
[17] J. Louie and J. F. Hartwig, *Tet. Lett.*, **36** (1995), 3609.
[18] A. S. Guram, R. A. Rennels and S. L. Buchwald, *Angew. Chemie*, **341** (1995), 1348.
[19] M. C. Harris, O. Greiss and S. L. Buchwald, *J. Org. Chem.*, **64** (1999), 6019.
[20] M. Barani, S. Fioravanti, M. A. Loreto, L. Pellicani and P. A. Tardella, *Tetrahedron*, **50** (1994), 3829.
[21] P. Griess, *Phil. Trans. R. Soc. Lond.*, **154** (1864), 683.
[22] L. Gattermann, *Ann.*, **347** (1906), 347.
[23] M. Gomberg and W. E. Bachmann, *J. Am. Chem. Soc.*, **46** (1924), 2339.
[24] R. Pschorr, *Ber.*, **29** (1896), 496.
[25] T. Sandmeyer, *Ber.*, **17** (1884), 1633.
[26] G. Balz and G. Schiemann, *Ber.*, **60** (1927), 1186.
[27] H. Bart, *Ann.*, **55** (1922), 429.
[28] H. Meerwein, *Ann.*, **405** (1914), 129.
[29] F. R. Japp and F. Klingemann, *Ber.*, **20** (1887), 3398.
[30] H. C. Yao and P. Resnick, *J. Am. Chem. Soc.*, **84** (1962), 3504 (and references therein).
[31] E. Nickel, *Die Farben Reactionen der Kohlenstoff Verbindungen* (Berlin: vonHermann Peters, 1890), p. 14.
[32] W. W. Hofmann and C. A. Martius, *Ber.*, **5** (1872), 720.
[33] E. W. Fischer, *Ber.*, **100** (1967), 2445.
[34] K. J. Orton, *J. Am. Chem. Soc.*, **95** (1909), 1465.
[35] C. Friedel and J. M. Crafts, *Comptes Rend.*, **84** (1877), 1392.

[36] Z. H. Skraup, *Ber.*, **13** (1880), 2086.

[37] O. Doebner and W. von Miller, *Ber.*, **16** (1883), 2464.

[38] A. Combes, *Bull. Soc. Chim. Fr.*, **49** (1888), 89.

[39] M. Conrad and L. Limpach, *Ber.*, **24** (1891), 2990.

[40] P. Friedlander and C. F. Gohring, *Ber.*, **16** (1883), 1883.

[41] A. W. Hofmann, *Ber.*, **13** (1880), 1224.

[42] F. Ullman, *Ann.*, **332** (1904), 38.

[43] A. W. Chapman, *J. Chem. Soc.* (1929), 569.

[44] C. Mannich and W. Krosche, *Arch. Pharm.*, **250** (1912), 647.

[45] H. Fujikawa, M. Ishii, S. Tokito and Y. Taga, *Mat. Res. Soc. Symp. Proc.*, **621** (2000), Q3.4.2.

[46] G. Meijer, G. Berden, W. L. Meerts, H. E. Hunziker, M. S. deVries and H. R. Wendt, *Appl. Phys. Lett.*, **69** (1996), 878.

4

Heterocyclic amines

4.1 An introduction to heterocyclic amines

A heterocyclic amine is defined, for the purposes of this book, as a molecule with a carbocyclic structure that contains at least one nitrogen atom and may, or may not, contain other additional heteroatoms such as oxygen, sulphur or even silicon or boron. There are many naturally occurring compounds containing heterocyclic amines; for example, in the vitamin B group, in naturally occurring penicillins, in nicotinamides and also in purines and nucleic acids, all of which will be examined in this chapter. In fact, the natural world is full of nitrogen-containing heterocycles including carbohydrates, many enzymes and also the amino-acid side chains of the DNA double helix.

Heterocyclic amines, like all heterocyclic compounds, may be either aromatic or non-aromatic in character. Many non-aromatic heterocycles closely resemble their aliphatic analogues in their physical and chemical properties, unless they possess a strained ring structure. The unique chemistry of heterocyclic amines derives mainly from the presence of ring strain for small-ring heterocycles such as aziridine and azetidine, or from aromaticity for heterocycles such as pyrrole, pyridine and the quinolines. For this reason, the emphasis in this chapter is on nitrogen-containing heterocycles with unique or unusual properties that set them apart from their aliphatic counterparts. Nitrogen can be sp^2 or sp^3 hybridized in heterocycles and can easily replace a carbon atom. It is actually technically correct to call an sp^2-hybridized nitrogen in a cyclic ring structure a cylic imine, but for the purposes of this chapter both cyclic imines and cyclic amines will be reviewed together. Once incorporated into a ring structure a nitrogen atom has four possibilities for using its lone pair of electrons, as follows:

(i) it can retain its lone pair pointing away from the ring, as in pyridine;
(ii) it can donate its lone pair of electrons into a ring system to create an aromatic molecule according to Hückel's $(4n+2)$ rule, as in pyrrole;

(iii) it can donate them to an external electrophile to form an adduct, as in the case of pyridine *N*-oxide; or

(iv) it can form a non-bonded complex with a metal atom or ion.

The term aromatic was once exclusively used as a classification for compounds that had the same properties as benzene or fused benzene compounds such as naphthalene (i.e. purely carbocyclic rings); however, this term is now applied to all compounds that obey the Hückel (4*n*+2) rule and have the correct number of π-electrons. For $n = 0$, there must be two π-electrons, for $n = 1$, there must be six π-electrons, for $n = 3$ there must be 14 electrons, etc. The *n* in the Hückel equation is actually just a mathematical integer (1, 2, 3, 4, . . . , etc.) and has no relation to the number of ring atoms.

Some heterocyclic molecules with ring nitrogen atoms have very low barriers to lone-pair donation and are able to flip between being non-aromatic and aromatic, depending upon the reaction conditions. An example of this is pyrrole, which is aromatic in character in basic solution but in the presence of acids forms a linear trimer with two aromatic rings and one central non-aromatic ring. The nitrogen lone pair in pyrrole is donated to the ring and so in basic or neutral conditions pyrrole has a degree of aromatic character (four electrons from the carbon atoms plus the nitrogen lone pair, giving six π-electrons in total, which complies with Hückel's law). However, in acid solution, protonation and trimerization take place and the aromatic character is lost. The extent of the resonance energy in pyrrole is estimated to be between 87 and 130 kJ mol^{-1} (c.f. benzene with a resonance energy of 263 kJ mol^{-1}).

In the case of pyridine the lone pair is not required to complete the six π-electrons required for aromaticity and it points away from the ring in a perpendicular p orbital in a region of localized electron density. This is why pyridine oxides and pyridinium salts such as the hydrochloride retain their aromaticity.

The extent of the ring resonance energy in pyridine is 96.3 kJ mol^{-1}. This partly explains why pyridine (p$K_a = 8.75$) is a significantly stronger base than pyrrole (p$K_a = -0.27$), which has to sacrifice the energy associated with resonance stabilization in order to donate its lone pair to another atom or molecule, but is a weaker base than piperidine (p$K_a = 2.88$).

The distinction between pyridine-type nitrogens (which are generally electron withdrawing) and pyrrole-type nitrogens (which are electron donating) is important and helps to explain the behaviour of nitrogen-containing heterocycles.

Table 4.1. *The level of aromaticity and resonance energy per π-electron for some nitrogen-containing heterocycles*

Heterocycle	Resonance energy/π-electron	% Aromatic character
Pyridine	0.058	82
Pyrazine	0.049	75
Pyrimidine	0.049	67
Pyrazole	0.055	61
Imidazole	0.042	43
Pyrrole	0.039	37
Isoindole	0.029	32

If the level of aromaticity in benzene is 100 then the aromatic character of nitrogen heterocycles, all of which have less aromatic character, can be expressed as a percentage as given in Table 4.1 together with the values of resonance energy per π-electron (total resonance energy divided by the number of π-electrons).

Mention should also be made of the ring numbering system used in heterocyclic chemistry. Sometimes this appears to defy logic, but it is a system that organic chemists have become used to over the decades and are reluctant to change. Heterocycles are usually numbered anti-clockwise from the heteroatom. Therefore pyridine (on the left) has nitrogen at position 1 and carbons at positions 2 to 6 as shown below (hydrogens not shown) and imidazole (on the right) follows this rule with nitrogens at positions 1 and 3.

This system works well for many heterocycles but there are some important exceptions: in particular, isoquinoline (below right) is numbered in the same way as quinoline (below left) so the nitrogen heteroatom occupies position 2.

This system is adapted slightly for acridine with the ring positions being numbered from the top right, as shown below. Some authors have extended the quinoline-type numbering system with the numbering starting from the bottom right. This system is not used in this book but may be encountered elsewhere by the reader. In this book, a figure has been added to the text in cases where the ring position could be unclear in a heterocycle.

4.2 Azirine

The simplest possible heterocyclic amine is azirine, which can exist in two isomeric forms, 1-azirine and 2-azirine.

1-Azirine   2-Azirine

Loss of a hydride ion from either of these isomers will create an azirinylium cation with two π-electons that should possess aromatic character.

The azirines possess a very high degree of ring strain, over 170 kJ mol^{-1} and are prone to ring-opening reactions. The unsubstituted azirines have only ever been detected in the gas phase as reactive intermediates from pyrolysis and thermolysis experiments, but several substituted azirines are known and have been isolated. Substituted 1-azirines can be prepared from the gas-phase thermolysis of vinyl azides or by the Neber Reaction from oximes with acidic hydrogens [1].

(Here Ts = tosyl.)

Substituted 2-azirines can be prepared by the gas-phase pyrolysis of triazoles at 750 °C in high vacuum but are unstable and can be detected only spectroscopically.

3,3-Dimethyl-2-phenyl azirine has been isolated and is surprisingly stable in hydrochloric acid solution. In fact 10 M hydrochloric acid is required to dissolve this substituted azirine completely. Electrophilic reagents (e.g. benzoyl chloride) attack this azirine at the nitrogen atom, as shown in Figure 4.1. It is also possible

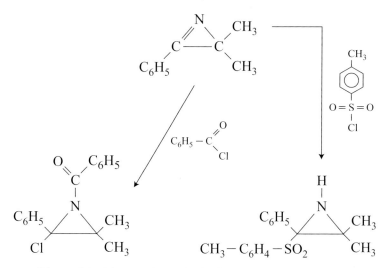

Figure 4.1. The reactions of 3,3-dimethyl-2-phenylazirine.

for other reagents such as 4-toluene sulphinic acid selectively to attack the carbon at position 2 in 1-azirines if the carbon at position 3 is disubstituted. Addition at position 2 is non-stereoselective, and both $(+)$ and $(-)$ enantiomers are obtained.

4.3 Aziridine (ethyleneimine)

Unlike azirine, which is unstable in its unsubstituted form, aziridine is a stable molecule under ambient conditions, and is supplied on a commercial scale as the pure substance and also as a stable alkaline aqueous solution. The main applications of aziridine (also known as ethyleneimine) are in the synthesis of hydroxylic polymers, in the inactivation of viruses and in the synthesis of 'nitrogen mustard' chemotherapy agents.

Pure aziridine is a colourless liquid under ambient conditions and has a boiling point of 56 °C. It behaves as a secondary amine with a pK_a of 7.98 but is less basic than dimethylamine, which has a pK_a of 10.9. The structure of aziridine is shown below [2]. Protonation of the aziridine ring produces the cyclic aziridinium cation, which is also formed in the Beckmann Rearrangement and in other rearrangement reactions involving amines.

An aliphatic C–N bond length is normally around 1.47 Å, which is comparable to the bond length found in aziridine. However, the C–C bond length in aziridine is 1.481 Å, which is shorter than the 1.54 Å found in alkanes. The ring-strain energy in aziridine is 96 kJ mol^{-1}, an indication of which may be made by measurement of ν_{C-H}(1475 cm^{-1}) and ν_{N-H}(1441 cm^{-1}) by infra-red spectroscopy. These absorption bands are shifted between 10 and 20 wavenumbers from where they would normally be expected (1465 cm^{-1} and 1460 cm^{-1}, respectively) [3].

The aziridine molecule can be *N*-substituted by several reagents but it is extremely unstable in acid conditions and will polymerize to form a linear polymer with explosive violence. Even the traces of carbon dioxide present in air can cause polymerization to occur.

From the structure of aziridine it might be predicted that it could easily be synthesized from ammonia and acetylene, but all attempts to date have failed.

The original synthesis of aziridine was made by the eponymous Gabriel Reaction by using 2-bromoethylamine in potassium hydroxide solution [4]. This synthesis is still used in the production of vaccines; 2-bromoethylamine hydrobromide is added to veterinary vaccine stock solutions after cultivation to generate aziridine *in situ*. This then inactivates any viruses present. (A better method of inactivation is to add an aqueous solution of aziridine directly to the stock solution but because of the dangers associated with handling aziridine many vaccine manufacturers prefer to use 2-bromoethylamine hydrobromide instead as it is much easier to handle.)

In addition to the Gabriel Reaction there are two other processes for the manufacture of aziridine described in the literature; these are the Wenker [5] and Dow [6] Processes. The Dow Process operated between 1963 and 1978 but has since been discontinued for large-scale manufacture. The Dow Process is based on the reaction between 1,2-dichloroethane and ammonia. Although the raw materials are inexpensive, the process produces a waste stream that has high disposal costs. Also, the use of a chlorinated raw material requires a corrosion-resistant manufacturing plant.

$$\text{Cl-CH}_2\text{-CH}_2\text{-Cl} + 3\text{NH}_3 \rightarrow \begin{array}{c} \text{CH}_2 \\ | \quad \diagdown \\ \quad \quad \text{N-H} \\ | \quad \diagup \\ \text{CH}_2 \end{array} + 2\text{NH}_4\text{Cl}$$

The Wencker Process, which was developed in 1935, uses a two-stage synthesis (see below) to generate aziridine from 2-aminoethanol with sulphuric acid and then sodium hydroxide. Although the raw material costs are higher than for the Dow Process, the plant-operating costs and waste-disposal costs are lower and the

Wencker Proces is the basis for most industrial production of aziridine worldwide today.

Stage 1

$$H_2N-CH_2-CH_2-OH + H_2SO_4 \rightarrow H_2N-CH_2-CH_2-OSO_3H + H_2O$$

Stage 2

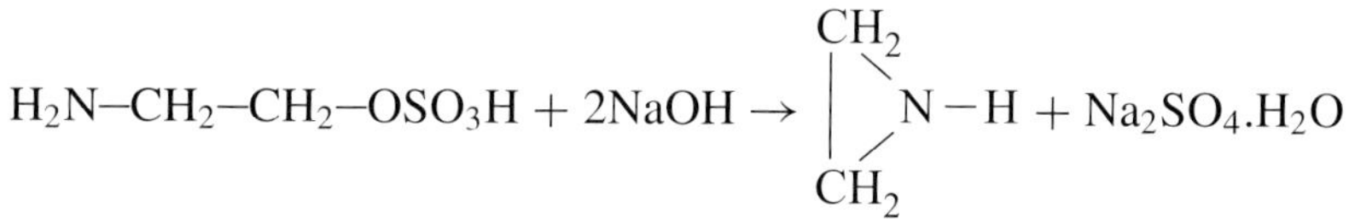

$$H_2N-CH_2-CH_2-OSO_3H + 2NaOH \rightarrow \begin{array}{c} CH_2 \\ | \\ CH_2 \end{array}\!\!\!N-H + Na_2SO_4 \cdot H_2O$$

Substituted aziridines can be prepared by the action of chloramine-T or bromamine-T on alkenes, as shown below (where X = Cl or Br and Ts = tosyl).

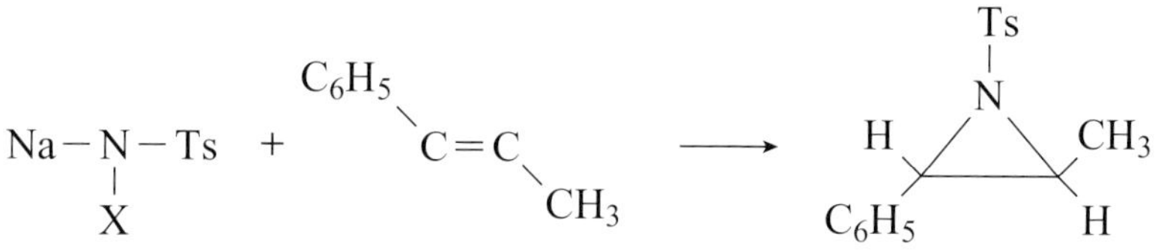

The reaction is catalysed by 5% copper(I) chloride and works best in acetonitrile solution. Bromamine-T and chloramine-T will also react with carbonyls to form aziridines in the same manner, but with both alkyl or aryl groups ending up on the same carbon.

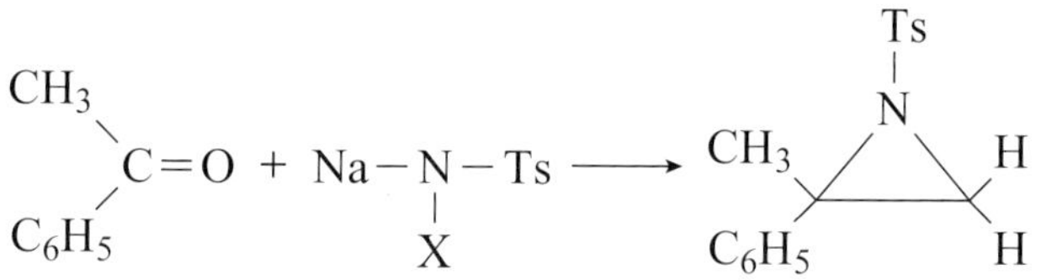

Another method of preparing substituted aziridines is from the reaction of alkyl or aryl magnesium halides with aryl alkyl ketoximes. The intermediate product is an azirine. Addition to the azirine in the second reaction step usually occurs onto the less-hindered side of the molecule.

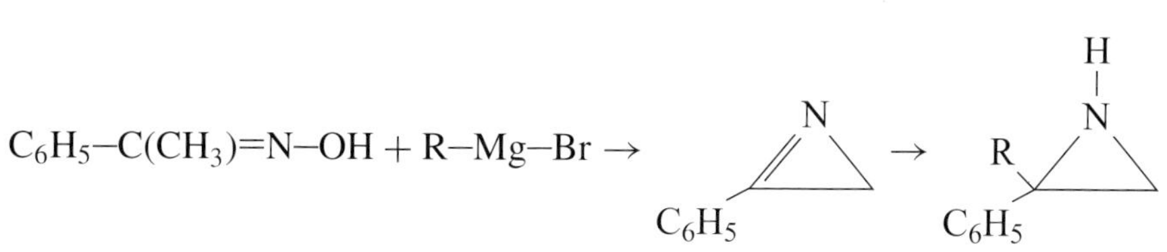

Many reagents, such as the Lwowski Reagent, 4-nitrobenzenesulphonoxy-urethane, will react with cyclic alkenes to form fused aziridines [7].

$$\bigcirc\!\!\!| + C_2H_5O-CO-NH-O-SO_2-C_6H_5-NO_2$$

78%   7%

Some other synthetic routes to substituted aziridines are given in Appendix 3 and also in Chapter 5, where metal-mediated nitrene reactions often generate aziridines both as reaction intermediates and as reaction products.

Aziridine is a secondary amine and does display some reactions typical of this class of amines, but because of its high ring strain the important reactions of aziridine are all characterized by ring-opening. These ring-opening reactions may be base catalysed, or acid catalysed, although acid-catalysed reactions are difficult to control and give mixtures of several products.

In even weakly acidic solutions the nitrogen atom on aziridine becomes protonated to form a cation and the ring then opens to form a carbonium ion. If the reaction is performed very carefully with hydrochloric acid it is possible to produce 2-chloroethylamine and also 2-chloroethylammonium chloride; however, this should not be attempted in the laboratory as the mixtures of acids and aziridine are liable to explode spontaneously.

The problem with acid-catalysed ring opening is that once the carbonium ion is formed it will react with any aziridine present and start to form linear polymer chains. However, if the ring opening is performed in anhydrous solution with hydrogen chloride or aluminium chloride then it is possible to isolate the reaction products. The 2,2-dimethylaziridine ring opens in the presence of hydrogen chloride and benzene to form a carbonium ion, which adds on a phenyl group as shown.

$$CH_3 \diagdown \underset{\underset{CH_3}{\diagup}}{\triangle}\overset{\overset{\overset{H}{|}}{N}}{} H + AlCl_3/C_6H_6 \rightarrow [(CH_3)_2-C^+-CH_2-NH_2]$$

$$\downarrow$$

$$CH_3-\underset{\underset{C_6H_5}{|}}{\overset{\overset{CH_3}{|}}{C}}-CH_2-NH_2$$

Mono-alkyl and mono-aryl substituted aziridines will also undergo ring opening in the presence of reagents such as cyanogen bromide and carbon disulphide to form adducts.

$$Br-CH_2-CH_2-NR-CN \xleftarrow{Br-CN} \underset{R}{\triangle}\overset{\overset{H}{|}}{N} \xrightarrow{CS_2} R-\underset{\underset{H}{|}}{\overset{H}{\underset{N}{\diagup}}}\overset{S}{\diagdown}\underset{H}{\overset{S}{\diagup}}$$

In contrast to acid-catalysed ring openings, the base-catalysed ring openings of aziridines are usually simpler and lead to the possibility of preparing resolvable chiral products.

The rings of substituted aziridines can be opened in the presence of aqueous ammonia and ethylamine. The reaction is rather slow and takes several weeks even at elevated temperatures. For unsymmetrical rings the attack occurs at the carbon atom with most hydrogens and Walden Inversions [8] usually take place. This has the effect that ring-opening reactions proceed in a *trans* manner.

$$\underset{H_3C}{\overset{\overset{\overset{R}{|}}{N}}{\triangle}}CH_3 + NH_{3(aq)}/C_2H_5NH_2 \rightarrow CH_3-CH-\underset{\underset{HNCR}{|}}{\overset{\overset{HNR}{|}}{CH}}-CH_3$$

(±, threo series)

Aziridine is able to participate in many *N*-substitution reactions where the ring structure is retained. For example, the reagent $[SbF_6][CH_3CO]$ will *N*-acetylate aziridine in the presence of sulphur dioxide at $-60\,°C$.

Aziridine wil also add to activated double bonds such as that in acrylonitrile.

$$\underset{\triangle}{\overset{\overset{\displaystyle H}{\underset{\displaystyle N}{|}}}{}} \; + \; CH_2{=}CH{-}CN \; \rightarrow \; \underset{\triangle}{\overset{\overset{\displaystyle CH_2{-}CH_2{-}CN}{\underset{\displaystyle N}{|}}}{}}$$

The anti-cancer agent Tetramin is prepared in a similar way from the reaction between 2-vinyloxirane and aziridine.

Aziridine can also be lithiated with methyl lithium and then used to prepare other *N*-substituted aziridines and even 1,1'-biaziridine. With nitrosyl chloride at $-60\,^{\circ}$C aziridine forms an unstable *N*-nitroso compound that decomposes to nitrous oxide and ethylene.

Although the aziridine ring is prone to ring-opening reactions, *N,N*-dialkyl halide salts of aziridine have been synthesized but starting from tertiary amines containing an *N*-(2-haloethyl) group rather than from aziridine or *N*-alkylaziridines. These dialkyl salts are difficult to isolate and can dimerize to form diquaternary piperazine salts or undergo ring opening with a hydrogen halide (HX) to form linear amines with a 2-haloethyl side chain.

$$R_2NCH_2{-}CH_2{-}Cl \rightleftharpoons Cl^- \; \underset{\triangle}{\overset{R\underset{+}{\diagdown}N\diagup R}{}} \; \xrightarrow[(HX)]{} R_2N{-}CH_2{-}CH_2{-}X$$

$$\Bigg[\underset{\underset{R\diagup N\diagdown R}{\overset{+}{N}}}{\overset{R\diagdown \overset{+}{N} \diagup R}{}} \Bigg] 2Cl^-$$

The main applications of aziridine are in the manufacture of paints, adhesives and paper, petroleum exploration and surface membranes. In all of these applications aziridine is used as a polymer. The polymerization of aziridine is usually performed in weakly acidic aqueous solution at 90–100 $^{\circ}$C in a stirred reactor. The polymers thus produced are typically between 10 000 and 20 000 in molecular weight and are spherical. The total amine content present in the polymer has been analysed by X-ray photoelectron spectroscopy and has been found to be roughly 30% primary amines, 40% secondary amines and 30% tertiary amines [9].

Polymeric aziridines are added to water-soluble paints to improve their wet adhesive properties. They are also added to paper during the manufacturing process to aid pigment dispersion and to improve the paper's wet strength. In the petroleum exploration industry, polyaziridines are blended with lignosulphonate and added to well cements to prevent fluid loss and are also used as demulsifiers for crude oil. In water treatment, polyaziridines can be used for the fabrication of reverse osmotic membranes and also as flocculants.

In addition to pure aziridine polymers, aziridine can also be polymerized with trifunctional acrylates to form new hydroxylic polymers for specific coating applications. The polymers produced with acrylates have a high degree of cross linkage and are used in the manufacture of coatings for laminates for exterior use. A typical reaction is shown below for 2-ethyl-2-(hydroxymethyl)-1,3-propane diol with aziridine.

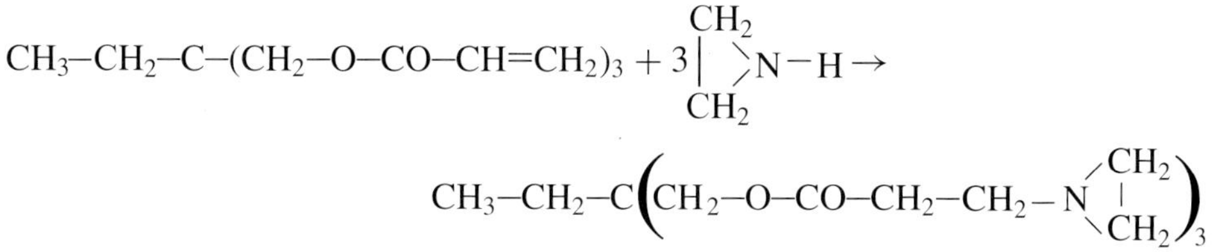

$$CH_3-CH_2-C-(CH_2-O-CO-CH=CH_2)_3 + 3 \left| \begin{array}{c} CH_2 \\ \\ CH_2 \end{array} \right\rangle N-H \rightarrow$$

$$CH_3-CH_2-C\left(CH_2-O-CO-CH_2-CH_2-N\begin{array}{c} CH_2 \\ | \\ CH_2 \end{array}\right)_3$$

It is difficult to get an impression of the total worldwide production of aziridine as it is mainly produced for captive use in polymer manufacture and is not isolated. The two main producers are BASF in Germany and Nippon Shokubai in Japan, and the total manufactured quantity is probably between 10 000 and 15 000 tonnes per year.

In addition to the industrial use of polymerized aziridines, aziridine is used for the production of animal vaccines and also for the synthesis of a range of anti-cancer agents, including Thiotepa; these uses are discussed in Chapter 8. However, these non-polymer applications account for less than 1 tonne per year of aziridine.

4.4 Diazirene

Diazirene is a stable colourless gas at ambient temperatures and was prepared for the first time in 1961 by Schmitz [10] from a carbonyl precursor and ammonia (see Appendix 3). In 1962, Graham [11] prepared disubstituted diazirenes from amidine hydrochlorides. The N=N double bond in diazirene is 1.288 Å long; this is comparable with that found in imines. Although the diazirene ring is highly strained it is chemically rather inert and is stable to attack by potassium metal dissolved in t-butanol. Diazirene can be monoalkylated with alkylmagnesium bromide Grignard Reagents to yield diaziridines.

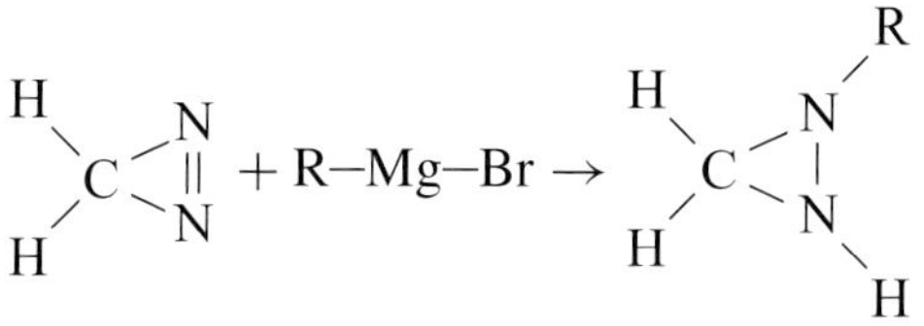

Diaziridines (like aziridines) are unstable to acid and decompose liberating hydrazines and, if the carbon group is dialkylated, giving a ketone. This reaction is reversible in some instances.

$$
\begin{array}{c}
R\!-\!\!\!\diagdown \\
\quad C\!\!-\!\!N\!\!-\!\!R'' \\
R'\!\!\diagup \quad \;\; N\!\!-\!\!H
\end{array}
\;+\; [H_3O]^+ \rightleftharpoons R''NH\!-\!NH_2 + RR'C\!\!=\!\!O
$$

The N–N single bond in diaziridine is 1.468 Å long. This is slightly longer than the N–N single bond in hydrazine, which is 1.449 Å. The difference in length reflects the ring strain.

Diaziridines can be thought of as intermediates to the more stable diazirenes, and can be easily converted to diazirenes with oxidizing agents such as chromium trioxide, sodium chlorate or iodine/triethylamine.

$$
R_2C\!\!\diagdown\!\!\begin{array}{c} N\!\!-\!\!H \\ N\!\!-\!\!H \end{array}
\quad + \quad [O] \quad \longrightarrow \quad R_2C\!\!\diagdown\!\!\begin{array}{c} N \\ \| \\ N \end{array}
$$

Diazirenes decompose slowly in sulphuric acid and more quickly thermally and by photolysis to yield nitrogen gas and hydrocarbon fragments formed from the carbenes generated. When dimethyldiazirene is decomposed, propene is a commonly encountered hydrocarbon fragment.

$$
(CH_3)_2C\!\!\diagdown\!\!\begin{array}{c} N \\ \| \\ N \end{array}
\quad \longrightarrow \quad N_2 \;+\; (CH_3)_2C\!: \;(\longrightarrow \quad CH_3\!-\!CH\!\!=\!\!CH_2)
$$

There are no commercial applications of diazirene or diaziridine, but their molecular structures are occasionally encountered as intermediates in metal-mediated amination reactions.

4.5 Azetidine

It might be thought that azetidine, a four-membered heterocyclic ring with a lower ring strain than three-membered rings, would be more easily formed from a cyclization reaction of a linear precursor than aziridine. However, in practice four-membered rings are highly difficult to prepare and many reagents that might be expected to cyclize to form azetidine instead prefer to form substituted aziridines, e.g. 2,3-dibromo-*N*-propylphenylsulphonamide, in alkaline solution.

This reluctance to form four-membered rings is thought to be partly explained by the reduction in mobility of four-membered linear chains and also by the relatively lower percentage of molecular orientations that can lead to four-membered-ring formation. Azetidine, which is also called trimethyleneimine and azacyclobutane, was first prepared in 1899 and is a colourless liquid with a boiling point of 61 °C [12].

$$H_2C \diagdown CH_2 \, \theta_2 \diagup N \diagdown H$$
$$\theta_1 \quad CH_2$$

$\theta_1 = 146.9°$
$\theta_2 = 88°$

Azetidine ($pK_a = 11.19$) is more basic than aziridine ($pK_a = 7.98$) and has a bent ring with plane angle dependent upon the ring substituents. The high ring strain in aziridine may inhibit the formation of quaternary nitrogen, which is less of a problem in a four-membered ring. The ring strain in azetidine is about one sixth of the ring strain in aziridine on the basis of molecular orbital calculations performed by the author assuming lowest-energy gas-phase conformations (see Appendix 1).

$$C_6H_5-SO_2-NH-CH_2-CHBr-CH_2-Br \longrightarrow X \rightarrow$$

$$\downarrow$$

$$C_6H_5-SO_2-N \diagup^{CH_2Br}$$

The cyclization of 3-aminopropyl halide salts generally gives poor yields of azetidines (aziridines are preferentially formed), but an alternative synthetic method from 3-aminopropanols via a Michael Addition with ethyl acetate gives good yields although it is a three-step process.

$$NH_2-CH_2-CH_2-CH_2-OH + CH_2=CH-CO_2C_2H_5 \rightarrow$$
$$HO-CH_2-CH_2-CH_2-NH-CH_2-CH_2-CO_2C_2H_5 \rightarrow CH_2-N-CH_2-CH_2CO_2C_2H_5$$
$$\text{(a)} \mid \qquad \mid$$
$$CH_2-CH_2$$
$$\text{(b)} \downarrow$$
$$CH_2-N-H$$
$$\mid \qquad \mid$$
$$CH_2-CH_2$$

(a) = $SOCl_2/CH_2Cl_2$
(b) = $KOH/250 °C$

Generally azetidine behaves like a secondary amine with the exception of ring-opening reactions. The ring is thermally stable up to around 360 °C, but azetidine will react with hydrogen chloride to form 3-chloropropylamine hydrochloride and also with hydrogen peroxide to form acrolein and ammonia. With nitrous acid a stable *N*-nitroso derivative is formed, which in the case of 1,3-dimethyl azetidine can be reduced with lithium aluminium hydride to form an *N,N*-disubstituted hydrazine.

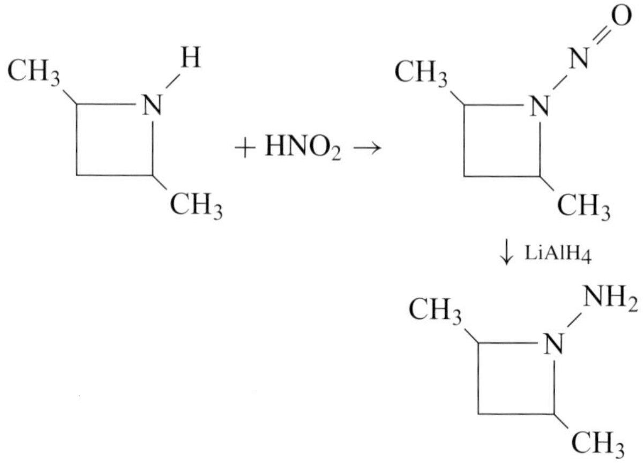

A range of alkyl and aryl *N*-substituted azetidines are known, as are *N,N*-diakyl-substituted quaternary salts, which have a tendency to polymerize. *N*-t-butyl-3-chloroazetidine reacts with potassium cyanide to form a 3-cyano derivative but when heated in vacuum rearranges to form *N*-t-butyl-1-chloromethyl aziridine.

   As well as alkyl-substituted azetidines, many alkyl substituted diazetidines are known. These are even more stable than azetidine. The four-membered ring in *N,N'*-dialkyl-2,4-diphenyl-1,3-diazetidines are exceptionally stable to heat, to refluxing in acid or alkali, and also to reducing agents such as lithium aluminium hydride. They can be formed from the photolytic dimerization of the corresponding *N*-alkylbenzylimine.

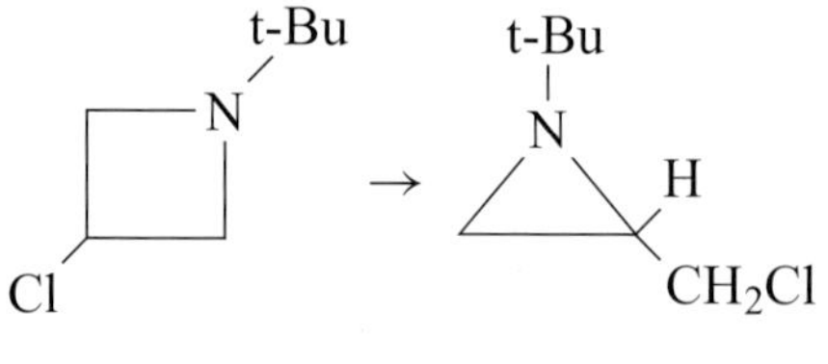

There are no commercial applications of azetidine but a derivative, azetidine-2-carboxylic acid, is present in the leaves of the plants *Polygonatum* and *Convallaria majalis*. It is thought that azetidine-2-carboxylic acid is formed from the biosynthesis of methionine [13]. In contrast to azetidine, the azetidin-2-ones or β-lactams have significant commercial application in penicillin and cephalosporin antibiotics, the chemistry of which has been extensively reviewed by Neu [14].

4.6 Pyrrole

Pyrrole was first isolated by Runge in 1834 from a distillation of coal tar, bone oil and a few other undefined protein derived fractions [15]. He called the unknown substance he obtained pyrrole (from the Greek word *purrhos*, meaning 'red'), as it turned a pine splint dipped in hydrochloric acid a red colour. In 1857 Anderson repeated Runge's experiment using pure bone oil, and in 1860 pyrrole was obtained as a synthetic chemical from the pyrolysis of ammonium mucate.

$$H_4NO_2C-(CHOH)_4-CO_2NH_4 \rightarrow \text{[pyrrole]} + 2CO_2 + NH_3 + [4H_2O]$$

Pyrrole is a colourless liquid (although it does turn brown if exposed to air), miscible with organic solvents and 6% soluble in water. It is both a weak base ($pK_a = -3.8$) and a weak acid ($pK_a = 17.5$) and displays intermolecular hydrogen bonding. The pyrrole molecule is completely planar and has a C–N–C bond angle of 108.9°. Pyrrole reacts readily with electrophilic compounds, can function as a 1,3-diene with activated dienophiles in Diels–Alder Reactions, and takes part in ring-opening reactions.

The structure of pyrrole is a resonance hybrid with six main contributing structures, shown as numbers 1 to 6 in Figure 4.2. There are another three tautomers that have been detected by Raman spectroscopy at low temperatures (7, 8 and 9) but which make only a minor contribution to the molecular properties of pyrrole [16].

The electron densities in pyrrole have been calculated based on possible resonance hybrids, and are illustrated in Appendix 1. They show slightly higher electron densities at ring positions 3 and 4 (1.0854) than at ring positions 2 and 5 (1.0848). In order to be aromatic the pyrrole ring has to have six π-electrons, as predicted by the Hückel rule. Four electrons can be provided by each of the C=C double bonds and the remaining two can be provided by the nitrogen atom if the molecule is positively charged. The consequence is that the nitrogen lone pair is not as freely available for donation as in a typical secondary amine and also that the positively charged nitrogen centre inhibits any approaching protons or cations. These two factors help to explain the weakly basic character of pyrrole. In weakly acidic and basic solutions, the nitrogen-attached hydrogen can be seen to exchange with hydrogen from the solvent by NMR (nuclear magnetic resonance) [17]. The nitrogen-attached hydrogen will also exchange with deuterium if heavy water (D_2O) is present. A molecular-orbital diagram for pyrrole showing five π-orbitals is presented in Figure 4.3, with white and shaded circles used to represent open and closed wave-functions respectively. The measured dipole moment for pyrrole is 1.8 D with the dipole arrow pointing away from the nitrogen atom towards the ring, indicating a net movement of electron density in this direction.

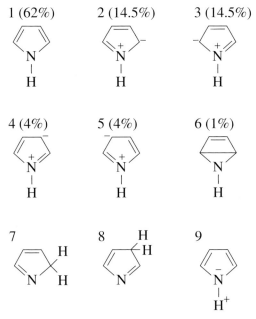

Figure 4.2. The resonance structures of pyrrole.

In many ways pyrrole is the most interesting nitrogen heterocycle from the point of view of its character as an amine, because donation of the nitrogen lone pair to the ring's π-orbitals bridges makes the molecule change from non-aromatic to aromatic in character. Pyridine is always aromatic, and azetidine and aziridine are always non-aromatic, so pyrrole forms the bridge between these two types of molecule.

4.6.1 The synthesis of pyrrole

There are several classical methods for the synthesis of pyrrole such as the Knorr, Paal–Knorr and Hantzsh Syntheses, all of which are described in Appendix 3. Commercially, pyrrole is manufactured by passing furan, ammonia (or a primary amine) and steam over a heated alumina catalyst [18].

$$\text{furan} \quad + \quad R-NH_2 \quad + \quad H_2O \quad \xrightarrow[120°C]{Al_2O_3} \quad \text{pyrrole}$$

Another good manufacturing process is to heat a primary amine with *cis*-but-2-ene-1,4-diol; this can give up to 95% yield in the presence of a dehydration agent and

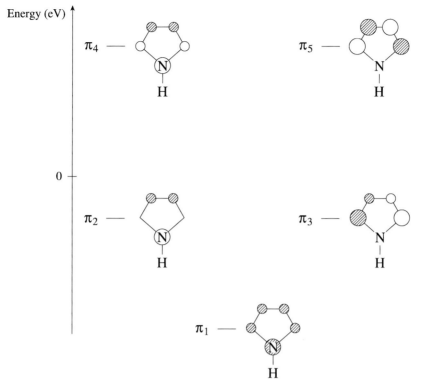

Figure 4.3. Molecular-orbital diagram for pyrrole.

a palladium catalyst.

$$\begin{array}{c} CH{=}CH \\ | \quad \ | \\ HO{-}CH_2 \ \ CH_2{-}OH \end{array} + R{-}NH_2 \xrightarrow[120\,°C]{Pd} \underset{R}{\boxed{N}} + 2H_2O + 2[H]$$

Pyrrole can also be prepared from a Diels–Alder Reaction between a diene and *N*-thionylaniline and from the reaction between diacetylene, ammonia and copper(I) chloride at 140–160 °C.

$$H_2C{=}CR{-}CR{=}CH_2 + C_6H_5{-}NSO \rightarrow \underset{C_6H_5}{\overset{R\qquad R}{\boxed{N}}}$$

$$R{-}C{\equiv}C{-}C{\equiv}C{-}R' + R''{-}NH_2 + CuCl \rightarrow R\underset{R''}{\boxed{N}}R'$$

Finally, pyrrole can be prepared by refluxing succinimide with zinc dust.

4.6.2 The reactions of pyrrole

As previously mentioned, pyrrole behaves as a weak acid and as a weak base. The nitrogen atom can be derivatized and the hydrogen atom replaced by acyl, alkyl or potassium. The pyrrole ring can be opened easily to form both linear and ring expansion compounds and can also be substituted. Pyrrole can be oxidized and reduced without ring cleavage.

(a) N-substitution

When pyrrole is heated with solid potassium hydroxide the imino hydrogen is replaced by the K^+ ion to form potassium pyrrole.

$$C_4H_4N-H + KOH \rightarrow [C_4H_2N]^-[K]^+ + H_2O$$

This reaction does not work with sodium hydroxide, but sodium pyrrole can be prepared from sodamide in liquid ammonia.

Potassium pyrrole reacts with acyl chlorides, acid anhydrides and alkyl halides (including methyl iodide, which will not react with pyrrole). It also gives the Kolbe–Schmidt Reaction to form 2- and 3-pyrrole carboxylic acids and the Reimer–Tiemann Reaction with alkaline chloroform and sodium ethoxide to yield pyrrole-2-aldehyde and 3-chloropyridine.

(Pyrrole-2-aldehyde is a deactivated aldehyde and does not give typical aldehydic reactions such as the Cannizzarro Reaction; the molecule is better thought of as a vinylogous amide instead of a true aldehyde.)

Potassium pyrrole is instantaneously hydrolysed by water and generally the dissociation of the pyrrole metal ionic bond is favourable under conditions leading to *N*-alkylation.

(b) With unsaturated hydrocarbons

Pyrrole will react with unsaturated hydrocarbons (e.g. with acetylene and acrylonitrile) to form *N*-alkyl derivatives in the presence of an alkaline catalyst such as potassium carbonate.

$$H-C{\equiv}C-H + C_4H_4N-H \rightarrow C_4H_4N-CH{=}CH_2$$
$$CH_2{=}CH-C{\equiv}N + C_4H_4N-H \rightarrow C_4H_4N-CH_2-CH_2-CN$$

(c) With Grignard Reagents

Pyrrole reacts with Grignard Reagents like ethyl magnesium bromide to form pyrryl magnesium bromide, as shown below. This derivative is blocked at the nitrogen position and electrophilic reagents usually react mainly at position 2, e.g. the reaction with methylchloroformate.

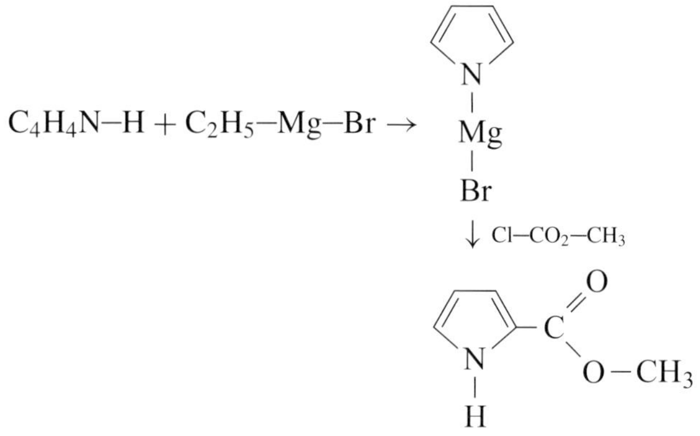

(d) Ring substitution

Pyrrole is attacked by electrophilic reagents at ring positions 2 and 5. Pyrrole is polymerized by sulphuric acid, and a mixture of tars is formed from its reaction with nitric acid but under very carefully controlled conditions it is possible to obtain (or to identify in an NMR experiment the presence of) 2-pyrrole sulphonate and 2-nitropyrrole respectively. Pyrrole reacts readily with halogens although the reaction products are not always stable. Pyrrole gives the Friedel–Crafts Reaction, which often works in the absence of a catalyst. However, unsubstituted pyrroles do not work in the Gattermann Reaction, although substituted pyrroles with deactivated rings give this reaction and the related Houben–Hoesch Reaction.

Pyrrole-2-carbaldehyde can be prepared in 90% yield by the Vilsmeier Reaction in the presence of excess phosphorous oxychloride, and pyrrole will also give the Mannich Reaction with formaldehyde and dimethylamine.

$$C_4H_4N{-}H + H_2C{=}O + (CH_3)_2N{-}H \rightarrow \underset{\underset{H}{|}}{\overset{}{\boxed{N}}}{-}CH_2{-}\underset{\underset{CH_3}{|}}{\overset{\overset{CH_3}{|}}{N}} + H_2O$$

Finally, pyrroles will react at ring position 2 with diazonium salts under mild acidic conditions to form coloured azo derivatives.

(e) Ring-opening reactions

There is marked increase in ring stability with pyrrole compared with azetidine, and aziridine, and the pyrrole ring is stable to most dilute acids and bases. The ring will, however, open on refluxing with ethanolic hydroxylamine hydrochloride to form succinaldehyde dioxime. Hydroxylamine acts to prevent polymerization of the reaction product and the ring-opening reaction is accelerated by the presence of alkyl groups on the ring, as these stabilize the protonation of pyrrole.

$$C_4H_4N{-}H + 2NH_2OH \rightarrow HO{-}N{=}CH{-}CH_2{-}CH_2{-}CH{=}N{-}OH$$

The pyrrole ring can be opened by treatment with silver nitrate solution in an ultrasonic bath, and also by the action of ozone.

(f) Ring oxidation and reduction

Pyrrole can be oxidized by chromium trioxide in acetic acid to form maleimide.

Oxidation with hydrogen peroxide (if carried out very carefully) produces a mixture of two pyrrolidinone tautomers.

In the presence of acidic hydrogen peroxide solution, a protonated pyrrolidinone is obtained. This is unstable in the presence of pyrrole and will immediately react to form dimers.

$$\text{pyrrole} + H_2O_2/[H]^+ \rightarrow \text{protonated pyrrolidinone}$$

$$\downarrow C_4H_4N\text{–}H$$

Pyrrole is partially reduced by zinc and acetic acid to 2,5-dihydropyrrole (pyrroline) and can be completely reduced with nickel and hydrogen at 200 °C to tetrahydropyrrole.

4.6.3 Applications of pyrrole

The pyrrole ring is found in nature in porphyrins and corrins; these will be discussed in Chapter 5, along with chlorophyll, vitamin B12 and haemoglobin. The pyrrole ring is used in many pharmaceuticals, and derivatives of pyrroles are present in plants (carrots and cacti) and also in a poison (2,4-dimethylpyrrole-3-carboxylic acid esters) used by a Colombian tree frog (*Phyllobates aurotaenia*). The pyrrole ring is also present in the two amino acids, proline and 4-hydroxyproline, which are discussed in Chapter 8. Fused pyrroles such as indole are discussed in Section 4.8, and have many applications in life science.

4.7 Pyrazole and imidazole

The term 'azole' is a suffix that is used for five-membered rings with a minimum of two heteroatoms, at least one of which is nitrogen. Both pyrazole and imidazole are based on five-membered rings, and may be thought of, like pyrrazole, as being based on the cyclopentadiene ring but with two ring nitrogen heteroatoms.

Pyrazole Imidazole

Both compounds have many similarities, such as their solubilities in water and their tautomeric structures, but there are also many important differences. For example, pyrazole is found only rarely in living systems, but imidazole derivatives are vital to living organisms and are present in vitamin B12, biotin, histidine and the pilocarpine alkaloids.

Imidazole was first synthesized in 1858 [19] from ammonia and glyoxal; more-recent syntheses have been developed from *para*-aldehyde and also from 2-bromomethyl-1,3-dioxolane, which on heating with formamide and ammonia yields imidazole.

However, the most widely used synthetic procedure for imidazoles is based on the condensation of α-amino aldehydes or α-amino ketone hydrochlorides with aqueous potassium thiocyanate. Desulphurization of the thione intermediate with nitric acid yields imidazole [20].

Imidazole is tautomeric as either of the ring nitrogen atoms can carry the hydrogen atom (structures 1 and 2 below) and charged resonance hybrids (structures 3 and 4) are also known.

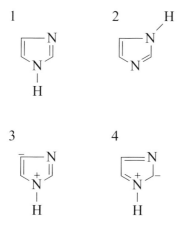

Imidazole has a pK_a of 7.2 and a pK_b of 14.5, and can function both as a weak acid and as a weak base. Imidazole forms strong intermolecular hydrogen bonds and associates of 20 or so molecules in benzene solution with imidazole molecules giving linear hydrogen-bonded chains.

Imidazole forms many stable salts and may be thought of as being between pyrrole and pyridine in its properties. Like pyrrole, imidazole forms a salt with potassium and will also produce a silver salt if mixed with ammoniacal silver nitrate.

Because of the contribution of charged resonance structures 3 and 4 shown above, electrophilic attack on the imidazole ring takes place primarily at ring position 4. Imidazole reacts with iodine and bromine and forms mono- and tri-substituted derivatives. Imidazole can also take part in diazo coupling reactions at ring position 2 (between the two nitrogens) to give red azo dyes. 2-Aminoimidazole can form diazonium salts. The antibiotic azamycin (2-nitroimidazole) is made from 2-aminoimidazole, nitrous acid and copper(II) sulphate.

The imidazole ring is stable to oxidation and strong acids, but opens readily in the presence of benzoyl chloride and sodium hydroxide.

The applications of imidazole derivatives in the personal-care industry are reviewed in Chapter 8. The imidazole ring was widely used in anti-fungal formulations, but these have declined in importance since reaching a maximum market penetration in the 1980s.

The first synthetic pyrazole was prepared in 1884 by Knorr [21] and was named antipyrene because of its antipyretic effect. Antipyrene was widely used as an anti-inflammatory drug (phenazone) until quite recently. A derivative called phenylbutazone was discovered in 1946 and is still used today, mainly for veterinary applications and occasionally for the treatment of gout in humans. (Phenylbutazone is permitted only for pleasure and leisure horses, i.e. those at riding schools and in retirement; its use for race horses is banned as it can make 'old nags' run like the wind and win races unfairly against better and younger competition.) In addition to antipyretics (or non-steroidal anti-inflammatory agents, to give them their full name) the pyrazole ring is found in the food colouring tartrazine and also in β-pyrazol-1-ylalanine, which is found in melon seeds and is one of the very few naturally occurring pyrazolones.

Antipyrene   Phenylbutazone

Pyrazole can be prepared from the reaction between acetylene and diazomethane in cold ether. This is an example of a 1,3-dipolar addition.

An alternative synthetic method is from 1,1,3,3-tetramethoxypropane and hydrazine in the presence of acid. Hydrolysis of 1,1,3,3-tetramethoxypropane gives malondialdehyde, which then reacts with hydrazine. In fact almost any

1,3-dicarbonyl will react with hydrazine to form pyrazole.

$$(CH_3O)_2-CH-CH_2-CH(OCH_3)_2 + [H]+ \rightarrow OHC-CH_2-CHO$$

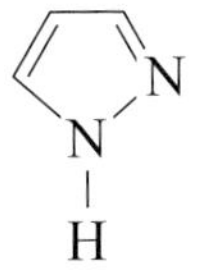

Like imidazole, pyrazole displays tautomerism and positions 3 and 5 are equivalent.

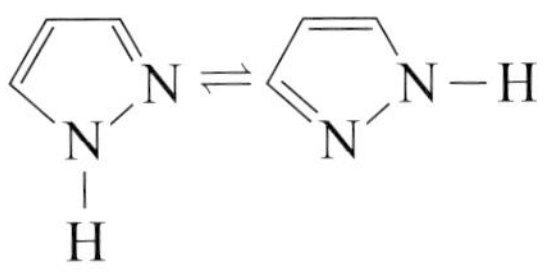

Unlike imidazole, which forms hydrogen-bonded linear aggregates in solution, pyrazole forms hydrogen-bonded dimers of the type shown below.

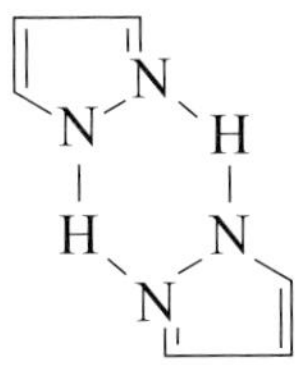

Although pyrazole will react with iodine at ring position 4 (like imidazole) it does not couple with aromatic diazonium salts. Pyrazole is also readily nitrated and sulphonated at position 4.

Two important differences between imidazole and pyrazole are their boiling points and their pK values.

	Boiling point at 760 mmHg	pK_a	pK_b
Imidazole	256 °C	7.2	14.0
Pyrazole	187 °C	2.5	14.5

The difference in boiling points can be explained by the differences in the extent of intermolecular hydrogen bonding, but the difference in pK_a values is slightly harder to explain. Both the imidazole and pyrazole rings contain two types of nitrogen, the imine sp^2 type (as found in pyridine, $pK_a = 5.14$) and the amine sp^3 type (as found in pyrrole, $pK_a = -3.8$). The sp^2 (pyridine-type) nitrogen in pyrazole has a pK_a of 2.5. However, the sp^2 (pyridine-type) nitrogen in imidazole is shielded from the sp^3

(pyrrole-type) nitrogen by a carbon atom and so it has a stronger basic character as its lone pair is more easily donated, and hence it has a higher pK_a of 7.2. This value is similar to that for 2-aminopyridine ($pK_a = 6.86$), which also has both types of nitrogen separated by a carbon atom.

4.8 Fused five-membered nitrogen heterocycles

There are three structural possibilities for a five-membered heterocyclic ring such as pyrrole to fuse onto a benzene ring, and these are represented by indole, isoindole and indolizine. The benzene-fused derivative of pyrazole is called benzopyrazole (or indazole) and the benzene-fused derivative of imidazole is called benzimidazole. Benzotriazole is also known, although the unfused, unsubstituted triazole ring is unstable.

Indole   Isoindole    Indolizine

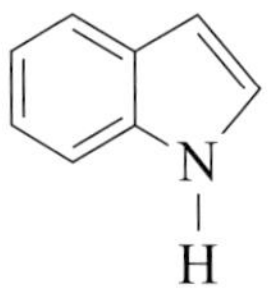

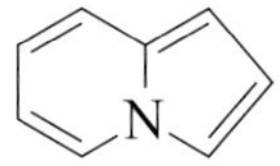

Benzopyrazole 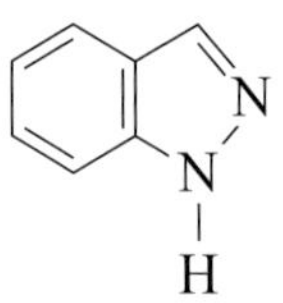   Benzimidazole 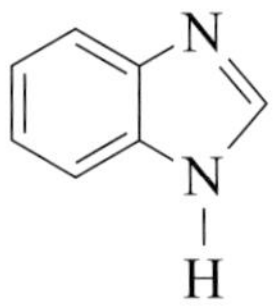   Benzotriazole

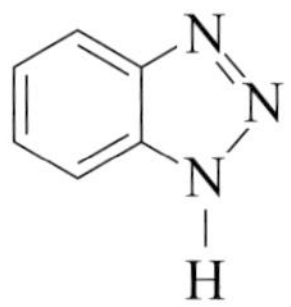

Of the above structures, the indole ring is probably the most important as it is found naturally in tryptophan and its derivatives, such as tryptamine, serotonin, auxins, and also in plant pigments (as indigo derivatives) and plant fragrances such as jasmine and orange blossom. Skatole, or 3-methylindole, will be familar to most readers as it is one of the unpleasant-smelling constituents of faeces.

Indole was prepared for the first time in 1866 from the distillation of oxindole with zinc dust [22], and the bicyclic structure of indole was elucidated for the first time by Baeyer in 1869 [23]. Indole is a colourless crystalline solid that melts at 52 °C and boils at atmospheric pressure at around 254 °C, although there is some decomposition. The resonance energy of indole is around 200 kJ mol^{-1}, which is a little less than the arithmetic sum of adding the resonance energy of pyrrole (105 kJ mol^{-1}) to the resonance energy of benzene (150 kJ mol^{-1}) but is in line with the increase in resonance energy in going from two benzene molecules to one molecule of naphthalene (resonance energy 254 kJ mol^{-1}).

= Indole

$105 \text{ kJ mol}^{-1} + 150 \text{ kJ mol}^{-1}$ (theory) $= 255 \text{ kJ mol}^{-1}$

Determined $= 200 \text{ kJ mol}^{-1}$

As in pyrrole, the electron density at ring position 3 in indole is higher than at ring position 2 (1.11 and 1.02 π-electrons respectively). However, electrophilic reagents prefer to attack indole at ring position 3 instead of at ring position 2, as found in pyrrole. The reason for this is that fusion with a benzene ring favours certain tautomers with extended resonance that create negative charge at ring position 3. In pyrrole, although ring position 3 has a slightly higher electronic charge, localization energies favour electrophilic attack at ring position 2. If ring position 3 in indole is blocked, reaction takes place at ring position 2. If both 3 and 2 are blocked the benzene ring is attacked at position 6.

Some classical synthetic methods for indole are given in Appendix 3 and these include the Fischer, Bischler, Reissert, Nenitzsecu and Madelung Reactions. Other methods of synthesis include the following.

(i) The reaction of 2-aminoarylketones with α-bromoketones.

(ii) The fusion of 2-nitrophenylacetic acid with sodium hydroxide in the presence of iron powder (this was the synthesis of Baeyer in 1869 [23]).

Like pyrrole, indole is easily *N*-protonated in solution and forms dimers and trimers. Electron-donating groups in the fused pyrrole ring assist *N*-protonation by stabilizing the positive charge, and electron-withdrawing groups have exactly the opposite effect. The N-hydrogen in indole is reasonably acidic and can be replaced with sodium or potassium.

Electrophilic reagents attack ring position 3 as previously explained, except for sodium hypochlorite, which yields the *N*-chloro derivative.

The indole ring is not susceptible to nitration and is decomposed by nitric acid, although 3-nitroindole can be prepared by refluxing with sodium methoxide and ethyl nitrate. The alkylation of indoles works only at high temperatures and occurs at ring position 3. The benzylation of indoles gives very low yields of mixed products and is generally not recommended as a synthetic procedure.

Indole-3-carbaldehyde can be prepared from the Vilsmeier–Haack Reaction with *N,N*-dimethylformamide/phosphorous oxychloride [24].

Indole-3-carboxaldehyde can be readily converted to tryptophan by the oxazolone method via condensation with *N*-acetylthiohydantoin and reduction with sodium/mercury amalgam. L-Tryptophan is an important amino acid and can be biosynthesized from serine pyrophosphate and indole by the mould *Neurospera crassa*. The chemistry of tryptophan is examined in Chapter 8.

Tryptophan

Indole gives the Mannich Reaction with formaldehyde and dimethylamine to yield gramine, which can then react with methyl iodide to form a quaternary trimethyl ammonium salt. When treated with potassium cyanide this salt forms 3-indoleacetonitrile, which can be easily hydrolysed to indole-3-acetic acid; this is better known as auxin, the plant-growth hormone.

(indole-3-acetic acid = auxin)

Unlike indole, which had been known since 1866, isoindole was first isolated only as recently as 1972 [25] from the thermal dissociation of a 1,4-amino-bridged tetralin derivative. Isoindole decomposes rapidly in air at room temperature, although 2-alkyl isoindoles are stable. The isoindole structure occurs in phthalocyanine dyes, and these are reviewed in Chapter 8. The resonance energy of isoindole is only 48.6 kJ mol^{-1}, which is less than a quarter of the resonance energy of indole.

Indolizine, which is occasionally called pyrrocoline in old chemistry books, is isolectronic with indole but is a much weaker base. Indolizines protonate at position

3, which is also the site of electrophilic attack. Indolizine is a colourless solid with a melting point of 74 °C and has no commercial applications but has interesting fluorescent properties that have been studied. Indolizine can be synthesized in good yield from 2-methyl pyridine.

$$+ \quad (i) \; Br-CH_2-CO-CO_2C_2H_5 \; \rightarrow$$
$$(ii) \; NaHCO_3$$
$$(iii) \; NaOH$$
$$(iv) \; \text{thermal decarboxylation}$$

Benzopyrazole can be prepared from cyclohexanone [26] and resembles pyrazole in its properties. It rearranges to form benzimidazole under photolytic conditions but to date it has not been the subject of extensive research unlike benzimidazole, which occurs in vitamin B12.

Benzimidazole can be prepared from the reaction between 1,2-phenylene diamine and formic acid. Benzimidazole has a high degree of aromaticity and is stable to mild oxidation and reduction. It is reduced by permanganate solution to form imidazole-4,5-dicarboxylic acids, and forms metal derivatives and salts with acids. In mixed nitric acid/sulphuric acid solution it nitrates at positions 5 and 6, and in alkaline halide solutions it is attacked at ring position 2. When 1,2-phenylene diamine is heated with mixed sodium nitrite and acetic acid, benzotriazole is formed instead of benzoimidazole.

$$(+CH_3CO_2H/NaNO_2) \downarrow \qquad \qquad \downarrow \; (HCO_2H/HCl)$$

Benzotriazole is useful as an amination reagent for the small-scale synthesis of amines [27] and its 1-hydroxy derivative has commercial application as a coupling agent for peptide synthesis [28]. Both of these applications occur by a similar mechanism and illustrate an unusual tautomeric exchange that is possible in benzotriazole. Benzotriazole-mediated amination was developed by Katritzky and Rachwal [27], and starts with the reaction between an aldehyde, a secondary amine and benzotriazole to form a tautomeric system. In this system it is not possible to specify which nitrogen atom the alkylamino group is attached to and there is rapid exchange. A similar tautomerism is displayed when 1-hydroxy benzotriazole is *N*-acylated and the carbonyl streching frequency shifts from 1815–1820 cm^{-1} down to 1730 cm^{-1} upon reaction with benzotriazole.

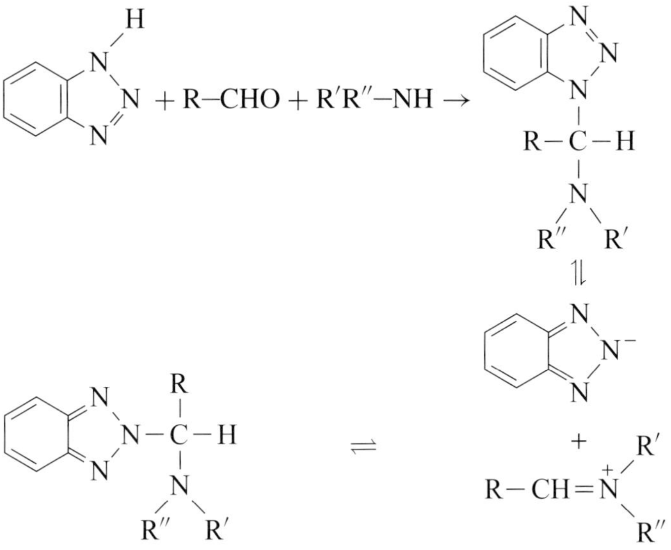

$$\text{benzotriazole–}H + R\!-\!CHO + R'R''\!-\!NH \rightarrow$$

$$R\!-\!\underset{\underset{\overset{\displaystyle N}{R''\quad R'}}{|}}{\overset{\displaystyle |}{C}}\!-\!H$$

$$\rightleftharpoons$$

benzotriazole$^-$ $+$ $R\!-\!CH\!=\!\overset{+}{N}\!\!\diagdown{\!R''}^{\!\!R'}$

The addition of a nucleophile, Nu, to the above tautomeric system forms a product with the following structure.

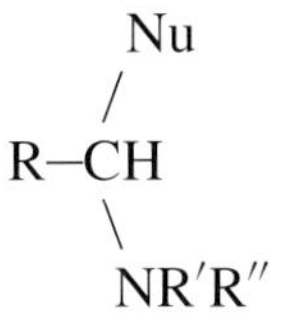

$$R\!-\!\overset{\displaystyle Nu}{\underset{\displaystyle NR'R''}{CH}}$$

When 1-hydroxy benzotriazole reacts with a carboxylic acid of the form $R\!-\!CO_2H$, which may also be an *N*-protected amino acid or an *N*-protected peptide, the tautomeric system is as follows.

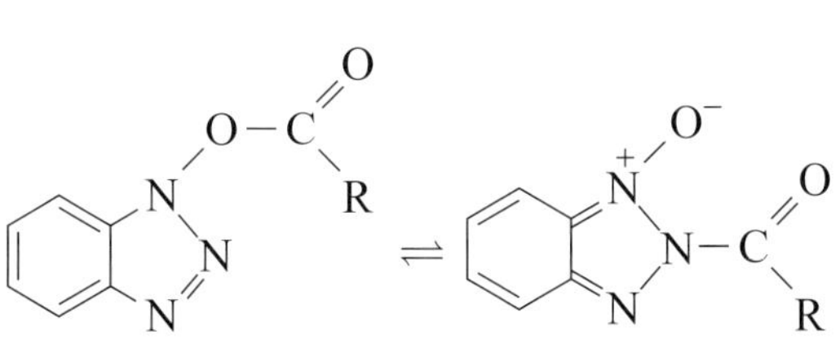

$$\text{(benzotriazole)}\!-\!O\!-\!\overset{\overset{\displaystyle O}{\|}}{C}\!\diagdown R \quad \rightleftharpoons \quad \text{(benzotriazole)}\!-\!C\!\diagdown R \ (O^-)$$

Addition of a primary amine $R'\!-\!NH_2$ to this system yields an amide of the form $R'\!-\!NH\!-\!COR$. This reaction works well in dimethylformamide at low temperatures without racemization, and gives peptide in high yields from amino acids with unreactive carboxyl groups.

4.9 Pyridine

Pyridine was first isolated as a pure compound in 1849 by Anderson [29], who obtained it from the thermal degradation of bones. Its structure was determined in 1864 by Körner (and also independently by Dewar), who noticed that although pyridine was basic it resembled benzene in many of its properties [30]. Pyridine has a resonance energy of 125.5 kJ mol^{-1} and because of its large dipole of 2.23 D it may be regarded as a resonance hybrid.

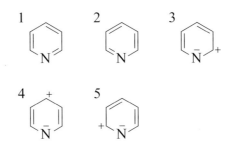

The nitrogen lone pair is not required for aromaticity in pyridine, and overall the nitrogen atom is electron withdrawing from the ring. For this reason resonance hybrids with positive nitrogen are not shown here and are not favourable for pyridine, as evidenced by the dipole moment, which is indicative of a concentration of negative charge on the ring nitrogen atom (see Appendix 1). Of the charged resonance structures shown above, 3 and 5 are most favoured as the separation of charge is less than for structure 4. Nucleophiles will therefore preferentially attack the pyridine ring at position 2 and electrophiles will preferentially attack at position 3.

Because the nitrogen lone pair is not required for aromaticity it might be expected that pyridine would be a strong base with a pK_a similar to an alkyl amine; however, the pK_a of pyridine is only 5.23. This value is a lot lower than that for the alkyl amines, which have pK_a values generally between 9 and 11.

In fact the pK_a of pyridine is nearer to that of the aryl amines (pK_a values generally between 1 and 7) than the alkyl amines. One explanation for this is that the sp^2-hybridized nitrogen in pyridine is more electrophilic than the sp^3 nitrogens in alkyl amines, and so its lone pair is more tightly bound and less available for donation. Unlike pyrrole, donation of the lone pair on the pyridine ring does not destroy the aromatic properties of the ring.

Theoretically pyridine may be considered to be a derivative of benzene in which one C–H unit has been replaced by a nitrogen. The aromatic character is essentially unchanged although the two paired π-molecular orbitals (π5 and π4, and π2 and π3) are no longer degenerate and display a small difference in energy. The molecular orbitals for pyridine are shown in Figure 4.4 with white and shaded circles representing open and closed wavefunctions respectively.

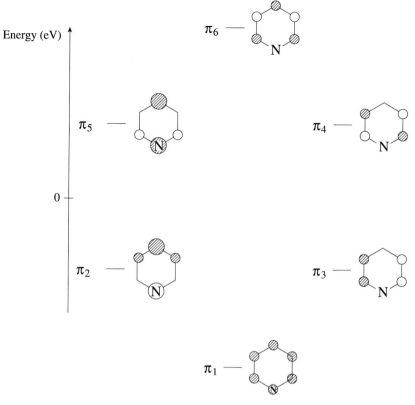

Figure 4.4. A molecular-orbital diagram for pyridine.

Pyridine can be synthesized by several classical methods such as the Hantzsh Synthesis and the Guareshi and Thorpe Reactions, which are given in Appendix 3. Some other methods of synthesis that can be used for pyridine include the following.

(i) From the reaction between glutaconic aldehyde and ammonia.

$$OHC-CH=CH-CH=CH-OH + NH_3 \rightarrow$$

(ii) From the thermolysis of pentamethylene diamine hydrochloride followed by catalytic dehydrogenation with palladium on a carbon support.

$$[H_3N-CH_2-CH_2-CH_2-CH_2-CH_2-NH_3]^{2+}2[Cl]^-$$

$$\rightarrow \quad + \quad H_2/Pd-C \rightarrow$$

(iii) From the reaction between coumarins or pyrones and ammonia.

Reactions of pyridine can occur at the nitrogen atom to form quaternary salts or cause ring opening, and also in the aromatic ring, which may be substituted by both nucleophilic and electrophilic reagents. However, before looking at the reactions of pyridine it is worth mentioning a few chemical reactions that either do not work well or do not work at all with pyridine. These are the Friedel–Crafts Reaction, the Diels–Alder Reaction and direct nitration and sulphonation reactions. The ring in pyridine may be considered to be deactivated compared with benzene. It is sometimes compared with nitrobenzene as both molecules behave similarly towards nucleophilic reagents. However, this comparison has its limitations and is not always valid, as nitrobenzene is a far weaker base than pyridine.

Pyridine reacts with halogens (X) at ambient temperature to form N-halogenpyridinium halides of the form $[C_5H_5N^+X][X]^-$. When heated with alkyl halides, $R-X$, quaternary salts of the form $[C_5H_5N^+-R][X]^-$ are formed. N-methyl pyridinium salts can also be prepared fron the reaction between dimethyl sulphate and pyridine. When N-methyl pyridinium halide salts are heated at 300 °C they undergo the Hofmann–Martius Rearrangement to form 2- and 4-methyl pyridines (see Appendix 3). When N-methyl pyridinium hydroxy salts are heated, anhydro bases are formed if there are 2- or 4-substituent alkyl groups present on the pyridine ring.

Pyridine also reacts with acyl chlorides to form N-acylpyridinium halides, which can then act as acylating agents for alcohols and other amino compounds.

$$C_5H_5N + R-CO-Cl \rightarrow [C_5H_5N^+-CO-R][Cl]^-$$
$$\downarrow R'-OH$$
$$R-CO_2R' + [C_5H_5N^+][Cl]^-$$

Chlorination of pyridine in the presence of aluminium chloride forms exclusively 3-chloropyridine. If the 2-chloro or 4-chloro isomers are required then these have to made starting from amino pyridines via the diazonium salt. The 2- and 4-chloropyridines are reactive and the chloro groups can be replaced easily by hydroxy, cyano or amino groups. Because pyridine forms complex ions with aluminium chloride and other Lewis Bases, the alkylation and acetylation of pyridine cannot be accomplished by the Friedel–Crafts Reaction.

Pyridine will react with nitric acid at 300 °C only in a mixture of concentrated sulphuric acid and fuming nitric acid. The product is 3-nitropyridine. However, if the pyridine ring contains a hydroxy or amino group then nitration with nitric acid proceeds normally in a similar manner to the nitration of benzene.

The reaction between pyridine and sulphuric acid also requires aggressive conditions and occurs at 220 °C in concentrated sulphuric acid containing 20% oleum to produce pyridine-3-sulphonic acid. Mercuric sulphate is often added to the reaction mixture as a catalyst. The sulphonyl group can easily be replaced by a hydroxy or a cyano group.

The nitrogen atom in pyridine can be lithiated, and alkylation takes place at position 2 if an alkyl lithium such as methyl lithium or butyl lithium is used. However, care must be taken not to add methyl lithium in excess as 2-methyl pyridine can be lithiated and by-products can be formed.

(Here $Y = CH_3$ or $-CH_2-CO_2C_2H_5$.)

Pyridine will react with strong nucleophiles such as sodamide in toluene and potassium hydroxide at 320 °C to form 2-amino or hydroxy or 2,6-disubstituted amino or hydroxy derivatives respectively.

Actually 2- and 4-hydroxy pyridines are really pyridones, as the $-OH$ group is present only in small quantities. For 3-hydroxy pyridine, where rearrangement to a pyridone is not possible, phenolic properties are observed.

Minor tautomer   Major tautomer

As mentioned earlier, pyridine can form a range of quaternary salts. When the substituent group attached to the nitrogen atom is a strong electrophile, the pyridine ring can easily be opened by nucleophiles. Good examples of substituents that facilitate ring opening are the phenyl group $-C_6H_5$, the cyano group $-C\equiv N$ and the sulphonate group $-SO_3$. Azulene can be easily prepared from N-phenyl pyridinium chloride by reaction with N-methylphenylamine followed by heating with sodium ethoxide in the presence of an auxiliary base.

$$[C_6H_5-N(CH_3)-CH=CH-CH=CH-CH=N(CH_3)-C_6H_5]^+[Cl]^-$$

Ring-opening reactions of pyridine can also take place photolytically, and this is the reason why pyridine turns slightly yellow in sunlight. However, the photochemistry of pyridine has been investigated only at low temperatures, where 2-azabicyclo[2.2.0]hexadiene can be formed. This compound is also known as Dewar pyridine [31].

The best-known N-substituted pyridine is probably pyridine N-oxide (also called pyridine-1-oxide), which can be easily made from the reaction between pyridine and hydrogen peroxide at 70–80 °C. This synthesis is not limited to hydrogen peroxide, and per acids such as peracetic acid and perbenzoic acid can also be used. Pyridine N-oxide exists as a resonance hybrid, which can act either to increase or decrease the electron density on ring atoms 2 and 4. The dipole moment of pyridine N-oxide (4.42 D) is lower than calculated from theory, indicating that resonance structures play a significant part in determining the reactivity of pyridine *N*-oxide. Pyridine *N*-oxide is a much weaker base than pyridine, but has been studied extensively as it can easily undergo many reactions that are difficult for pyridine as it is more reactive to both electrophiles and nucleophiles than pyridine alone. For example, pyridine *N*-oxide can be nitrated at $100\ °C$ in mixed sulphuric and nitric acids, and once the 4-nitropyridine *N*-oxide has been formed the pyridine ring can be regenerated in phosphorous trichloride and an inert solvent such as dichloromethane. Unfortunately the reaction between pyridine *N*-oxide and sulphuric acid is not markedly improved on from the reaction between pyridine and sulphuric acid. Substitution reactions with pyridine *N*-oxide usually occur at the 4-position; however, methyl iodide attacks the oxygen and forms *N*-methoxypyridinium salts, which can react with nucleophiles predominantly at the ring position 2. 4-Aminopyridine *N*-oxide, which can be prepared by the reduction of 4-nitro-pyridine *N*-oxide, is interesting because although it might be thought that it would form a tautomeric system it does not, and the amino group can be converted into a diazonium salt.

Pyridine *N*-oxide can be used as an oxidation reagent and can convert the α-bromo derivatives of carboxylic acids to aldehydes in a toluene or xylene solvent with the loss of a carbon. Pyridine *N*-oxide can also react with acid anhydrides (or carboxylic acids in the presence of acetic anhydride to convert them to anhydrides) to form aldehydes in an analogous manner, but with the formation of by-products.

$$n\text{-Pr}-\text{CH(Br)}-\text{CO}_2\text{H} \xrightarrow{\textit{(xs pyridine N-oxide/toluene)}} n\text{-Pr}-\text{CHO}$$

$$\text{C}_6\text{H}_5-\text{CH}_2-\underset{\underset{\text{C}_6\text{H}_5}{|}}{\text{C}}-\text{CO}_2\text{H} \rightarrow \text{C}_6\text{H}_5-\text{CH}_2-\text{CO}-\text{C}_6\text{H}_5 + \text{C}_6\text{H}_5-\text{CH}_2-\text{CRR}'-\text{C}_6\text{H}_5$$

(Here $R = CH_3-CO_2-$ and $R' = H$ or CH_3-CO_2-.)

There are several methods for converting pyridine *N*-oxide back to pyridine. The use of phosphorous trichloride has already been mentioned, but also nickel or iron in the presence of acetic anhydride have been reported to work well for 4-chloro- or nitro-substituted pyridine *N*-oxides, and sulphur dioxide and hydrogen with palladium supported on carbon can also be used.

The total worldwide production of pyridine (and pyridine bases) is around 90 000 tonnes per year. This amount is split between pyridine obtained from natural

Figure 4.5. The structures of nicotinic acid (a) and nicotinamide (b), both of which are called niacin.

sources and synthetic pyridine. Pyridine can be obtained from coal and oil shale, and roughly 1 t of coal will contain 0.1–0.03 kg of pyridine-containing residues from which pyridine can be obtained by distillation in a yield of about 55%. Synthetic pyridine is produced by a vapour-phase process from the reaction between formaldehyde, acetaldehyde and ammonia over zeolites at 350 °C [32]. The product from this reaction is a mixture of pyridine and 3-methyl pyridine, which then has to be separated by distillation. A variation of this process is to replace both formaldehyde and acetaldehyde by acrolein. However, the acrolein process produces a greater percentage of 3-methyl pyridine in the product mixture. Pyridine is widely distributed in nature as nicotinamide and as nicotinic acid (see Figure 4.5). Biochemists refer to both of these compounds as niacin as both are precursors to the nicotinamide nucleotide coenzymes NAD (nicotinamide adenine dinucleotide) and NADP (nicotinamide adenine dinucleotide phosphate). In older biochemistry books niacin is sometimes referred to as vitamin B3 but this name is incorrect and is no longer used for niacin. The role of niacin in nutrition has been reviewed by Bender [33].

The synthesis of NAD, NADP and another coenzyme, $CH_3CO-SCoA$, is the main fate of dietary tryptophan (with the exception of the small amount of tryptophan used for protein synthesis and for the neurotransmitter 5-hydroxy tryptophan), which take place by a six-step, seven-step and eight-step biosynthesis, respectively.

Another well-known pyridine derivative is the alkaloid nicotine, which occurs in the leaves of the tobacco plant.

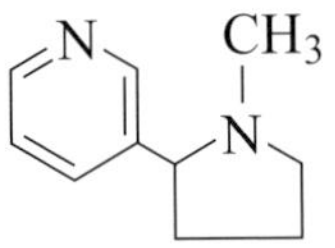

Nicotine

Derivatives of nicotine have been developed in Japan as pharmaceuticals to 'kick start' the hearts of patients who have suffered cardiac arrest. Another pharmaceutical application of nicotine is in the manufacture of nicotine transdermal patches and gums to assist people who wish to stop cigarette dependence.

(a)   (b)   (c)

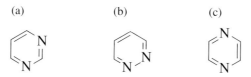

Figure 4.6. The structures of pyrimidine (a), pyridazine (b) and pyrazine (c).

The pyridine ring is is also present in the structure of vitamin B6. This is also called pyridoxine and it is synthesized by intestinal flora. Dietary defficiencies of vitamin B6 are rare, but in the case of gross clinical deficiencies its absence can result in skin lesions, ulceration and convulsions.

About half of the pyridine produced annually is used in the manufacture of agrochemicals, particularly insecticides. These will be described in Chapter 8. Typical agrochemicals produced from pyridine include Diquat, Chloropyrifod and Paraquat. Other important uses of pyridine are as an industrial solvent, in the manufacture of the anti-dandruff shampoo additive zinc pyrithionate, and in many pharmaceuticals. The two most important pyridine derivatives are 2-hydroxy, 3,5,6-trichloropyridine, for agrochemical production, and 2-vinyl pyridine, which is used in terpolymer latexes.

4.10 Pyrimidine, pyrazine and pyridazine

Six-membered rings with two nitrogen atoms are known collectively as the diazines, of which there are three isomers, pyrimidine, pyridazine and pyrazine, as shown in Figure 4.6.

Of the three diazines, the most studied is pyrimidine. This ring structure is found in many naturally occurring nucleic acids such as cytosine, 5-methylcytosine, thymine and uracil. The pyrimidine ring is also common in the structures of bacterial folate antagonists such as the pharmaceuticals trimethoprim, sulphondiazine and sulphacytine.

Pyridazine is a weak base, with a pK_a of 2.33. It will accept only a single proton and even the strongest alkylation agents will produce only monoalkylated derivatives. Pyridazine is unreactive with electrophilic reagents, but pyrazine *N*-oxide will react with electrophiles in an analogous manner to pyridine *N*-oxide. Pyridazine can be prepared from maleic anhydride and hydrazine via a five-stage

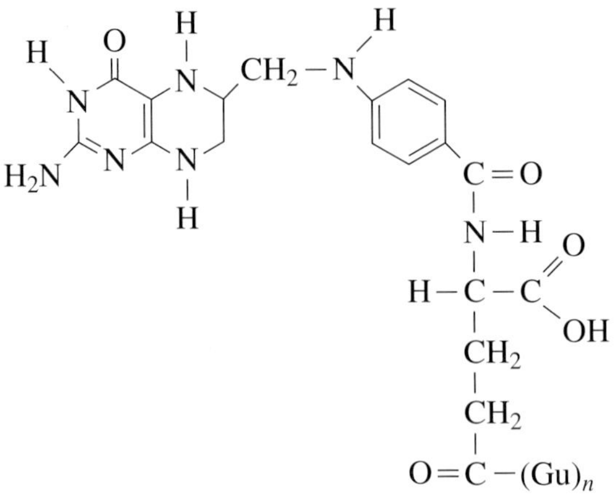

Figure 4.7. The structure of folic acid.

synthesis that involves a tautomeric intermediate.

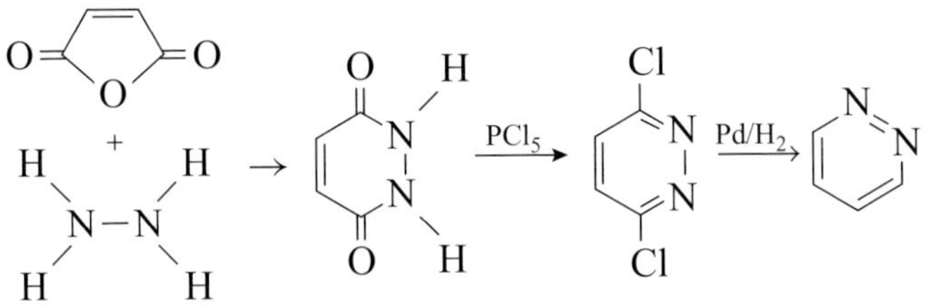

Pyrazine is an even weaker base than pyridazine, with a pK_a of 0.6. Unlike pyridazine it can be monoalkylated by normal reagents such as alkyl halides, and dialkylated by strong alkylators such as triethyloxonium tetrafluoroborate. Pyrazine is not attacked at all by electrophiles and the ring is stable to ring opening by electrophiles. Pyrazine can be chlorinated at 400 °C to give 2-chloropyrazine and reacts with sodamide to form 2-aminopyrazine.

Pyrazine can be prepared from the vapour-phase dehydrogenation of piperazine, and substituted pyrazines can be prepared fron the self-condensation of α-amino compounds, as shown below.

$$4R-CO-CHR-NH_2 + O_2 \rightarrow 2H_2O + 2 \text{ (pyrazine ring)}$$

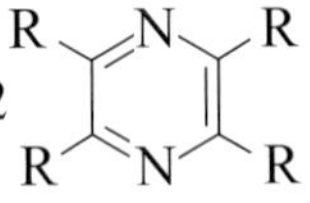

The mould *Aspergillus* produces the antibiotic aspergillic acid. This occurs naturally and contains a pyrazine ring, but the pyrazine ring is more widespead in nature in its fused form as pteridine derivatives such as folic acid (see Figure 4.7) and riboflavin (vitamin B2).

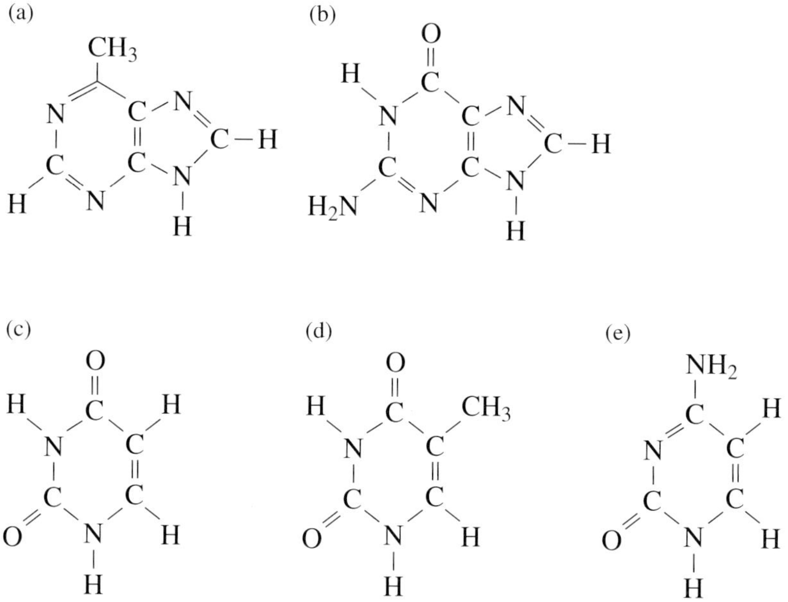

Figure 4.8. Structures of pyrimidine bases: (a) adenine, (b) guanine, (c) uracil, (d) thymine and (e) cytosine.

In contrast to pyridazine and pyrazine, the pyrimidine ring (which was discovered in 1845 but not synthesized in its pure form until 1900 [34]) is widely distributed in living systems and is found in vitamins B1 and B2, purines, uric acid and coenzymes with uracil and cytosine side chains that are essential to the biosynthesis of lipids and complex carbohydrates. The pyrimidine ring is also found in the structures of several pharmaceuticals including barbiturates, antimalarials, antimetabolites and antibacterials. As the chemistry of life has evolved over several millennia it is worth looking at the properties of pyrimidine to understand why it is found in so many biologically important molecules that are essential to life. The pyrimidine molecule is planar and has good potential for forming hydrogen bonds. In nucleic acids, pyrimidine is usually present as uracil (which has two carbonyl groups), as thymine (which has two carbonyl groups and a methyl substituent) or as guanine or cytosine (which have one carbonyl group and one amino group each in their molecular structure) (see Figure 4.8).

All of these molecules have a good capacity for the formation of hydrogen bonds at the substituent amino groups in adenine, cytosine and guanine, and at the ring amino groups in positions 1 in adenine and guanine, and position 3 in uracil, thymine and cytosine. The oxygen atoms at position 2 in cytosine, position 4 in thymine and uracil, and position 6 in guanine, are all very electronegative and form strong hydrogen bonds. These five molecules are the most important nitrogen-containing bases found in nucleotides and are found in all living cells. Watson and Crick identified the hydrogen-bonded pairs adenine–thymine (A–T) and guanine–cytosine (G–C) as being important in maintaining and regulating the stacking of helical DNA [35]. The dependence of DNA on the pyrimidine ring is thought to be a consequence both of the molecule's size and also its ability to form hydrogen bonds. Pyrimidine is easily available to living systems from the molecular building blocks carbamoyl phosphate (from the urea cycle) and aspartic acid, which are coupled by the enzyme aspartate transcarbomoylase.

$$[H_2N\!-\!CO\!-\!OPO_3]^{2-} + HO_2C\!-\!CH(NH_2)\!-\!CH_2\!-\!CO_2H \rightarrow$$

Carbamoyl aspartate cyclizes to form orotic acid; this then undergoes another five bio-reactions before forming cytidine triphosphate, which adds to ribose.

Pyrimidine is only slightly soluble in water and is weakly basic, but will form salts with acids. Pyrimidine is best regarded as a resonance hybrid of the following four structures and, like pyridine, is a deactivated ring.

Ring position 5 has the greatest π-electron density (1.103) and this is the site of electrophilic attack. Although the ring is relatively resistant to attack by nucleophiles, any nucleophilic attack is expected at positions 2, 4 and 6, which have π-electron densities of 0.776, 0.825 and 0.825 respectively. When a hydroxy or amino group is present in positions 2, 4 or 6 the aromaticity of the pyrimidine ring is lowered, as the keto and imino tautomers are known to exist in biological systems. Amino groups at positions 2 and 4 are easily replaced by hydroxy groups by heating with water. The carbon–carbon bond lengths in pyrimidines have been measured by X-ray diffraction and are in the range of 1.35–1.40 Å; this is comparable to benzene (1.40 Å) but is longer than found in ethene (1.33 Å) [36]. The resonance energy of pyrimidine (109 kJ mol^{-1}) is lower than found in pyridine (129.5 kJ mol^{-1}) and benzene (150.5 kJ mol^{-1}). Pyrimidine can best be considered as being a pyridine derivative and to a lesser extent as a cyclic amidine. It is a much weaker base than pyridine and imidazole, and will accept two protons under strongly acidic conditions (pK_a = 1.3 and −6.9). Protonation does not, however, affect the ring resonance energy.

Electrophilic substitution at ring position 5 is similar to ring substitution at ring position 3 in pyridine. Similarly nucleophilic substitution at positions 2, 4 and 6 in pyrimidine correspond to substitution at ring positions 2 and 4 in pyridine.

The earliest synthesis of a pyrimidine ring is that of barbituric acid in 1879 from urea, malonic acid and phosphorous oxychloride [37].

$$HO_2C{-}CH_2{-}CO_2H + H_2N{-}CO{-}NH_2 \xrightarrow[(POCl_3)]{}$$

In 1900 Gabriel treated barbituric acid with zinc powder and hot water and produced unsubstituted pyrimidine [34]. Another synthetic route to pyrimidine uses 1,1,3,3-tetraethoxypropane and formamide in the presence of aluminium oxide at 200 °C.

$$(C_2H_5O)_2{-}CH{-}CH_2{-}CH(OC_2H_5)_2 + 2HCONH_2 \rightarrow$$

The pyrimidine ring is generally stable to most reagents (with the exception of hot sodium or potassium hydroxide solutions) but can be attacked by sodamide and Grignard Reagents such as phenylmagnesium bromide to give a mixture of 2-, 4- and 6-substituted products. If there are activating groups present on the ring (such as amino or hydroxy substituents) at positions other than 5, then this ring position can be attacked by electrophilic reagents. It is possible to prepare diazonium salts in this way. Finally, 5-nitro pyrimidine displays an unusual addition reaction and forms 5-nitro,6-hydroxy-1,6-dihydropyrimidine on mixing with water.

4.11 Quinoline, isoquinoline and quinolizine

There are three heterocyclic compounds that could be formed from the replacement of a single carbon in the molecular structure of naphthalene with a nitrogen. These three compounds are quinoline, isoquinoline and the quinolizinium cation, as shown below as (a), (b) and (c) respectively.

(a)   (b)   (c)

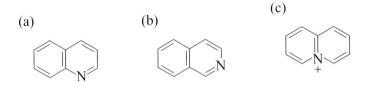

Quinoline is a colourless liquid at room temperature and is only slightly soluble in water (0.7% at 25 °C). The cyclic structure of quinoline was elucidated by Körner [38]. The molecule exists as a resonance hybrid with a resonance energy 198 kJ mol^{-1}, with the most prevalent resonance hybrid being the one normally shown with symmetrical double bonds. The π-electron densities for quinoline are as shown below, from which nucleophilic attack can be predicted at positions 2 and 4 and electrophilic attack at positions 1, 3, 6 and 8.

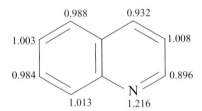

Like pyridine, quinoline forms acid salts and also an N-oxide. Quinoline has a pK_a value of 4.94, which is similar to that of pyridine (p$K_a = 5.23$), and a dipole of 2.10 D. Quinoline is sulphonated at 220 °C to give quinoline-8-sulphonic acid and also reacts with nitric acid at 0 °C to give a mixture of 5- and 8-nitroquinolines. With elemental bromine, quinoline initially forms a π-complex, but on heating or in the presence of an auxiliary base, 3-bromoquinoline is formed. Nucleophilic reagents such as sodamide (the Chichibabin Reaction), akyl or aryl lithiums and potassium hydroxide will attack quinoline preferentially at position 2, but if this position is already occupied then position 4 will be attacked.

Quinolines with methyl groups in position 2 (quinaldine) and position 4 (lepidine) are very active and these groups can be converted to quinoline-based dyes such as Ethyl Red and Pinacyanol and undergo metallation reactions similar to the 2- and 4-methylpyridines. Finally, ring opening of quinoline can be conducted with sodium/mercury amalgam (Emde's Method) and also by von Braun Reactions.

Quinoline occurs naturally in coal tar, and its derivatives are present in many plant alkaloids. In addition to being extracted from coal tar, quinoline can also be manufactured by preparative reactions such as the Skraup, Friedlander, Conrad–Limpach, Doebner–Miller and Pfitzinger Syntheses, which are all given in Appendix 3.

Quinoline finds commercial application in the manufacture of reprographic chemicals and in the synthesis of pharmaceuticals. It is not used as widely as pyridine, but the quinoline ring structure is present in pharmaceuticals such as primaquine, quinine, mefloquinine and chloroquine.

Like quinoline, isoquinoline was first isolated from coal tar, and it has also been isolated from crude petroleum. The isoquinoline ring is numbered is exactly the same way as quinoline, so in isoquinoline the nitrogen atom occupies position 2. It is a colourless solid at room temperature and, like quinoline, is slightly soluble in water. Isoquinoline is a slightly stronger base than quinoline ($pK_a = 5.14$) and has a larger dipole of 2.60 D. Both of these properties of isoquinoline can be explained by the fact that charged resonance hybrids such as the two shown below seem to have a greater effect in defining the properties of isoquinoline than in quinoline.

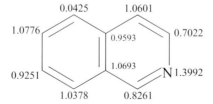

Nucleophilic attack on isoquinoline occurs preferentially at ring position 1, which is the most electron-deficient position on the ring. Electrophilic attack occurs mainly at ring positions 5 and 8 and halogenation seems to give either 5- or 4-substituted derivatives, depending on whether aluminium chloride is used as a catalyst. With aluminium chloride, which must coordinate with the nitrogen atom, halogenation gives the 5-halogeno product, followed by the 5,8-dihalogenated product and then the 5,7,8-trihalogenated product. If halogenation is carried out under neutral conditions in the absence of aluminium chloride, the 4-halogeno derivative may be obtained. The π-electron densities of isoquinoline are shown below.

Like quinoline, isoquinoline forms quaternary salts and can be oxidized to form an *N*-oxide. The isoquinoline ring is found in many naturally occurring alkaloids such as papaverine, morpholine and emetine, and also in the structures of a few synthetic pharmaceuticals such as Famotine and Esproquin. The tetrahydroisoquinoline ring is present in neuromuscular blocking agents such as atracurium and tubocurramine.

Like quinoline and isoquinoline, the quinolizinium cation is isoelectronic with naphthalene. The quinolizinium cation can be attacked at position 4 by nucleophilic reagents such as lithium hydride, sodium borohydride in ethanol and Grignard Reagents. However, many other reagents cause the ring of the quinolizium ion to open and form derivatives of pyridine. Quinolizine can be prepared from 2-cyanoaniline by treatment with (a) 3-ethoxypropylmagnesium bromide followed by (b) refluxing in hydrobromic acid and then (c) a second reflux in acetic anhydride. It is interesting that bromine does not substitute the quinolizinium ion and it is stable to bromine in acetic acid.

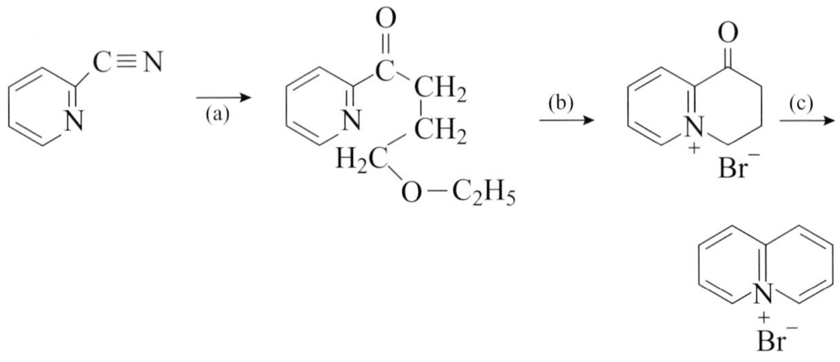

The quinolizinium ion is found in nature incorporated into several alkaloids and has the π-electron densities shown below. Unlike quinoline and isoquinoline, the quinolizinium ion is symmetrical and the π-electron densities are the same in both rings at equivalent positions.

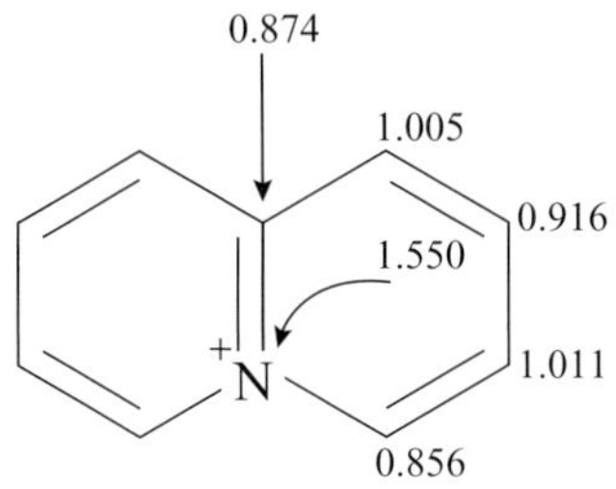

4.12 Nitrogen-containing heterocycles with large rings

Nitrogen-containing heterocycles with seven-, eight- and nine-membered rings are known, and generally all exhibit high reactivity and little or no aromatic character as they either do not comply with Hückel's $(4n+2)$ rule or are distorted too far from planarity to allow electronic overlap and the formation of a π-orbital over the entire molecular ring system. The one possible exception to this is the nine-membered heterocycle azonine, which could contain ten π-electrons if the nitrogen lone pair was taken into account. When the N-carboxyethyl derivative of azonine is allowed to react with potassium t-butoxide in tetrahydrofuran at −20 °C, a stable potassium salt is formed. This salt does display some aromatic character; however, *N*-derivatization with alkyl halides destroys any partial aromatic character as the substituted nitrogen adopts a tetrahedral configuration and the two bonds formed to the azonine ring are not coplanar.

Of the large-membered nitrogen heterocycles the most important, and the most studied, is the seven-membered azepine ring known as azepan-2-one or caprolactam (literally 'goat's milk'), which is used to produce Nylon 6. The first commercially developed form of nylon was Nylon 6.6, made by Dr Wallace H. Carothers of Dupont in the 1930s from adipoyl chloride and hexamethylene diamine [39]. (Carothers in fact made many different nylon formulations from 1935 to 1938, but the best formulation was Nylon 6.6, which was commercialized – hence the name.) However, on 28th January 1938, Dr Paul Schlack discovered how to spin fibres of Nylon 6.6, and this stimulated the market for nylon clothing and stockings as an affordable alternative to silk. Schlack called his spun fibre Perlan [40]. Caprolactam was developed as a way of producing nylon without infringing the Dupont Nylon 6.6 patent, and nylon made from caprolactam is called Nylon 6 to distinguish it from nylon made from adipoyl chloride (or adipic acid) and hexamethylene diamine.

The azepine ring can be formed from the reaction of a 1,6-difunctionalized hexane with either ammonia or a primary amine, or from a ring-expansion reaction from benzene with a singlet nitrene.

The product from this synthesis is 1-ethoxycarbonylazepine, which was produced for the first time in 1963 [41]. Unsubstituted azepine is very reactive and is hard to isolate except at very low temperatures. It is thought that 1H-azepine isomerizes spontaneously to the 3H isomer, which is also very hard to isolate. For this reason most work on the azepine ring has been performed using the 1-ethoxycarbonyl derivative, which acts both as a diene and also as a dienophile in Diels–Alder Reactions. Other synthetic routes to azepines are known from the cyclization of di-nitriles [42] and Diels–Alder Reactions between substituted cyclopentanones and azetidines [43].

Caprolactam is manufactured in three main ways, two of which start from cyclo-hexane. The first route catalytically oxidizes cyclohexane to cyclohexanone and then forms a cyclic seven-membered ether (6-hexanolactone or ε-caprolactone) by reaction with peroxyacetic acid that reacts with ammonia to form caprolactam. The second route uses nitrosyl chloride to convert cyclohexane into an oxime, which then undergoes an acid-catalysed Beckmann Rearrangement to give caprolactam. The third route is the Kvaener Process, which is used at the Caproleuna works, with phenol as a raw material.

Caprolactam is polymerized by heating with about 5% to 10% water at 250 °C, and the polymerization reaction can be either base or acid catalysed.

$$nC_6H_{11}NO + H_2O \rightarrow -[-CO(CH_2)_5NH-]_n-$$

The polymerization reaction proceeds via a gem diol formed from caprolactam and water. The ring of the gem diol opens to allow reaction with a second caprolactam polymer and then this repeats to form a linear polyamide (silk is a naturally occurring polyamide).

$$H_2N-CH_2-CH_2-CH_2-CH_2-CH_2-C(=O)-NH-CH_2-CH_2-CH_2-CH_2-CH_2-CO_2H$$

$+ xs$ caprolactam $\rightarrow$ Nylon 6 polyamide

The total worldwide production of caprolactam is around 115 million tonnes per year and the total worldwide production of Nylon 6 and Nylon 6.6 is about 4.5 million tonnes of fibre and 1 million tonnes of resin.

The azepine ring structure is common in pharmaceuticals, particularly in those such as carbamazepine and imipramine used for the treatment of sleep disorders. The well-known pharmaceutical family of diazepams contain a seven-membered ring with two nitrogen heteroatoms as shown below (where B = H or OH).

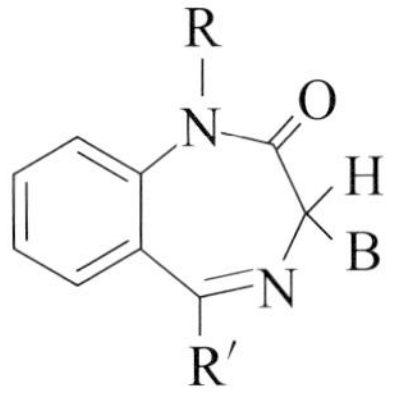

Finally, the eight-membered nitrogen heterocycle azocine was isolated for the first time in 1968 in the form of a 2-methoxy derivative. Like the nine-membered ring azonine, azocine also forms a potassium salt that appears to possess slight electron delocalization. The problem with azocines is that they are prone to the formation of methylene-bridged bicyclic tautomers that become more stable with increasing temperature. To date this molecule has been the subject of academic research only, and has no commercial applications.

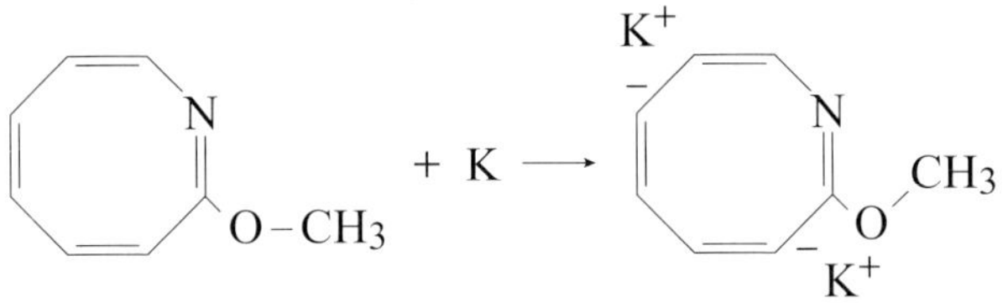

References

[1] D. J. Cram and M. J. Hatch, *J. Am. Chem. Soc.*, **75** (1953), 33.
[2] B. Bak and S. Skaarup, *J. Mol. Struct.*, **10** (1971), 385.
[3] R. M. Acheson, *An Introduction to the Chemistry of Heterocyclic Compounds* (New York: J. Wiley, 1997), p. 12.
[4] S. Gabriel, *Ber.*, **21** (1888), 1049.
[5] H. Wencker, *J. Am. Chem. Soc.*, **57** (1935), 2338.
[6] US Patent 3326895 (1967, assigned to Dow Chemical).
[7] M. Barani, S. Fioravanti, M. A. Loreto, L. Pellacani and A. Patardella, *Tetrahedron*, **50** (1994), 3829.
[8] P. Walden, *Ber.*, **28** (1895), 1287.
[9] R. Bahulekar, N. R. Ayyangar and S. Ponrathanan, *Enzyme Microb. Tech.*, **113** (1991), 858.
[10] E. Schmitz, R. Ohme and D. R. Schmidt, *Ber.*, **95** (1962), 2714.

[11] W. H. Graham, *J. Am. Chem. Soc.*, **84** (1962), 1063.

[12] P. Huszthy, *J. Hetero. Chem.*, **30** (1993), 1197.

[13] E. Leete, *J. Am. Chem. Soc.*, **86** (1964), 3162.

[14] H. C. Neu in *The Chemistry of the β-lactams*, ed. M. I. Page (London: Blackie Academic, 1992), p. 29.

[15] F. Runge, *Ann. Phys.*, **31** (1834), 67.

[16] R. C. Lord and F. A. Miller, *J. Chem. Phys.*, **10** (1942), 328.

[17] G. P. Bean, *J. Chem. Soc. Chem. Commun.* (1971), 421.

[18] N. V. Sidgwick, *The Organic Chemistry of Nitrogen*, 3rd revision by I. T. Miller and H. D. Sprigham (Oxford: Clarendon Press, 1966).

[19] H. Debus, *Ann.*, **107** (1858), 204.

[20] DE Patent 0012371 (1982, assigned to BASF).

[21] L. Knorr, *Ber.*, **16** (1883), 2587.

[22] L. van Order, *Chem. Rev.*, **30** (1942), 69.

[23] A. Baeyer and A. Emmerling, *Ber.*, **2** (1869), 679.

[24] G. Smith, *J. Chem. Soc.* (1954), 3842.

[25] R. Bonnett and R. F. C. Brown, *J. Chem. Soc. Chem. Commun.* (1972), 393.

[26] C. Ainsworth, *J. Am. Chem. Soc.*, **79** (1957), 5242.

[27] A. R. Katritzky and S. Rachwal, *Tetrahedron* (1991), 2683.

[28] W. König and R. Geiger, *Ber.*, **103** (1970), 788.

[29] T. Anderson, *Trans. R. Soc. Edin.*, **20** (1885), 251.

[30] J. B. Cohen, *J. Chem. Soc.*, **127** (1925), 2978.

[31] K. E. Wilzbach and D. J. Rausch, *J. Am. Chem. Soc.*, **92** (1970), 2178.

[32] DE Patent 2703069 (1978, assigned to Degussa AG) and NL patent 7809552 (1978, assigned to Stamicarbon N.V.)

[33] D. A. Bender, *The Nutritional Chemistry of the Vitamins* (Cambridge: Cambridge University Press, 1992).

[34] S. Gabriel, *Ber.*, **33** (1900), 3666.

[35] J. D. Watson, *The Double Helix* (New York: Atheneum, 1968).

[36] P. J. Wheatley, *Acta. Cryst.*, **13** (1960), 80.

[37] A. Baeyer, *Ann.*, **127** (1863), 199.

[38] L. Dobbin, *J. Chem. Educ.*, **11** (1934), 596.

[39] W. H. Carothers, *Chem. Rev.*, **8** (1931), 353.

[40] US Patent 2241321 (1941, assigned to I. G. Farben Industrie A.G.)

[41] W. Lwowski, T. J. Merrich and T. W. Mattingly, *J. Am. Chem. Soc.*, **85** (1963), 1200.

[42] W. A. Nasutavicus, S. W. Tovey and F. Johnson, *J. Org. Chem.*, **32** (1967), 3325.

[43] A. Hassner and D. J. Anderson, *J. Org. Chem.*, **39** (1974), 3070.

5

Inorganic amines, hydrazine, hydroxylamine and amine ligands

5.1 Introduction to the inorganic chemistry of the amines

There are ten possible acyclic compounds which can be formed between hydrogen and nitrogen. Five of these have been isolated: ammonia, hydrazine, diazine (N_2H_2), tetrazine (N_4H_4) and hydrazoic acid (HN_3). Another four have been identified as short-lived intermediates: triazene (N_3H_3), triazane (N_3H_5), tetrazane (N_4H_6) and nitrene (NH). The tenth compound, tetrazete (N_4H_2) has only recently been isolated [1]. The N_4^{2-} ion is square planar and possesses aromatic character. In addition to the organic amines that have been examined in the previous two chapters, there are also many examples of inorganic amines where the hydrogen atoms on ammonia are replaced with non-carbon-containing substituents. Three of these compounds – hydroxylamine, hydrazine and hydrazoic acid – have significant industrial and commercial importance (although hydrazoic acid is not strictly an amine). Most inorganic amines are prepared from ammonia, and the Raschig Synthesis is important here in substituting the hydrogen atoms of ammonia for other groups such as halides, hydroxyl groups or other nitrogen-containing fragments.

Like carbon atoms, nitrogen atoms have a tendency towards catenation. However, this tendency is less pronounced than in carbon for two reasons: firstly because of the presence of non-bonding electron pairs on the smaller nitrogen atoms, which (relative to carbon) act to weaken nitrogen–nitrogen bonds (and attack by electrophiles); and secondly because of the larger bond energy of the N_2 molecule, which makes the free energy of formation of nitrogen chains less negative and so reduces their overall stability. Chains of up to eight nitrogen atoms have been identified, and many of them include carbon-based amine-group substituents. The following molecules are examples.

CH$_3$—NH—N=NH   with three nitrogens
(CH$_3$)$_2$—N—N=N—N(CH$_3$)$_2$ with four nitrogens
C$_6$H$_5$—N=N—N(CH$_3$)—N=N—C$_6$H$_5$ with five nitrogens
C$_6$H$_5$—N=N—N(C$_6$H$_5$)—N=N—N(C$_6$H$_5$)—N=N—C$_6$H$_5$ with eight nitrogens

All of these molecules are of academic interest only, and are unstable. In 1957, Huisgen and Ugi confirmed the existence of a five-nitrogen ring in phenylpentazole [2]. Although the pentazole ring is also unstable the tetrazole ring with four nitrogens and one carbon has since found application in the sartan class of pharmaceuticals used to treat hypertension [3].

Pentazole   Tetrazole

Tetrazoles are formed from the reaction of sodium azide with aryl cyanides or with 2-substituted quinolines. However, because sodium azide is explosive, tributyl tin azide is usually the preferred reagent for this synthesis. Silver azide can be formed from the reaction between silver metal and aqueous ammonia and it, too, is very explosive.

(*Author's note*. I once caused a small explosion in a fume hood when trying to measure the redox potential of an aqueous-ammonia solution with a silver electrode. The silver electrode was in contact with the ammonia solution for less than 20 min and cannot have generated more than 1 g of silver azide, but the explosion sent the electrode clean across the laboratory! When I owned up to my supervisor what had happened, I was told 'Oh yes, everyone does that because this reaction is not taught to students at universities anymore'. From that day on I always treated azides with the utmost respect and fortunately have never had another 'incident' when handling them.)

The replacement of hydrogen atoms in ammonia with halides creates a new family of compounds called amine halides. Although many amine halides are known, only three are of interest to the synthetic chemist: chloramine; its bromo equivalent, bromoamine; and trichloramine. Other amine halides such as dichloramine, fluoro-dichloramine, chlorodifluoramine and difluoroamine have been synthesized but are of academic interest only and will not be discussed here.

Trifluoroamine is of interest here because the central nitrogen atom appears to have no donor properties. Although the geometry of the molecule is still only just pyramidal (see Appendix 1), which is indicative of the presence of a lone electron pair, there is no localized area of negative electronic charge on the molecule's isosurface. Trifluoroamine is a stable molecule and forms an amine oxide with oxygen in the presence of an electric arc. Unlike other amine oxides, it is thought that the oxygen atom is covalently double bonded to nitrogen here and that the molecule exists as an ion pair stabilized by resonance with the three fluorine atoms sharing a negative charge.

$$\left[\begin{array}{c} F \\ \diagdown \\ \diagup N = O \\ F \end{array}\right]^{+} \left| F^{-} \right.$$

From Pauling electronegativities the amount of ionic character of the N–F bond should be $1 - e^{-1/4(4-3)} = 22\%$ ionic.

The resonance energy of nitrogen trifluoride oxide is sufficient to stabilize the molecule with respect to dissociation into its constituents. A similar effect is also observed for the molecule nitryl chloride, NO_2Cl, which has a resonance-stabilized form $[NO_2]^+Cl^-$.

Trifluoroamine, or nitrogen trifluoride as it is also known, is widely used in the electronics industry for cleaning chemical vapour deposition (CVD) reactors. Perfluoroethane was used for this purpose in the past, but because it is a greenhouse gas its use has been discontinued. Interestingly, nitrogen trifluoride is also classified as a greenhouse gas, but it is easier to handle and to destroy after use. Nitrogen trifluoride reacts with any deposits in the CVD reactors and converts them to volatile fluorides, which can be easily flushed out from the reactors. Nitrogen trifluoride is also used as a source of fluorine for the plasma etching of polysilicon wafers and of silicon nitride, tungsten silicate and tungsten films. Its use in 2002 was around 1.8 million tonnes and is growing by about 50% per year. Nitrogen trifluoride can be manufactured from the reaction of ammonia with fluorine or from the electrolysis of ammonium fluoride.

5.2 Chloramine

In its pure state chloramine, NH_2Cl, is violently explosive and very difficult to handle even on a small scale in the laboratory. However, in the gas phase, in the form of a dilute solution in water or as its tosyl derivative, it can be handled safely and is used as an intermediate in a number of industrial processes. Chloramine can be prepared by the action of sodium hypochlorite on ammonia at 273 K or alternatively from ammonia and chlorine gas [4].

$$NH_3 + NaOCl \rightarrow NH_2Cl + NaOH$$

$$2NH_3 + Cl_2 \rightarrow NH_2Cl + NH_4Cl$$

The second reaction works well, with yields of over 80%, and requires only a slight stoichiometric excess of ammonia. Any excess ammonia after reaction can be removed by passing the gas stream through anhydrous copper(II) sulphate. Because pure chloramine has never been isolated its freezing point and boiling point are matters for speculaton, but the freezing point is thought to be around 207 K. To the present author's knowledge, no-one has succeeded in measuring the boiling point of chloramine as it is far too unstable and explodes before boiling. Chloramine has a similar trigonal-pyramidal structure to ammonia but the bond angle Cl−N−H is only 102° (4° less than the angle in ammonia), and the molecule is less polar than ammonia because the polarity of the N−Cl bond opposes the polarity of the unpaired electrons on the nitrogen atom and the N−H bonds [5]. Chloramine is a very weak acid ($pK_a = 14$) and base ($pK_b = 15$), and its aqueous solutions are neutral.

In aqueous solution, chloramine exists in equilibrium with dichloramine and nitrogen trichloride. The major component present is dependent upon the pH of the solution.

$$H_3O^+ + 2NH_2Cl \rightleftharpoons HNCl_2 + NH_4^+ + H_2O$$

$$H_3O^+ + 3HNCl_2 \rightleftharpoons 2NCl_3 + NH_4^+ + H_2O$$

At pH < 3 nitrogen trichloride is the more-stable chemical in solution, whereas at pH > 8 chloramine is the predominant species (up to pH11, at which point the hypochlorite ion is formed). At intermediate pH, in the range 3–5, dichloramine becomes the more-stable species and at pH 5–8 there is a mixture of dichloramine and chloramine in solution.

Unlike chloramine, dichloramine is extremely unstable. It has never been isolated in the solid or liquid phase, and has only ever been studied in aqueous solution or in the gas phase. At high pH chloramine reacts with hydroxyl ions to form hypochlorites.

$$NH_2Cl + NaOH \rightleftharpoons NH_3 + NaOCl$$

Chloramine is a weaker base than ammonia and in most of its reactions it behaves as an oxidizing agent.

$$[NH_2Cl] + 2H^+ + 2e^- \rightleftharpoons [NH_4]^+ + Cl^-$$

Chloramine acts as an aminating agent and reacts with donor bases (**B**) to form adducts with the elimination of chlorine. When the base is ammonia the reaction product is hydrazine.

$$\mathbf{B} + NH_2Cl \rightarrow [\mathbf{B}-NH_2]^+ + Cl^-$$
$$NH_3 + NH_2Cl \rightarrow [NH_3NH_2]^+ + Cl^-$$
$$[NH_3NH_2]^+ + NH_3 \rightarrow NH_2-NH_2 + [NH_4]Cl$$

The reaction between chloramine and ammonia was first carried out by Raschig in aqueous alkaline solution, but he experienced difficulty trying to remove the water, so the reaction is often best carried out in a non-aqueous solvent to eliminate this problem. Ammonia also works well as a solvent for this reaction.

If an amine is used instead of ammonia in this reaction, a substituted hydrazine is obtained. For example, dimethylamine gives *N,N*-dimethylhydrazine, which is used as a rocket propellant.

$$(CH_3)_2N-H + H_2N-Cl \rightarrow (CH_3)_2N-NH_2 + [(CH_3)_2NH_2]Cl$$

It is best to use secondary amines for this reaction because primary amines can give a complicated mixture of products that are impossible to separate. Also, chloramine can react with the dialkyl-substituted hydrazine produced from the reaction to yield by-products such as $(CH_3)_2N-N=N-N(CH_3)_2$ and $(CH_3)-N-N=CH_2$ via the formation of the intermediate dimethyl diazine $(CH_3)_2-N-N$. Quaternary hydrazinium salts are obtained with tertiary amines.

$$R_3N + NH_2Cl \rightarrow [R_3N-NH_2]Cl$$

However, chloramine can also react with hydrazines; in the above reactions short reactor times are required to minimize the formation of by-products such as triazanium salts.

$$NH_2-NH_2 + NH_2Cl \rightarrow [H_2N-NH_2-NH_2]^+Cl^-$$

Triazanium chloride is unstable, but alkyl derivatives such as 1,1-dimethyl-triazinium chloride ($[NH_2-N(CH_3)_2-NH_2]Cl$) are stable crystalline solids.

In general chloramine will react with most electron donors, such as trialkyl and triphenyl phosphines and arsines, as well as with their mono-chloro-substituted derivatives. However, when chloramine reacts with a phosphine bonded to a nitrogen atom, as in the case of $(Ph)_2P-N(Et)_2$, the adduct is formed on the phosphorus atom rather than on the adjacent nitrogen atom.

$$(Ph)_2P-N(Et)_2 + NH_2Cl \rightarrow [(Ph)_2P(NH_2)(N(Et)_2]Cl$$

Chloramine reacts with alkoxides to form hydroxylamines.

$$RONa + NH_2Cl \rightarrow RO{-}NH_2 + NaCl$$

Chloramine also gives oxidative-addition reactions and can donate Cl^+ from its protonated form NH_3Cl to electronegative reagents (where Nu is a nucleophile). (Note that the Cl^+ ion is relatively uncommon, but can also be generated from the hypohalous acid HOCl.)

$$[Nu]^- + [Cl{-}NH_3]^+ \rightarrow Nu{-}Cl + NH_3$$

With very electropositive reagents, a hydrogen may be cleaved from chloramine and replaced, for example in a chloramidation reaction (where E is an electrophile).

$$E^+ + NH_2Cl \rightarrow E{-}NH{-}Cl + H^+$$

Dichloramine may be prepared in this way from the reaction between hypochlorous acid and chloramine.

When chloramine is exposed to ultra-violet light or high temperatures (above $350\,^\circ C$) it is cleaved to form free radicals, which can chlorinate alkenes and the terminal hydrogens of alkane chains. The reaction is difficult to control and can give a mixture of different reaction products, as is often the case with free-radical reactions.

$$NH_2Cl \rightleftharpoons {\bullet}NH_2 + {\bullet}Cl$$
$$(CH_3)_3C{-}H + 2{\bullet}Cl \rightarrow (CH_3)_3C{-}Cl + (CH_3)_2CH{-}CH_2Cl$$
$$(CH_3)_2C{=}CH_2 + 3{\bullet}Cl \rightarrow CH_2{=}C{-}CH_2{-}Cl + (CH_3)_2CCl{-}CH_2Cl$$

5.3 Nitrogen trichloride

Nitrogen trichloride is an unstable yellow oil under ambient conditions. Its usefulness as a synthetic reagent is limited by the fact that it is spontaneously explosive in the pure state and is best handled as a dilute solution in a chlorinated, organic solvent. Although the N–Cl bonds in nitrogen trichloride are normal covalent bonds and are similar to Cl–Cl and N–N single bonds, nitrogen trichloride is unstable with respect to the nitrogen–nitrogen triple bond to an extent of $229\ kJ\ mol^{-1}$, which accounts for its unstable behaviour. As previously mentioned, chloramine rearranges in aqueous solution at pH < 3 to form nitrogen trichloride (also called trichloroamine). However, nitrogen trichloride has a very low solubility in water and has a tendency to form droplets that often explode.

Because of nitrogen trichloride's explosive nature, great care must be taken during the Raschig Hydrazine Synthesis not to use an excess of chlorine as the explosive nitrogen trichloride could be produced instead of chloramine.

$$NH_3 + 3Cl_2 \rightleftharpoons NCl_3 + 3HCl$$

A better synthetic route to nitrogen trichloride is to use ammonium chloride in a mixture of water and a non-miscible chlorinated solvent [6] into which chlorine gas can be introduced. With this method the produced nitrogen trichloride can be extracted into a dilute non-aqueous solution where it is relatively non-hazardous (nitrogen trichloride is more soluble in apolar solvents than in polar solvents). Any residual nitrogen trichloride left behind in aqueous solution can be destroyed by the addition of excess alkali, which will convert it to hypochlorite ions.

$$NH_4Cl + 3Cl_2 \rightleftharpoons NCl_3 + 4HCl$$
$$NCl_3 + 3NaOH \rightarrow NH_3 + 3NaOCl$$

Nitrogen trichloride is useful in organic synthesis as, like chloramine, it can aminate alkanes, arylalkanes and cycloalkanes at terminal carbons in the presence of aluminium chloride at 0–10 °C [7]. The reaction mechanism is S_N1 and the leaving group is the hydride ion $[H]^-$.

$$NCl_3 + AlCl_3 \rightarrow [Cl_2N-AlCl_3]^- Cl^+$$
$$R_3C-H + Cl^+ \rightarrow [R_3C]^+ + HCl$$
$$[R_3C]^+ + [NCl_2]^- \rightarrow R_3C-NCl_2(\rightarrow R_3C-NH_2)$$

This reaction can be used to make 1-aminoadamantane and 1-amino-1-methyl-cyclopentane in good yield from adamantane and 1-methylcyclopentane respectively.

Although nitrogen trichloride is unstable and prone to decompose explosively to chlorine and nitrogen, it is more stable than either chloramine or dichloramine. The decomposition of nitrogen trichloride is accelerated by light and by silver halides or arsenic halides. Nitrogen trichloride also reacts with ammonia to form hydrogen chloride and nitrogen.

$$NCl_3 + NH_3 \rightarrow N_2 + 3HCl$$

Although the three chlorine atoms in the nitrogen trichloride molecule are electron withdrawing, the molecule is still a tetragonal pyramid with a Cl–N–Cl bond angle of 107.1° [8]. It is possible to add chlorine to nitrogen trichloride to create the tetrachloroammonium ion $[NCl_4]^+$ as found in $[NCl_4][AsF_4]$. However, nitrogen trichloride does not react with bromine or iodine to give nitrogen tribromide or nitrogen triiodide; both of these are best prepared by different methods.

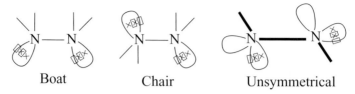

| Boat | Chair | Unsymmetrical |

Figure 5.1. The conformations of hydrazine.

5.4 Hydrazine

Hydrazine (or diazine) was first prepared by Curtius in 1889 [9] but it had few practical or commercial applications until the Second World War when it was used as a propellant for rockets. This application stimulated interest in hydrazine as a high-energy source of fuel and led to further research into this interesting chemical.

At first sight the hydrazine molecule might be expected to exhibit free rotation about the nitrogen–nitrogen bond in a similar manner to the rotation about the carbon–carbon bond in ethane. However, the presence of two nitrogen lone pairs on adjacent nitrogen atoms provides a 25–41 kJ mol^{-1} barrier to free rotation, and at room temperature three conformations of hydrazine are known. These are the boat, chair and unsymmetrical (gauche) forms shown in Figure 5.1. The unsymmetrical form accounts for nearly 100% of the molecular configuration at room temperature.

The nitrogen atoms are $2s^2 2p^3$ hybridized and exhibit similar tetrahedral geometries to the carbon atoms in alkenes. In fact, hydrazine is structurally very similar to hydrogen peroxide. At room temperature pure hydrazine is a colourless oily liquid with a density of 1 g cm^{-3} and a similar odour to ammonia. In the pure state hydrazine boils at 113.5 °C and melts at 2 °C, and forms strong intermolecular hydrogen bonds. It has a dielectric constant of 51.7 and a dipole moment of 1.84 D.

Hydrazine forms a stable monohydrate, $N_2H_4.H_2O$, which is used in many commercial applications especially if hydrazine is required in an aqueous solution. Hydrazine hydrate is a viscous liquid that which fumes in air and has a characteristic fish-like amine smell.

The main commercially used method for the manufacture of hydrazine is based on the Raschig Process, which was developed by Raschig in 1907 [10]. In this process, or in its subsequent modifications, sodium hypochlorite is allowed to react with aqueous ammonia at pH 8–12. A 30-fold stoichiometric excess of ammonia is usually used in the commercial process.

In the first stage of the process, chloramine is formed by the nucleophilic substitution of a hydroxy group bound to a chlorine atom.

$$ClO^- + H_2O \underset{(-NH_3)}{\overset{(+NH_3)}{\rightleftharpoons}} Cl-OH \rightleftharpoons H_2O + Cl-NH_2 + OH^-$$

Once formed chloramine reacts with ammonia to form hydrazine.

$$H_3N + Cl-NH_2 \rightarrow H_2N-NH_2 + HCl$$

Unfortunately hydrazine also reacts with chloramine and competes with ammonia in the second stage of the reaction.

$$N_2H_4 + 2NH_2Cl \rightarrow N_2 + 2NH_4Cl$$

In the Raschig Reaction hydrazine decomposes 18 times faster than it is formed, which is why a high excess of ammonia is added to the reaction mixture. Two other ways in which the overall balance of the reaction can be tipped towards the production of hydrazine are (i) to increase the reaction temperature as the rate of reaction for hydrazine synthesis increases more rapidly with increasing temperature than does the reaction for its conversion to nitrogen, and (ii) to add a complexing agent to the reaction mixture to remove any metal ions present that could catalyse the decomposition reaction. Typically gelatin is added to the reaction mixture to complex metal ions, particularly the copper(II) ion as this is very effective at decomposing hydrazine (even at p.p.m. levels). Although other complexing agents such as ethylenediaminetetraacetic acid have been tried in the Raschig Process, they do not work as well and it is thought that gelatin has a beneficial catalytic effect on the reaction as well as acting as a sequestrant. In a typical manufacturing process, chlorine gas is added to aqueous sodium hydroxide solution at $0\,^{\circ}C$ in a molar ratio of 1:2. The reaction to form sodium hypochlorite is exothermic, and cooling must be applied to the reaction vessel to maintain the temperature. Once the required amount of chlorine has been added, a stoichiometric quantity of ammonia sufficient to form chloramine is added.

At this stage, more ammonia (in large excess) is added under pressure and as the addition of ammonia to water is exothermic the heat generated is sufficient to raise the heat of the reaction solution to over $120\,^{\circ}C$. Under these conditions hydrazine is formed in a few seconds in about 75% yield. Distillation of the excess ammonia from the reaction mixture (it is usually recycled) leaves behind a crude solution of hydrazine hydrate. This can be purified by distillation to form commercial-grade hydrazine hydrate, which is sold as 64% hydrazine solution. This is the major problem with the Raschig Synthesis: it is possible to obtain only a weak aqueous solution of hydrazine from the reaction, and this must be concentrated thereby incurring additional cost. If anhydrous hydrazine is required (for example, for use as a rocket fuel) different synthetic strategies must be employed. Hydrazine hydrate may be dehydrated by heating with anhydrous sodium hydroxide. This produces 98% hydrazine, which may then be treated with anhydrous barium oxide to yield pure anhydrous hydrazine.

Three other processes for the manufacture of hydrazine are the Bergbau–Bayer Process, the Pechiny–Ugine–Kuhlmann Process and the Sisler Process.

The Bergbau–Bayer (or ketazine) Process [11] uses ammonia, acetone and sodium hypochlorite at 35 °C. The advantage of adding acetone is that as soon as hydrazine is formed it reacts with the acetone carbonyl group to produce ketazine and so prevents its side reaction with chloramine.

$$H_2N-NH_2 + 2O=C(CH_3)_2 \rightarrow (CH_3)_2C=N-N=C(CH_3)_2 + 2H_2O$$

If insufficient acetone is added, acetone hydrazone, $(CH_3)_2C=N-NH_2$, is produced. Acetone hydrazone and ketazine are both hydrolysed by water to produce hydrazine hydrate and acetone. The overall process is exactly the same as for the Raschig Process, i.e. $2NH_3 + Cl_2 \rightarrow N_2H_4 + 2HCl$, and hydrazine is obtained as the monohydrate. It is also possible to use urea in this process instead of acetone, but this method is not commercially viable. A slightly modified Bergbau–Bayer Process with methyl ethyl ketone used instead of acetone in first stage is called the Whiffen Process, but this is no longer used commercially.

The Pechiny–Ugine–Kuhlmann Process [12] uses hydrogen peroxide instead of hypochlorite as the oxidising agent but otherwise is similar to the Bergbau–Bayer–Whiffen Process in that acetone is used to produce ketazine.

$$2NH_3 + H_2O_2 + 2(CH_3)_2C=O \rightarrow (CH_3)_2C=N-N=C(CH_3)_2 + 4H_2O$$

The processes described so far produce aqueous solutions of hydrazine as the reaction products. The process developed by Sisler [13] uses the gas-phase reaction between chlorine and ammonia to produce anhydrous hydrazine.

$$2NH_3 + Cl_2 \rightarrow NH_2Cl \rightarrow NH_4Cl$$

$$Cl-NH_2 + 2NH_3 \rightarrow N_2H_4 + NH_4Cl$$

However, the Sisler Process has not found widespread application in industry to date.

Another method of producing water-free hydrazine is to precipitate crystalline hydrazine sulphate from aqueous solution, as this salt is relatively insoluble in water. The Olin Modification of the Raschig Process also produces anhydrous hydrazine. In the Olin Modification sodium hypochlorite solution is added to a three-fold excess of ammonia at 5 °C to produce chloramine, which is then added to a 30-fold excess of ahydrous ammonia at 20–30 MPa at 130 °C. An azeotropic distillation of the reaction mixture with aniline yields anhydrous hydrazine [14].

The chemical properties of hydrazine are defined by the observations that it is a powerful reducing agent and also an endothermic molecule that is unstable to decomposition to nitrogen and ammonia.

Care must be taken when heating hydrazine as it disproportionates explosively especially in the presence of platinum and tungsten (which can also catalytically decompose ammonia to produce hydrogen and nitrogen).

$$3N_2H_4 + \text{heat} \rightarrow 4NH_3 + N_2$$
$$3N_2H_4 + \text{heat} + \text{catalyst} \rightarrow 3N_2 + 6H_2$$

Both hydrazine and hydrazine hydrate act as strong reducing agents.

Several reaction pathways are present depending upon the nature of the oxidizing agent used in the reaction, the pH and the temperature.

In weak acid solution three reactions are possible:

$$N_2H_4 + [H]^+ \rightleftharpoons [NH_4]^+ + \tfrac{1}{2}N_2 + H^+ + e^-$$

or
$$\rightleftharpoons \tfrac{1}{2}[NH]^+ + \tfrac{1}{2}NH_3 + \tfrac{5}{2}[H]^+ + 2e^-$$

or
$$\rightleftharpoons N_2 + 5H^+ + 4e^-.$$

In alkaline solution another three reactions are possible:

$$N_2H_4 + [OH]^- \rightleftharpoons NH_3 + \tfrac{1}{2}N_2 + H_2O + e^-$$

or
$$N_2H_4 + \tfrac{5}{2}[OH]^- \rightleftharpoons \tfrac{1}{2}NH_3 + \tfrac{1}{2}[N_3]^- + \tfrac{5}{2}H_2O + 2e^-$$

or
$$N_2H_4 + 4[OH]^- \rightleftharpoons N_2 + 4H_2O + 4e^-.$$

Most reactions are not quantitative but if the last reaction has iodine as an oxidant, the redox potential is $E^{\ominus} = -1.16$ V at pH 14.

Hydrazine reacts with oxygen, and its aqueous solutions can become oxidized in air unless protected with a layer of paraffin. Two reaction mechanisms are known and they may operate together although the final products are the same.

The first mechanism produces hydrogen peroxide as an intermediate:

$$N_2H_4 + 2O_2 \rightarrow N_2 + 2H_2O_2$$

followed by
$$2H_2O_2 + N_2H_4 \rightarrow N_2 + 4H_2O$$

giving overall
$$2N_2H_4 + 2O_2 \rightarrow 2N_2 + 4H_2O.$$

The second mechanism produces di-imine as an intermediate:

$$2N_2H_4 + O_2 \rightarrow 2N_2H_2 + 2H_2O$$

followed by
$$2N_2H_2 + O_2 \rightarrow 2N_2 + 2H_2O$$

giving overall
$$2N_2H_4 + 2O_2 \rightarrow 2N_2 + 4H_2O.$$

When ignited in air, hydrazine burns rapidly with great evolution of heat. It was the combusion properties of hydrazine and its derivatives that interested rocket engineers in the potential of hydrazine as a propellant.

$$N_2H_4 + O_2 \rightarrow N_2 + 2H_2O \quad \Delta H^{\ominus} = 622 \text{ kJ mol}^{-1}$$

Hydrazine reacts as a reducing agent with halogens, oxygen and sulphur to form N_2 and the respective binary non-metal hydride.

$$N_2H_4 + 2X_2 \rightarrow N_2 + 4HX$$

The reaction between sulphur and hydrazine can be used to desulphurize petroleum.

$$N_2H_4 + 2S \rightarrow 2H_2S + N_2$$

The reactions between hydrazine and halogens are often explosive and spontaneous.

Hydrazine reacts with many metal oxides and chlorides, and will reduce them back to metals if they are present in alkaline solution. Effluent from metal and electroplating works is often treated with hydrazine prior to discharge to precipitate out any dissolved metals, which can then be filtered off before entering the sewer. This process is of special importance in hydrometallurgy as it can be used to recover valuable Group VIII metals from dilute aqueous solution.

However, as the metal particulates produced by hydrazine reduction are usually very small and will pass through most filters it is normal practice to add a flocculant to encourage precipitation. Hydrazine derivatives are also used in the treatment of boiler water and rocket fuels; these applications are described in Chapter 8.

Hydrazine is used in organic chemistry as a reducing agent for carbonyl groups in the Wolf–Kishner Reaction [15] and also for the formation of hydrazones and azones. For example, hydrazine reacts with one carbonyl group to form a hydrazone.

$$\begin{array}{c}\setminus\\C=O + H_2N-NH_2 \rightarrow -\overset{|}{\underset{OH}{C}}-NH-NH_2 \rightarrow \overset{\setminus}{\underset{/}{C}}=N=NH_2 + H_2O\\/\end{array}$$

If the hydrazone then reacts with a second carbonyl group, an azine is formed.

$$\begin{array}{c}\setminus \quad\quad /\\C=N-N=C\\/ \quad\quad \setminus\end{array}$$

If semicarbazide is used in the reaction, semicarbazones are produced.

$$NH_2-CO-NH-NH_2 + O=\overset{/}{\underset{\setminus}{C}} \rightarrow H_2N-CO-NH-N=\overset{/}{\underset{\setminus}{C}} + H_2O$$

If alkyl or aryl derivatives of hydrazine are used, azine formation is avoided and crystalline hydrazones can be obtained; the best known is probably phenyl hydrazone. Most hydrazones can be converted back to the original carbonyl by refluxing in dilute hydrochloric acid. However, this reaction does not work so well for alkyl- and aryl-substituted hydrazones, and regeneration of the carbonyl is best accomplished in these case's with acetyl acetone in very dilute aqueous acid.

When hydrazones or semicarbazones are heated with sodium ethoxide (or potassium hydroxide) at 180 °C, nitrogen is produced together with a methylene group in place of the original carbonyl group. This is the Wolf–Kishner Reaction.

$$\underset{R}{\overset{R'}{\diagdown}}C{=}N{-}NH_2(+NaOC_2H_5) \rightarrow \underset{R}{\overset{R'}{\diagdown}}CH_2 + N_2(+H_2O)$$

The yield from the Wolf–Kishner Reaction is generally better than from the Clemmensen Reaction (with zinc amalgam and hydrochloric acid) and works best with carbonyl groups without excessive steric crowding as this inhibits formation of the hydrazone.

If the reduction of sterically crowded ketones is required, the Huang–Minlon [16] Modification of the Wolf–Kishner Reaction can be used at higher temperatures with diethylene glycol as a solvent. The Wolf–Kishner Reaction works well at room temperature if dimethyl sulphoxide is used as a solvent. Potassium tertiary butoxide also works well as a base.

If hydrazine is added to the reaction mixture too quickly in dimethyl sulphoxide the exothermic reaction can overheat the reaction mixture and the product obtained will be a polymeric tar.

(*Author's note*. It is often best to add the hydrazine to the reaction via a dropping funnel at a very slow rate and to be patient. It is better, in my experience, for a Wolf–Kishner Reaction to take 5 h to complete and produce the desired product in good yield than for it to be completed in 1 h with zero yield – especially as my first attempt at the Wolf–Kishner Reaction was the final step in an eight-stage process!)

The key to the Wolf–Kishner Process is the formation of an intermediate between the hydrazone and a base (**B**) via proton transfer.

$$\underset{R'}{\overset{R}{\diagdown}}C{=}N{-}NH_2 + \mathbf{B} \rightleftharpoons [\underset{R'}{\overset{R}{\diagdown}}C{=}N{-}NH \rightleftharpoons \underset{R'}{\overset{R}{\diagdown}}C{-}N{=}NH]^- + [\mathbf{B}H]^+$$

$$\rightarrow [\underset{R'}{\overset{R}{\diagdown}}CH]^- + N_2 + [\mathbf{B}H]^+ \rightarrow \underset{R'\ H}{\overset{R\ H}{\diagdown\diagup}}C + \mathbf{B}$$

(Here R = an alkyl or aryl group, R′ = an alkyl or aryl group or hydrogen.)

The Wolf–Kishner Reaction is well suited to the reaction of those carbonyl-containing chemicals that are acid sensitive, and is very useful in the production of alkyl-substituted aryls. For example, benzene can be acylated with R−CO−Cl in a Friedel–Crafts Reaction and can then be reduced to an alkyl-substituted aryl via the Wolf–Kishner Reaction.

$$R{-}CO{-}Cl + C_6H_6 \rightarrow R{-}CO{-}C_6H_5 + HCl$$
$$R{-}CO{-}C_6H_5 + N_2H_4\ (+\mathbf{B}) \rightarrow R{-}CH_2{-}C_6H_5 + N_2 + H_2O$$

A problem with the Wolf–Kishner Reaction is that if unsaturated carbonyl compounds are reduced the C=C double bond will often migrate and unexpected products may be obtained from the reaction. This effect is especially pronounced with α,β-unsaturated carbonyls. In some instances this type of carbonyl-containing compound can be reduced by using potassium butoxide at low temperatures (below 30 °C) without migration of the double bond, but in other cases migration will occur. If a literature search does not make this clear for the particular compound under investigation a different reducing agent should be used.

When the carbonyl-containing molecule is a carboxylic acid ester or an acyl chloride rather than an aldehyde or ketone, the carbonyl group is not reduced and instead an acid hydrazide is produced.

$$R{-}CO{-}X + N_2H_4 \rightarrow R{-}CO{-}NH{-}NH_2 + HX$$

(Here X $= -$Cl or $-$O$-$R or $-$O$-$Ar.)

When phosgene is used, a dihydrazide is obtained.

$$COCl_2 + 4N_2H_4 \rightarrow O{=}C(NH{-}NH_2)_2 + 2N_2H_5Cl$$

Acid hydrazides resemble acid amides but are less susceptible to hydrolysis. Like hydrazine, acid hydrazides are good reducing agents.

When treated with nitrous acids, acid hydrazides form acid azides, which when boiled with an alcohol react to form *N*-alkyl urethanes in good yield via the Curtius Rearrangement. They can then be decarboxylated to yield primary amines. The route is a little convoluted but has the advantage that the overall yield is good and the produced primary amine is uncontaminated by secondary and tertiary amines.

$$R{-}CO{-}NH{-}NH_2 + HNO_2 \rightarrow R{-}CO{-}N_3 + 2H_2O$$
$$R{-}CO{-}N_3 + R'OH \rightarrow R{-}NH{-}CO_2{-}R'$$
$$R{-}NH{-}CO_2{-}R' + NaOH \rightarrow R{-}NH_2$$

When tosyl hydrazides are heated with lithium aluminium hydride the carbonyl group is reduced and then the molecule decomposes to produce a hydrocarbon.

$$R-CO-NH-NH-Tosyl + LiAlH_4 \rightarrow [R-CH_2-NH-NH-Tosyl]$$
$$\downarrow$$
$$Tosyl-H + N_2 + R-CH_3$$

As well as being reduced, acyl hydrazides are able to be oxidized by metal salts. Reaction with copper(II) acetate or copper(II) chloride generates the acyl cation, which can then react with a nucleophile.

$$R-CO-NH-NH_2 \xrightarrow[Cu\ salt]{} [R-CO]^+ + N_2 \xrightarrow[Nu^-]{} R-CO-Nu$$

Typical nucleophiles in this reaction are alkoxides, amines and water, which yield esters, amides and carboxylic acids respectively.

The reaction between acyl hydrazides and lead tetraacetate is interesting as there are two possible products. Oxidative cleavage with lead tetraacetate alone proceeds by an intramolecular mechanism to yield a *cis*-diazo product, but if another nucelophile is added (such as a heterocyclic amine) a *trans*-diazo product is obtained.

$$R-CO-NR'-NHR' + Pb(OAc)_4 \rightarrow$$

(i) intramolecular reaction → $\underset{N=N}{R'} + R-CO-OAc$

(ii) intermolecular reaction → $\underset{R'}{\overset{R'}{N=N}} + R-CO-Nu$

Hydrazine and substituted hydrazines can be used to prepare heterocyclic compounds with adjacent nitrogen atoms by reaction with dicarbonyls in the presence of acid. For example, pyrazole can be prepared from most 1,3-dicarbonyl compounds and hydrazine or monosubstituted hydrazine. Typically this reaction is performed between 1,1,3,3-tetramethoxypropane, which hydrolyses in acid solution to form malondialdehyde. This then reacts with hydrazine to form pyrazole.

Similarly, maleic anhydride and hydrazine can be used to form pyridazine, and 2-chloronitrobenzene and hydrazine can be used to produce 1-hydroxybenzotriazole via an intramolecular oxidation/reduction reaction.

The most important substituted hydrazine is phenyl hydrazide. This can be manu-factured from aniline by first forming a diazonium salt, which can then be reduced with tin(II) chloride and hydrochloric acid.

$$C_6H_5-NH_2 + HNO_2 + HCl \rightarrow [C_6H_5-N_2]^+Cl^- + H_2O$$

$$[C_6H_5-N_2]^+Cl^- + SnCl_2 + HCl \rightarrow C_6H_5-NH-NH_2.HCl$$

Alkyl hydrazines can be prepared by adding primary amines such as methyl amine into the second stage of the Raschig Synthesis in place of ammonia. For more complicated alkyl amines, hydrazine can be alkylated but the problem is that because hydrazine is a symmetrical molecule alkylation reactions can take place at either end and di- and polysubstituted products may be obtained. A better synthetic strategy is to use a protected hydrazine such as BOC-hydrazide or tosyl-hydrazide, which can be monoalkylated at one end and then deprotected to yield the required alkyl amine [17]. This technique is discussed further in Chapter 7.

Phenyl hydrazine is oxidized if exposed to air and has basic properties including the formation of stable salts (i.e. stable to atmospheric oxidation). Like hydrazine itself, phenyl hydrazine is a powerful oxidising agent and will reduce Fehling's Solution. Phenyl hydrazine is reduced with hydrochloric acid and zinc metal to give aniline and ammonia.

$$C_6H_5-NH-NH_2 \xrightarrow[\text{Zn/HCl}]{} C_6H_5-NH_2 + NH_3$$

The most important reaction of phenyl hydrazine is with carbonyls to form hydra-zones. With sugars, phenyl hydrazine reacts to form osazones. Osazones are well-defined crystalline solids and are used in analytical chemistry to identify sugars. When treated with zinc and acetic acid phenyl hydrazones are reduced to form primary amines.

$$C_6H_5-NH-N=CHR + Zn/AcOH \rightarrow C_6H_5-NH_2 + R-CH_2-NH_2$$

For sugars and carbonyls that do not produce isolatable hydrazones and osazones, 2,4-dinitrophenylhydrazine is often the reagent of choice and will usually produce crytallizable derivatives.

5.5 Hydroxylamine

If hydrazine may be thought of as two amide groups joined together (NH_2-NH_2), then replacement of one of these groups by a hydroxy group would produce hydroxylamine and replacement of both hydroxy groups would produce hydrogen peroxide. Unfortunately there is no known reagent that will bring about this transformation from hydrazine to hydrogen peroxide but the concept of sequential replacement of amide groups is useful in explaining the properties of hydroxylamine, as these are in the most part somewhere between hydrazine and hydrogen peroxide. The amide group is a better proton acceptor than the hydroxy group, but the hydroxy group is a better oxidant than the amide group. Hydroxylamine is therefore a poorer reductant but better oxidant than hydrazine and a better reductant and poorer oxidant than hydrogen peroxide.

(*Author's note.* Is is difficult to think of hydrogen peroxide as a reducing agent because its oxidizing properties are so well known. However, I have used hydrogen peroxide in the laboratory to reduce platinum-metal salts in aqueous solution back to platinum metal. This procedure was developed by Bond in the 1960s and works well [18].)

Hydrazine	**Hydroxylamine**	**Hydrogen peroxide**
H_2N-NH_2	H_2N-OH	HO–OH
Strong reductant	$\rightarrow$	*Weak reductant*
Weak oxidant	$\leftarrow$	*Strong oxidant*

In the pure state at ambient conditions, hydroxylamine occurs as white crystalline needles or flakes. These are hygroscopic and unfortunately very unstable, and are prone to spontaneous explosive decomposition.

Although hydroxylamine melts at 30.2 °C it must be stored below −30 °C as it is prone to decomposition. It is soluble in water and also in methanol and ethanol. It is sparingly soluble in organic solvents such as benzene and carbon tetrachloride. However, despite being soluble in water, hydroxylamine is rapidly protonated and is stabilized by protonation.

$$NH_2OH + H_2O \rightleftharpoons [NH_3OH]^+ + [OH]^- \quad K_{298} = 6.6 \times 10^{-9}$$

Hydroxylamine is usually prepared *in situ* in solution just before use, or a stable salt such as hydroxylamine sulphate or hydroxylamine hydrochloride is used. This can lead to great confusion as many chemical texts use 'hydroxylamine' as an inclusive term for hydroxylamine and its salts and may refer to chemical reactions with hydroxylamine when in fact hydroxylamine sulphate or hydroxylamine hydrochloride are used. Hydroxylamine is used as a nucleophile in aromatic substitution reactions, as a reducing agent, and for the conversion of aldehydes to nitriles. The problem with hydroxylamine is that it has two reactive functions with different pK_a values (5.8 and 13.7) and can react at nitrogen and at oxygen (often leading to unwanted side products). Hydroxylamine can also be converted into hydroxylamine-*O*-sulphonic acid (HOSA), which is a good aminating agent.

Hydroxylamine can be prepared by the reduction of hydroxylamine hydrochloride by sodium butoxide in solution or from the oxidation of nitrogen monoxide or the hydrolysis of nitroparaffins. Alternatively, if hydroxylamine sulphate is added to liquid ammonia this will precipitate ammonium sulphate, which can be filtered off leaving a solution of hydroxylamine in ammonia. The resulting solution can either be used directly (provided that liquid ammonia is a suitable solvent for the desired synthesis), or the ammonia can be evaporated to leave free hydroxylamine.

Hydroxylamine is produced commercially as its sulphate or hydrochloride salt (see below) from the Raschig Reaction (i), from the electrolytic reduction of nitric acid between lead electrolytes in the presence of sulphuric acid (ii), or from the catalytic hydrogenation of nitric oxide over carbon-supported platinum in sulphuric acid (iii).

$$(i) \ NH_4NO_2 + 2SO_2 + NH_3 + H_2O \rightarrow (NH_4)_2[HON(OSO_2)_2]$$

$$(ii) \ HNO_3 + 6H^+ + 6e^- \rightarrow NH_2OH + 2H_2O$$

$$2NH_2OH + H_2SO_4 \rightarrow [NH_3OH]_2^+ \ [SO_4]_2^-$$

$$(iii) \ 2NO + 3H_2 + H_2SO_4 \rightarrow [NH_3OH]_2^+ \ [SO_4]_2^-$$

In the solid state hydroxylamine adopts a *trans* configuration, but *cis* and *gauche* configurations are also known.

Hydroxylamine reacts with Michael acceptors to form β-amino acids and their esters [19].

$$X-CH=CH-CO_2H \xrightarrow[NH_2OH/NaOEt/EtOH/H_2O]{} X-\overset{\overset{\displaystyle NH_2}{\displaystyle |}}{C}H-CH-CO_2H$$

Hydroxylamine reacts with aldehydes and ketones to form oximes.

$$O=CH-CO_2CH_3 \xrightarrow[NH_2OH.HCl/NaHCO_3/H_2O]{} HO-N=C-CO_2CH_3$$

Nitriles are produced instead of oximes if hydroxylamine is refluxed with aldehydes in the presence of formic acid. This reaction can also be performed by refluxing hydroxylamine with pyridine and then adding toluene to remove any water azeotropically.

$$C_6H_5-CHO + NH_2OH \rightarrow C_6H_5-C\equiv N + 2H_2O$$

With 1,3-dicarbonyl compounds or 1-carbonyl-3-nitriles or 1-carbonyl-3-enes, hydroxylamine reacts to form isoxazoles.

(Here X = SR$_2$.)

When 4,4-dimethyl-3-oxopentanenitrile is used in this reaction with hydroxylamine the product is 3-amino-5-t-butylisoxazole.

If hydroxylamine is reacted with oleum at room temperature or heated with chlorosulphonic acid then hydroxylamine-*O*-sulphonic acid (HOSA) is formed. This is a useful reagent for amination and for converting alkenes to primary amines.

$$NH_2-OH + H_2S_2O_7 \rightarrow H_2N-O-SO_2OH + H_2SO_4$$

HOSA is able to react both as an electrophile and as a nucleophile. Under basic conditions it can aminate carbon, nitrogen and sulphur nucleophiles including pyridine, trimethylamine, quinoline, purines, β-diketones, aromatic rings, thioamides and thioethers.

$$Ph-NH_2 + HOSA \rightarrow Ph-NH-NH_2$$

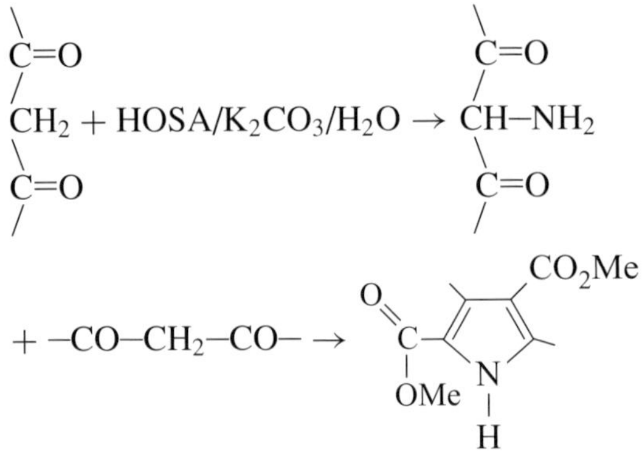

$$
\begin{array}{ccc}
\backslash \\
C{=}O & & \backslash \\
/ & & C{=}O \\
CH_2 + HOSA/K_2CO_3/H_2O \rightarrow & & / \\
\backslash & & CH{-}NH_2 \\
C{=}O & & \backslash \\
/ & & C{=}O \\
& & /
\end{array}
$$

$$+ \ {-}CO{-}CH_2{-}CO{-} \ \rightarrow$$

Alkenes may be converted into amines via the boronation technique developed by Brown and previously described in Chapter 2. Alkenes are reacted with diborane formed *in situ* from the reaction between sodium borohydride and boron trifluoride etherate in diglyme and then HOSA is added to form the amine.

Carboxylic acids may also be converted into amines by using HOSA but the yields are usually poor. The reaction is best carried out in polyphosphoric acid at $115\,^{\circ}C$, but mineral oil may also be used at $160\,^{\circ}C$.

$$p\text{-}CH_3-C_6H_4-CO_2H + HOSA \rightarrow p\text{-}CH_3-O-C_6H_4-NH_2$$

Interestingly, if HOSA is refluxed with an amine under basic conditions in the presence of potassium, sodium hydroxide or pyridine, deamination can take place; primary alkyl amines can be converted to alkanes and aminobenzoic acids can be converted to benzoic acid.

$$o\text{-}H_2N-C_6H_4-CO_2H + HOSA/NaOH/H_2O \rightarrow C_6H_5-CO_2H$$

Like hydroxylamines, HOSA will also convert aldehydes to nitriles and ketones to oximes.

5.6 Nitrogen-containing donor ligands

Ammonia and many amines are able to form complexes with transition metal ions. When ammonia acts as a ligand the resulting complexes are known as ammines; when the ligand is $[NH_2]^-$ the complex formed is an amino complex. The ammines of cobalt and chromium were first investigated in the early twentieth century, for which work Alfred Werner was awarded the Nobel Prize in 1913 [20]. Transition-metal ions generally form octahedral, hexasubstituted ions $[MX_6]^{n+}$, where X is a neutral or anionic monodentate ligand such as H_2O or Cl. For Group VIII metal ions with d^8 configurations, such as platinum(II), square-planar configurations are also known. Addition of ammonia or a nitrogen-containing ligand to the metal ion in solution (usually methanol or ethanol but liquid ammonia can also be used) can result in the displacement of X and the formation of ammines. The ions $[MX(NH_3)_5]^{n+}$ and $[MX_2(NH_3)_4]^{n+}$ are the most common, but complete mono-tonic series are known for Cr(III) and Co(III) ions from $[MX_{6-y}(NH_3)_y]^{(y-3)+}$, where $y = 0$ to 6.

One problem with this method of synthesis when ammonia is added to a methanolic aqueous solution of a metal salt such as rhodium trichloride or ruthenium trichloride is that a mixture of isomers and different ammines are formed that are impossible to separate or characterize.

Coordinated ammonia complexed to a metal cation can generally undergo oxidation and condensation reactions, but these are really of academic interest only unless as part of a template synthesis for the manufacture of phthalocyanines and porphyrin rings.

The most important transition-metal ammine complexes are the square-planar Pt(II) anti-tumour complexes of which the best known are *cis*-platin, otherwise known as *cis*-dichlorodiammineplatinum(II), and carboplatin, which has the full name of *cis*-diammine(1,1-cyclobutanedicarboxylato) platinum(II). *Cis*-platin and carboplatin are successful at treating cervical, testicular and small lung cancers, and have also been used for neck and head cancers.

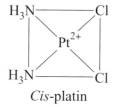

Cis-platin

Cis-platin and carboplatin are both administered as intravenous solutions and it is thought that the chloride ions are replaced by hydroxide groups at physiological pH once the uncharged molecule penetrates the cell walls bound to blood plasma. Interstitial-cell electrolyte contains 4 mmol of $[Cl]^-$ compared with about 100 mmol ouside the cell. This low chloride-ion concentration facilitates hydrolysis of both *cis*-platin and carboplatin to form *cis*-$[Pt(NH_3)_2(H_2O)_2]^{2+}$. The loss of the chloride ions increases the bioactivity of the platins as $[OH]^-$ is a better leaving group than Cl^-, and so increases its reactivity once inside the cancer cells. The stereochemistry

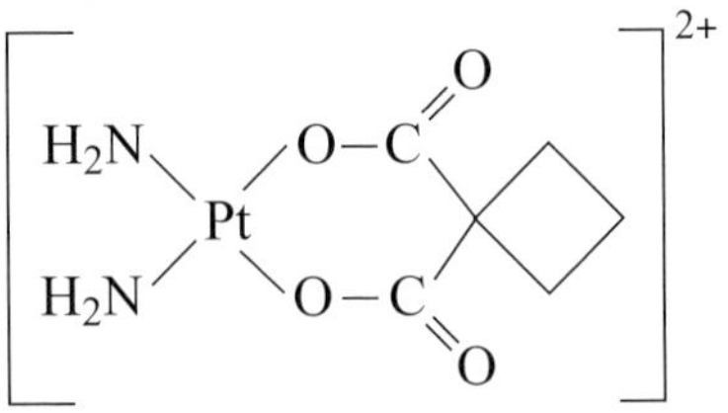

Figure 5.2. The structure of carboplatin.

of the square-planar Pt(II) allows the molecule to bond to two N-7 DNA positions and form an intrastrand cross-link between two adjacent deoxyguanylates. This link across two adjacent sites is sufficient to prevent DNA replication and is cytotoxic [21]. Typically a 1,2 intrastrand *cis*-platin linkage distorts the DNA helix by 34° and unwinds it by 13°, and so blocks replication. The *trans*-isomers do not exhibit this effect as they form interstrand links that do not inhibit DNA replication and so are ineffective as anti-cancer agents. The structure of carboplatin is shown in Figure 5.2. It has improved water solubility over *cis*-platin and is less nephrotoxic and causes less nausea to the patient.

Replacement of ammonia ligands with amines offers greater possibilities, and systems with six-coordinated alkyl amines such as $[Co(NH_2CH_3)_6]^{3+}$ and $[Ru(NH_2-CH_2-CH_2-NH_2)_3]^{2+}$ are known.

However, with six alkyl amines around a single metal centre the level of steric crowding is high and it is more common to find systems with five amines, such as $[MX(NH_2R)_5]^{2+}$, where M = Co, Rh and Cr.

Because of steric crowding, primary alkyl amines with no α-substituents coordinate best although a few systems with α-substituted primary amines, secondary amines and tertiary amines have been reported. Surprisingly aziridine complexes well and many be synthesized *in situ* from coordinated 2-bromoethylamine [22].

$$[Co(en)_2(Br)(NH_2-CH_2-CH_2-Br)]^{2+} \rightarrow [Co(en)_2(OH)(HNC_2H_4)]^{2+} + 2HBr$$

(Here en = ethylene diamine.)

Ruthenium tris(ethylenediamine) complexes can be reduced to form α-diimine complexes.

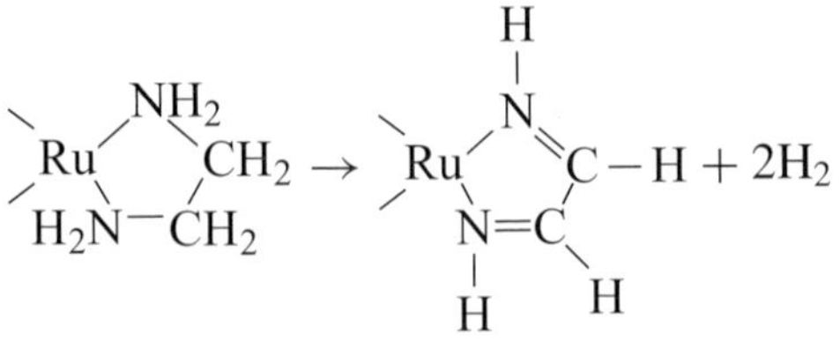

Also, cyclization reactions are known for systems involving glycine ester ligands, such as those with cobalt(III) ions.

For pyridine the maximum packing around a metal centre seems to be four, for example in $[RhCl_2(py)_4]^+$ with the relatively large rhodium(III) ion. The synthesis of rhodium–pyridine complexes is strongly dependent upon the choice of solvent. Reaction of Na_3RhCl_6 with pyridine in aqueous solution produces only the tris-pyridine derivative, but in a mixed water/ethanol solution the tetrapyridine derivative is obtained. It is thought that in the tetrapyridine salt the four pyridine ligands occupy the equatorial positions with the chlorine anions in the axial positions.

Other nitrogen heterocycles such as pyrazole, imidazole, pyrimidine, pyridazine, 2′,2-bipyridine, 1,10-phenanthroline and piperazine are also known as ligands [23]. Stronger metal ligand interactions are generally found for ligands with more delocalized amines such as bipyridine and *o*-phenanthroline, where delocalization over an extended aromatic frame lowers the energies of the π-HOMO and π-LUMO orbitals, making them weaker π-donors but better π-acceptors.

Two molecules of interest are the tris-bipyridine complexes with ruthenium and chromium, $[Ru(bipy)_3]^{n+}$ and $[Cr(bipy)_3]^{n+}$. The ruthenium tris-bipyridyl complex is used as a sensitizer in photolytic systems and the chromium tris-bipyridyl complex is used for probing excited-state photochemistry.

Of greater interest is the use of nitrogen-containing ligands in homogeneous asymmetric catalysis. The Trost Ligand [24] shown in Figure 5.3 contains two amine groups and is finding increasing application for asymmetric hydrogenation. It will also catalyse reactions such as the Michael Reaction and the Diels–Alder Addition.

Asymmetric catalysis with complexed transition-metal ions provides a convenient method of introducing a chiral centre into an achiral substrate. During the

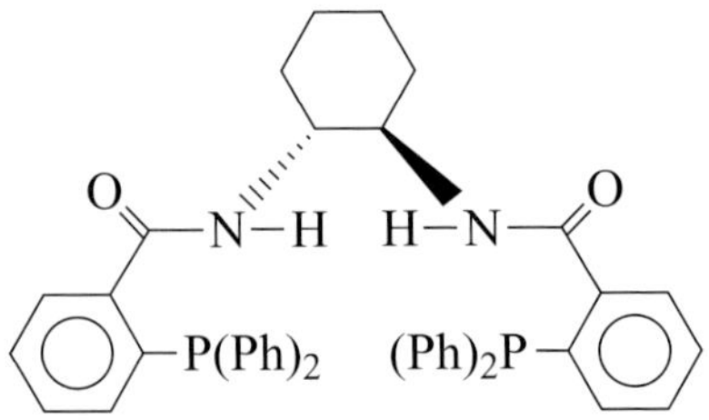

Figure 5.3. The Trost Ligand.

reaction the achiral substrate is fixed onto a sterically crowded metal centre so that only one face is available to the attacking species.

The ligands typically used to prepare catalysts for asymmetric reactions are very bulky and many are based on the substituted triphenyl phosphines.

Of course nitrogen itself can be a ligand; the symbiotic nitrogen fixing bacteria living in the root nodules of peas, beans, clover and aquatic ferns are examples. To date, many metal-based systems have been developed where coordination to a central metal ion means that the dinitrogen $N\equiv N$ triple bond is weakened and the derivatization can take place. The best known of these systems are based on molybdenum, manganese, titanium and tungsten, with diphos or cyclopentadienyl ligands. Nucleophilic attack on the coordinated nitrogen is facilitated by methyl or phenyl lithium, and electrophilic attack is facilitated by reagents such as alkyl halides and acyl chlorides.

$$(\text{diphos})_2 W(N_2) + C_6H_5{-}CO{-}Cl \rightarrow (\text{diphos})_2 ClW(N{=}N{-}CO{-}C_6H_5) + N_2$$

(where diphos $= (C_6H_5)_2P{-}CH_2{-}CH_2{-}P(C_6H_5)_2$).

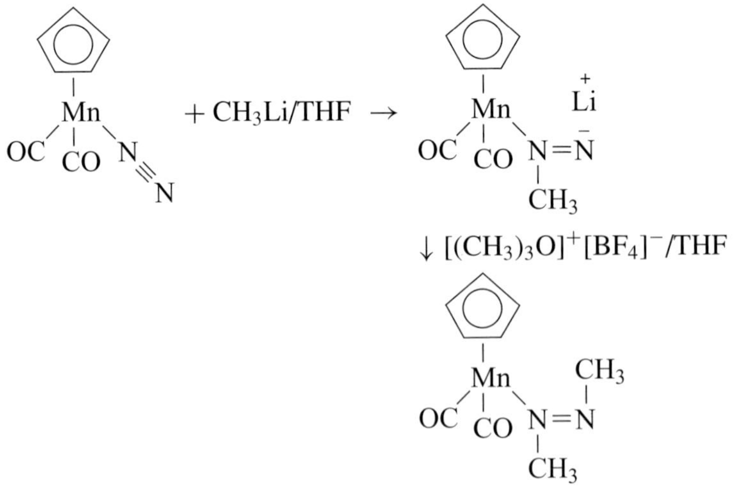

The nitride ion, N^{3-}, can also form coordination complexes and, like dinitrogen-coordinated nitrides, can take part in amination reactions.

The metal–nitride bond is very short, typically only 1.16 Å in length, and the nitride ion is one of the strongest π-donors known when complexed to metals.

Nitrides can form four main types of coordination complex.

(i) Complexes with a terminal nitrogen, such as

$$F_3S{\equiv}N \text{ and } N{\equiv}MN(TPP), \text{ where } TPP = \text{tetraphenylporphryin.}$$

(ii) Complexes with bridging nitrogen, like the one shown below.

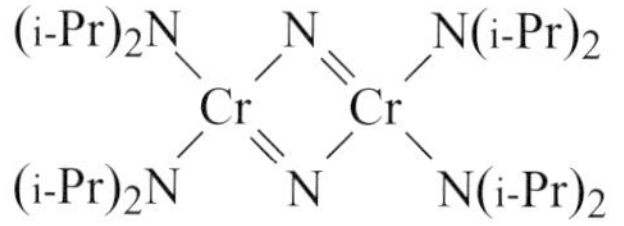

$$\begin{array}{ccc} (i\text{-Pr})_2N & N & N(i\text{-Pr})_2 \\ & \diagdown \diagup \diagdown \diagup & \\ & Cr \quad Cr & \\ & \diagup \diagdown \diagup \diagdown & \\ (i\text{-Pr})_2N & N & N(i\text{-Pr})_2 \end{array}$$

(iii) Complexes with tetrahedral N^{3-}, such as $[N(HgCH_3)_4]^+[Cl]^-$.

(iv) Triangular N-centred complexes such as $[Ir_3N(SO_4)_6(H_2O)_3]^{6-}$, where a central nitride ion is surrounded by three coordinated iridium ions in triangular geometry.

Coordinated terminal nitrides with magnesium can be used as reagents for amination reactions. The nitride ions can be generated from the reaction of tertiary butyl isocyanate with a metal oxide or from the reaction between active metals and urea via a metastable intermediate.

$$t\text{-Bu}{-}N{=}C{=}O + M{=}O \rightarrow M{\equiv}N + H_2C{=}C(CH_3)_2$$

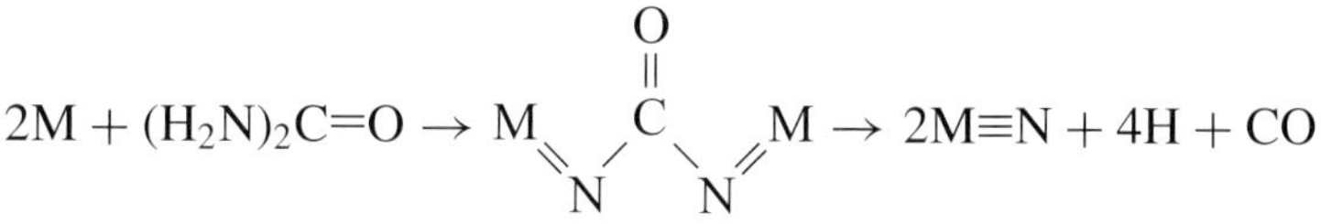

$$2M + (H_2N)_2C{=}O \rightarrow M \underset{N \quad N}{\overset{\overset{O}{\|}}{\diagdown \underset{}{C} \diagup}} M \rightarrow 2M{\equiv}N + 4H + CO$$

The first reported use (in 1983) of coordinated nitride ions as amination reagents was by Groves and Takashi, who used a nitride complex with manganese and 2,4,6-trimethylphenyl ligands (TMP), which aminated cyclooctene in the presence of trifluoroacetic acid [25].

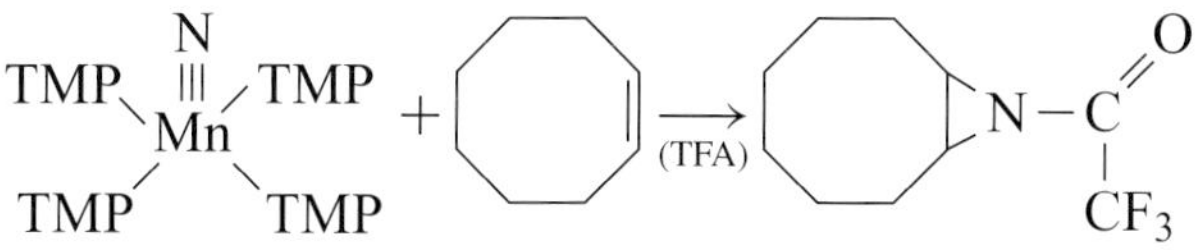

More recently (in 1996), Carreira and co-workers developed the Saltmen Ligand, shown in Figure 5.4. When complexed to manganese and a nitride ion, this ligand forms an effective reagent for the amination of styrenic double bonds [26].

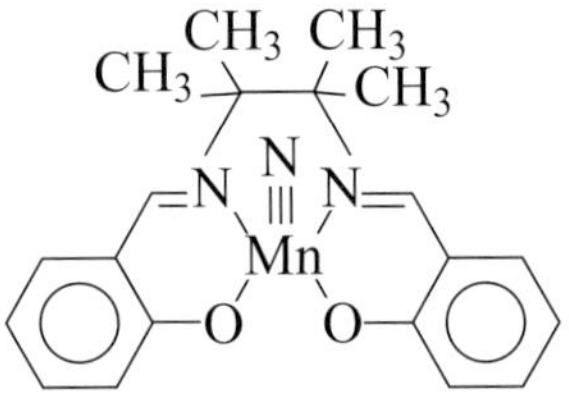

Inorganic amines

$$C_6H_5-C=C' + (Saltmen)MnN \xrightarrow[(a)+(b)]{} C_6H_5-CH(OH)-CH_2-NH-CO-CF_3$$

(a) $(CF_3-CO_2)_2O/CH_2Cl_2$ followed by
(b) aqueous $NaHCO_3$/tetrahydrofuran

Figure 5.4. The Saltmen Ligand complexed to manganese and nitrogen.

5.7 Nitrogen-containing macrocyclic ligands

The naturally occurring porphin ring is vital for the preservation of life on Earth. The porphin-ring system and its adducts with metals (called porphyrins) are found in haemoglobin, myoglobin, chlorophyll and vitamin B12 (vitamin B12 actually contains a corrin ring, which has one methine group less than the porphyrin ring but is otherwise similar). These biochemicals are vital to the chemistry of life as they are responsible for the use of solar energy to produce oxygen and for the transport of oxygen in animals. The porphin-ring system is a macrocyclic tetrapyrrole system with conjugated double bonds, and this system can exist in the neutral form or as doubly charged positive and negative ions (from the gain or loss of two electrons). When complexed with metal ions, two ring nitrogen atoms donate two protons to form the doubly negative anion. The metal-to-nitrogen bond distance is usually around 200 pm and the square-planar geometry favours ions such as Ni^{2+}, Cu^{2+}, Co^{2+}, Fe^{2+} and Zn^{2+}.

Many porphyrin and corrin rings are planar and contain delocalized electrons with aromatic character. For example, the inner ring of porphin (see Figure 5.5) contains 16 atoms, each contributing one electron. Therefore when two electrons are added from the dianionic form the total number of electrons equals 18, which conforms to Hückel's $(4n+2)$ rule for stable aromatic systems. Addition of a dipositive metal ion therefore produces a stable neutral species. Because the porphin ring is planar (or very nearly planar) there is considerable scope for electronic resonance and stabilization of charge.

The porphin ring is stable to concentrated sulphuric acid and can exhibit both acidic and basic properties. For example, alkoxides can deprotonate the inner nitrogen atoms ($pK_a = 16$) to form the porphyrin dianion, and strong acids such as trifluoroacetic acid can protonate the ring nitrogens ($pK_b = 9$). Porphyrins display characteristic UV–visible spectra (see Figure 5.6) that consist of a strong band at around 400 nm called the Soret Band together with weaker bands from 450 to

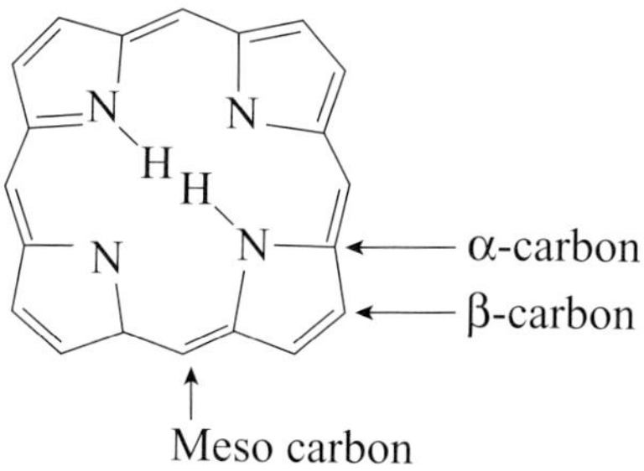

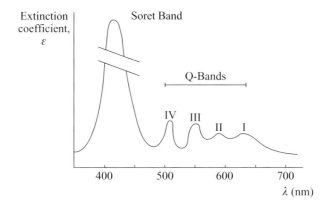

Figure 5.5. The structure of the inner ring of the porphyrin system.

Figure 5.6. The UV–visible spectrum of a typical metallo-porphyrin showing the Soret and Q-Bands.

700 nm called Q-Bands. The carbons in the porphin-ring system can be classified into three main types: α-carbons (adjacent to nitrogen), β-carbons (one carbon away from nitrogen), and meso-carbons (on the methine bridges). The corrin-ring system, which contains one methine group less than the porphin-ring system, contains only six double bonds and is not fully delocalized but still allows four possible resonance structures to stabilize the ring structure.

When the porphyrin ring is complexed with metal ions such as cobalt, iron and magnesium the resulting compounds are respectively vitamin B12, haem and chlorophyll. Collectively these are called metallo-porphyrins. They are important in biological systems because changing substituents on the porphyrin ring or the complexed metal ion can change their activity. They are also able to act as redox intermediates in electron-transfer reactions and so help change the oxidation state of the complexed metal ions. For example, in haem (Figure 5.7) the oxidation state of the central iron atom can be either +2 or +3 depending upon whether oxygen is complexed or not.

$$\text{Haem } (Fe^{2+}) + O_2 \rightarrow \text{Hematin } (Fe^{3+})$$

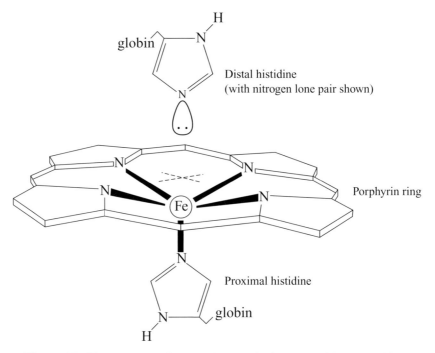

Figure 5.7. The structure of haem (the prosthetic group of haemoglobin).

Haemoglobin transfers oxygen from the lungs (or gills) to the body cells where it is to be used and transferred to myoglobin for use in respiration. Haemoglobin is a tetramer of myoglobin and has a molecular weight of about 64 500. It can absorb four molecules of oxygen. The bonding of oxygen to iron in oxyhaemoglobin and oxymyoglobin is interesting and is the opposite of what might be expected in a simple chemical system. For example, haemoglobin and myoglobin contain high-spin iron(II), which changes to low-spin iron(III) on complexation with oxygen. Molecular oxygen is not normally a strong-field ligand and would not be expected to make this type of electron spin pairing with iron. For this reason the porphyrin ring is thought to be important in controlling this complexation reaction. It is not an inert ligand and instead accepts a negative charge. The iron(II) ions complexed to the porphin ring actually sit just below the plane of the ring but when complexed to oxygen they are oxidized to iron(III), which is slightly smaller and fits perfectly into the ring. The exact mechanism of oxygen complexation is unclear in haemoglobin and it is thought to bind end-on to the Fe(III) d_{xz} and d_{z^2} orbitals.

   It is thought that as soon as one iron site in haemoglobin binds to molecular oxygen it results in complexation being triggered at the other sites. The mechanism for this is not clear but it may involve stress in the protein chains. Haemoglobin is also able to bind carbon dioxide by forming carbamates with the free amino groups

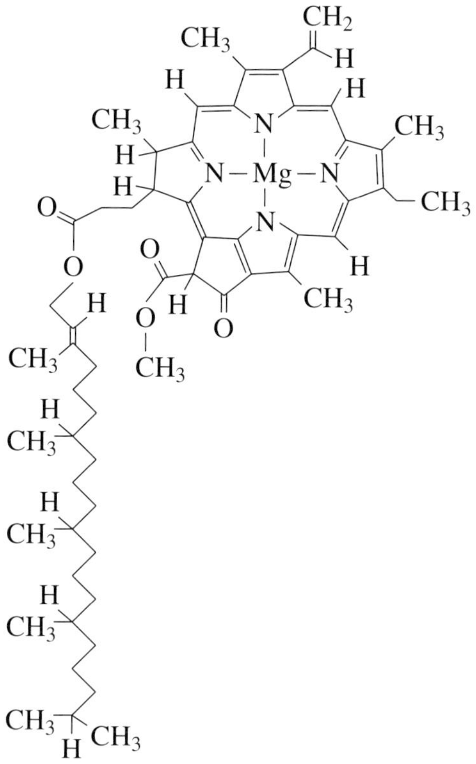

Figure 5.8. The structure of chlorophyll.

on the α- and β-globin chains; for example, with arginine 141 the reaction is as follows, where Hb— represents the globin chain [27].

$$Hb{-}NH_2 + CO_2 \rightarrow [Hb{-}NH{-}CO_2]^- \, [H]^+$$

Haemoglobin binds carbon dioxide when it delivers oxygen and releases carbon dioxide when it binds oxygen. The reciprocal exchange of carbon dioxide and oxygen is controlled by pH. Physiological pH is higher in the lung capillaries than in the tissue capillaries and red blood cells. This effect is known as the Bohr Effect [28] and defines the respiratory cycle. It should be noted that only about 10% of the carbon dioxide produced by respiring tissues is transported as a carbamino complex with haemoglobin. The balance is transported in blood plasma as bicarbonate produced from the reaction between water, carbon dioxide and the enzyme carbonic anhydrase.

Chlorophyll (shown in Figure 5.8) enables the photosynthetic process in plants. Overall this may be thought of as the reduction of carbon dioxide with hydrogen

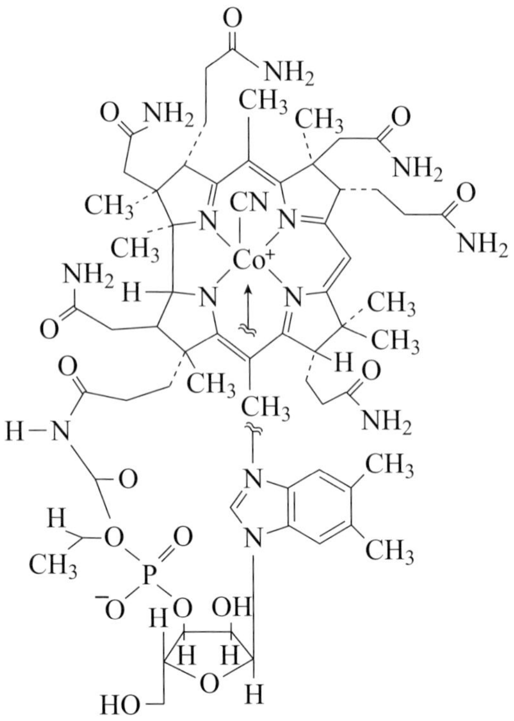

Figure 5.9. The structure of vitamin B12.

from water.

$$2H_2O + CO_2 \rightarrow O_2 + H_2O + CH_2O$$

The energy required to drive this process is provided by a photon hitting a chlorophyll molecule and starting a series of redox reactions. Free hydrogen is not actually present in the cells of plants but the energy from the photon is able to reduce water complexed to the magnesium centre so that it can combine with carbon dioxide.

$$(Mg_{centre}-2H_2O)_{oxidized} \rightarrow (Mg_{centre})_{reduced} + O_2 + 4H^+$$

Vitamin B12 (see Figure 5.9) contains a corrin ring with a cobalt(III) ion. The cobalt centre may be reduced by either one or two electrons to form Co(II) or Co(I). Vitamin B12 is an anti-pernicious-anaemia factor and is found in the liver and blood streams of animals, although interestingly vitamin B12 can only be manufactured by certain microorganisms and not by animals or plants. In these microorganisms, vitamin B12 is used to catalyse a number of methylation reactions. In mammals vitamin B12 catalyses three reactions (with specific enzymes): methylation (methionine synthetase), the catabolism of valine (methylamalonyl CoA mutase) and the isomerization between leucine and β-leucine (leucine amino mutase).

Many cobalt porphyrin-ring systems have been made synthetically but they do not catalyse these biochemical reactions *in vitro* as they cannot be reduced from cobalt(III) to cobalt(I) in aqueous media. Therefore it is the corrin ring system which gives vitamin B12 its bioactivity.

Given the structural similarities between haem, vitamin B12 and chlorophyll it is not surprising that there exists a common precursor for their biosynthesis. This precursor is called uroporphyrinogen III and is synthesized by the reaction between succinyl-CoA and glycine to form δ-aminolaevulinic acid.

Eight molecules of δ-aminolaevulinic acid are needed to form uroporphyrinogen III (two molecules of δ-aminolaevulinic acid condense to form phorphobilinogen and four molecules of phorphobilinogen condense to form uroporphyrinogen III with the loss of four molecules of ammonia).

$$CoA-S-CO-CH_2-CH_2-COOH + H_2N-CH_2-COOH \rightarrow$$
$$\text{Succinyl-CoA} \qquad \qquad \text{Glycine}$$

$$HOOC-CH(NH_2)-CO-CH_2-CH_2-COOH(-CO_2) \rightarrow$$
$$\alpha\text{-Amino-}\beta\text{-oxoadipic acid}$$

$$NH_2-CH_2-CO-CH_2-CH_2-COOH + \delta\text{-aminolaevulinic acid} \rightarrow$$
$$\delta\text{-Aminolaevulinic acid}$$

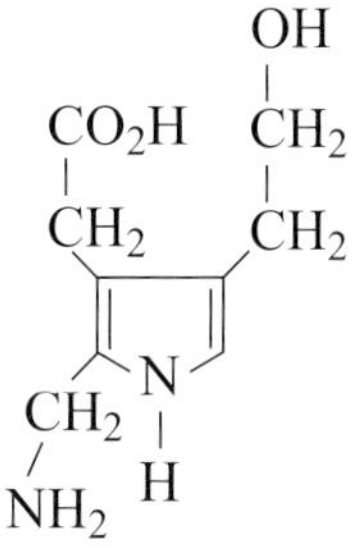

Porphobilinogen

$4 \times$ Porphobilinogen $\rightarrow$

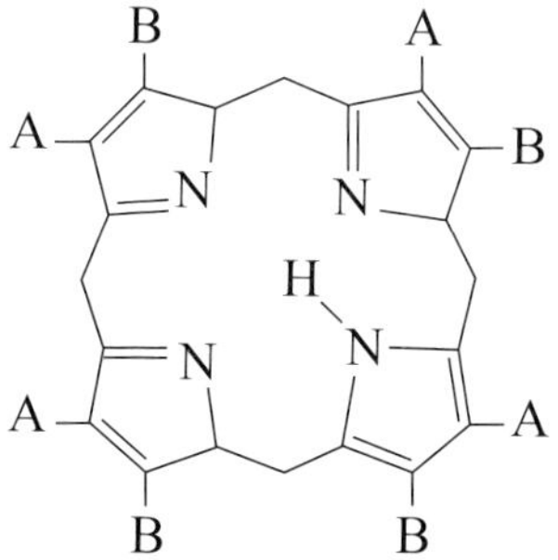

Uroporphyrinogen III

(Here A $= -CH_2-COOH$ and B $= -CH_2-CH_2-COOH$.)

In the laboratory, porphins may be conveniently synthesized by heating pyrrole-2-aldehyde with formic acid and ethanol, or by heating pyrrole with formaldehyde in methanol and pyridine [29].

The porphyrin-ring and corrin-ring systems in vitamin B12, haem and chlorophyll are all planar and their metal ions are in octahedral symmetry. In haem, one axial position is occupied by a histidine molecule bound to a peptide chain and the other is occupied by a water molecule; in vitamin B12, a dimethylbenzimidazole nucleotide occupies one axial position and a water molecule occupies the other; and in chlorophyll, both axial positions are occupied by bridging water molecules.

In vitamin B12 and haem, an incoming group is able to displace a water molecule from one of the axial positions and bond to the central metal ion resulting in a change of the metal's oxidation state. In chlorophyll, photon absorption leads to an activated state between the metal ion and the water molecules already bonded to the molecule so that it can reduce carbon dioxide.

The overall reaction schematics for vitamin B12 and haem are as follows. (Note that in the case of chlorophyll the overall reaction is the same but water is transferred from the ground state to an activated state without desorption.)

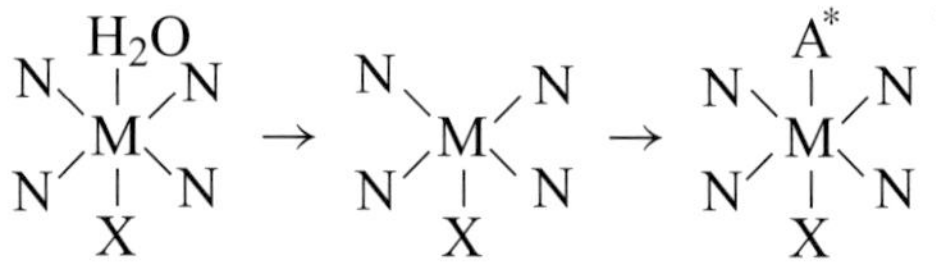

$$
\begin{array}{ccccc}
& \overset{\displaystyle H_2O}{\underset{\displaystyle X}{\;\;N\!\diagdown\!\underset{M}{|}\!\diagup\!N\;\;}} & \rightarrow & \underset{\displaystyle X}{\;\;N\!\diagdown\!\underset{M}{|}\!\diagup\!N\;\;} & \rightarrow & \overset{\displaystyle A^*}{\underset{\displaystyle X}{\;\;N\!\diagdown\!\underset{M}{|}\!\diagup\!N\;\;}}
\end{array}
$$

A^* is the absorbed species: in the case of chlorophyll this is water in a state of high energy, in the case of haem this is molecular oxygen, and in the case of vitamin B12 this is a methyl-containing group.

Chlorophyll is not a haem because the metal ion is manganese (which actually sits above the ring plane). Haems are also necessary for photosynthesis (in addition to chlorophyll) and are present in plant chloroplasts and mitochondria, which are necessary for animal and plant respiration. Haem groups form the active centres of cytochromes, which consist of a porphyrin ring bound to an iron centre with a polypeptide chain bound to the iron centre via a histidine nitrogen in one axial position and a methionine sulphur in the other axial position, as shown below.

Cyctochromes are fully coordinated and react indirectly by electron transfer to reduce oxygen and to release energy in respiration. In evolutionary terms, cytochrome c is one of the oldest chemicals involved in living processes and is found in yeasts, higher plants, insects, animals and man. Despite some differences in the protein chains in different species, cyctochrome c from plants will react in animal systems and vice versa; this may be indicative of a common evolutionary history.

Although for many years porphyrins were all thought to be planar and aromatic, several researchers have recently synthesized non-planar porphyrins whose structures have been verified by X-ray crystallography [30].

It is interesting that porphyrins are planar in the first instance as normally nitrogen adopts a tetrahedral configuration rather than a trigonal-planar configuration. In porphryins there are four nitrogens in the inner ring, two of which are deprotonated by the central metal ion with the overall gain of two electrons to the ring, and this completes the 18 electrons required for aromaticity.

If we assume that two nitrogens are deprotonated and two are not, then even with the many resonance structures possible in the delocalized porphyrin ring it is reasonable to assume that there must be a degree of ring strain. In planar porphyrins the energy gained by aromaticity is greater than the energy of the ring strain, but in non-planar porphyrins the opposite is true.

If the porphyrin ring is thought of as four nitrogen-to-metal bonds connecting a carbocyclic system to a metal-cation centre, then it can be seen that the role of the nitrogen bonding is the critical factor in defining the properties and geometry of the porphyrins.

Three different types of metal-to-nitrogen bonds have been identified by X-ray crystallography, and these define the geometry of the porphyrin ring and whether or not it is planar or non-planar (aromatic or non-aromatic) [31]. These are:

(i) short metal-to-nitrogen bonds (less than 1.95 Å, e.g. Ni(II)),
(ii) medium metal-to-nitrogen bonds (around 2 Å, e.g. Zn(II)),
(iii) long metal-to-nitrogen bonds (over 2.1 Å, e.g. Ta(II)).

The hole in the porphin ring has a radius of about 2.04 Å, which is adequate to accept most ions from the first transition series as, from covalent-bond radii, the length of a metal-to-nitrogen bond should be around 2 Å.

For metal–porphyrin complexes the nitrogen lone pairs on the ring are able to interact with the 'doughnut' on the d_{z^2} and the four lobes on the $d_{x^2-y^2}$ orbital.

When the central metal ion changes oxidation state the π^*-orbitals are able to accept or donate an electron as required via the overlap between the four nitrogen lone pairs and the d orbitals of the central metal ions.

Thus the three important factors by which haem, vitamin B12 and chlorophyll are able to operate biochemical processes are the combination of their unique chemical structures, which incorporate:

(i) a central metal atom with a variable oxidation state or coordination;
(ii) an aromatic system able to absorb, stabilize and release electronic charge as required; and
(iii) four nitrogen atoms with axial lone pairs that are able to act as electron bridges between the aromatic system and the coordinated metal ion for the transfer of electronic charge.

The stability of the metal–porphyrin system is the reason for its widespread use by living organisms ranging from simple yeasts to plants, animals and humans. It is strange that although oxygen is regarded as necessary for respiration and photosynthesis it is the action of nitrogen in the form of a complex amine that makes these processes operate.

Another macrocyclic ring system related to the porphyrins is the phthalocyanines. The phthalocyanines are isoelectronic with the porphyrins but are totally synthetic and unknown in nature. The chemistry of this class of compunds is discussed in Chapter 8, which covers the application of amines in reprography and photography.

5.8 Sulphur–nitrogen compounds

Tetrasulphurtetranitride (S_4N_4) was discovered in 1835 by Gregory [32], and remained of only academic interest until comparatively recently when it was discovered that the polythiazyl polymer $(SN)_x$, which can be prepared by the sublimation and vapour-phase deposition of tetrasulphurtetranitride, displays superconducting properties below 0.3 K [33]. One of the unachieved goals of the physical sciences remains the development of ambient-temperature superconductors, so the electronic properties of polythiazyl were extensively studied for many years to try to gain an understanding of the factors giving rise to superconductive behaviour. The discovery of superconducting mixed-metal oxides at liquid nitrogen temperatures in the mid 1980s has shifted the primary research focus away from polythiazoles because liquid nitrogen is readily available in most chemistry and physics laboratories but liquid helium is less accessible and much more problematic to handle.

(*Author's note*. In the mid 1980s I was briefly involved in research into superconducting mixed-metal oxides and helped to develop a europium-oxide-based system that was superconducting in liquid nitrogen and displayed the Meissner Effect [34].)

The bonding in tetrasulphurtetranitride is difficult to represent by conventional line drawings as there is significant interaction between adjacent sulphurs and also between sulphur centres separated by nitrogen. However, a molecular-orbital approach works well.

In the diagram below, the U-shaped S_4N_4 molecule, which has D_{2d} symmetry, is shown (the four nitrogen atoms occupy the corners of a square) with broken lines (---) representing interatomic interactions. These fall short of single bonds but nevertheless represent regions of detectable electron density.

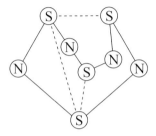

Tetrasulphurtetranitride reacts with S_2Cl_2 in ammonia at 0 °C to form an imide S_7–NH, which can also be prepared from sodium azide on cyclooctasulphur in dimethylformamide solution.

However, S_7–NH is not a typically imide and is appreciably acidic with a pK_a of around 5 and forms a mercury salt $Hg(S_7N)_2$.

Tetrasulphurtetranitride reacts with tin(II) chloride in ethanol to form $(SNH)_4$, which has a tiara structure not unlike cyclooctasulphur [35].

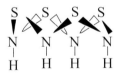

A series of cyclic derivatives of cyclooctasulphur has been prepared, ranging from S_7NR to $S_4(NR)_4$, where R can be H or CH_3. All have puckered eight-membered rings as shown above and have shorter N–S bonds than might be expected (0.06 Å shorter than normal N–S single bonds), which may be indicative of N→S π-bonding. The nitrogen atoms in these ring structures are in nearly trigonal-planar geometry and the bond angles are around 119°.

A recent study of these and related molecules by X-ray charge density analysis has demonstrated the presence of π-bonding with delocalization of the nitrogen lone pairs [36]. In this unusual family of sulphur–nitrogen ring systems, the amine nitrogen atoms are not unlike the pyrrole-type nitrogens described in Chapter 4 that demonstrate electron-releasing effects into heterocyclic ring systems.

References

[1] Q. S. Li and L. P. Cheng, *J. Phys. Chem.*, **107**: 16 (2003), 2882.
[2] R. Huisgen and I. Ugi, *Ber.*, **90** (1957), 1946.
[3] W. H. Birkenhager and P. W. de Leeuw, *J. Hypertens.*, **17**: 7 (1999), 873.
[4] S. R. M. Ellis, G. V. Jeffreys and G. T. Wharton, *IEC Process Design Rev.*, **3** (1964), 18.
[5] H. B. Bürgi, *J. Mol. Struct.*, **10** (1971), 31.
[6] T. M. Chapin, *J. Am. Chem. Soc.*, **51** (1929), 2112.
[7] J. Jander, *Adv. Inorg. Chem. Radiochem.*, **19** (1976), 2.
[8] H. B. Burgi, *J. Mol. Struct.*, **10** (1971), 31.
[9] T. Curtius, *Ber.*, **23** (1890), 3023.
[10] F. Raschig, *Schwefel und Stickstoffstudien* (Berlin: Verlag, 1924).
[11] DE Patent 1082889 (1960, assigned to Bayer AG).
[12] P. Tellier, *Bull. Soc. Chim. Fr.* (1972), 2954; *Chemische Rundshau* (October 1974), 635.
[13] H. H. Sisler, G. M. Omietanski and B. Rudner, *Chem. Rev.*, **57** (1957), 1021.
[14] US Patent 2935451 (1955, assigned to Olin Industries).
[15] L. Wolf, *Ann.*, **394** (1912), 86.
[16] H. Minlon, *J. Am. Chem. Soc.*, **68** (1946), 2487.
[17] S. A. Lawrence, *Chimica Oggi* (May 2000), 20.
[18] G. C. Bond and H. Connor, *Plat. Metals Rev.*, **9** (1965), 14.
[19] E. G. Rozantzev and M. B. Niemann, *Tetrahedron*, **20** (1964), 131.
[20] A. Werner, Nobel Lecture, 11 December, 1913.
[21] B. K. Keppler, *Metal Complexes in Cancer Chemotherapy* (New York: VCH, 1993).
[22] D. A. House, *Coord. Chem. Rev.*, **23** (1977), 223.
[23] S. A. Cotton, *The Chemistry of Precious Metals* (London: Blackie Academic, 1997), p. 121.
[24] B. M. Trost, *J. Am. Chem. Soc.*, **105** (1983), 2073.
[25] J. T. Groves and T. Takashi, *J. Am. Chem. Soc.*, **105** (1983), 2073.
[26] J. DuBois, J. Hong, E. M. Carreira and M. W. Day, *J. Am. Chem. Soc.*, **118** (1996), 915.
[27] M. Perrella, D. Bresciani and L. Rossi-Bernardi, *J. Biol. Chem.*, **250** (1975), 5413.
[28] W. E. Fransworth, *Med. Hypotheses*, **48** (1997), 229.
[29] J. S. Lindsey, I. C. Scheimann, H. C. Hsu, P. Kearney and A. M. Marguerettaz, *J. Org. Chem.*, **52** (1987), 877.
[30] T. E. Clement, L. T. Nguyen, R. G. Khowy, D. J. Nurco and K. M. Smith, *Heterocycles*, **45** (1997), 651.
[31] D. J. Nurco, C. J. Medforth, T. P. Forsyth, M. M. Olstead and K. M. Smith, *J. Am. Chem. Soc.*, **118** (1996), 10 918.
[32] W. Gregory, *J. Pharm. Chim.*, **21** (1835), 315.
[33] H. W. Roesky, *Angew. Chemie Int. Ed.*, **18** (1979), 91.
[34] M. S. Vong, P. A. Sermon and S. A. Lawrence, *Synth. Metals*, **33**: 2 (1989), 123.
[35] R. L. Sass, *Acta Crystallogr.*, **11** (1958), 497.
[36] H. W. Roesky, *Adv. Inorg. Chem. Radiochem.*, **22** (1972), 286.

6

Small-scale syntheses and analytical methods for amines

6.1 Introduction

The following synthetic methods are presented to give the reader an understanding of how reactions with amines may be performed in the laboratory. Readers of this book must not attempt any of these syntheses without first obtaining qualified supervision and also studying the material safety data sheets for each reagent, intermediate, final reaction product and by-product beforehand in order to take the necessary safety precautions. All waste streams and any leftover chemicals from these experiments should be disposed of in accordance with the prevailing national and local regulations. All of these chemical reactions are best performed in an inspected fume cupboard with an air extractor and with the experimenter wearing appropriate skin, eye and breathing protection, all of which should be used as described in the material safety data sheets.

Although the present author can vouch that the experiments work, if properly conducted, neither he nor the publisher can be held responsible for accidents that may take place in the course of experimentation, and the reader is asked to keep the above safety advice firmly in mind!

6.2 Synthetic methods for alkyl and aryl amines

Although most alkyl amines are made on an industrial scale by the reaction between ammonia and alcohols or alkyl chlorides and fractionation of the resulting product mixture, and most aryl amines are produced from the reduction of nitrated aryls, there are several other methods of synthesis that can be used in the laboratory and on a pilot scale to prepare specific amines in good yield. Some examples of these reactions will be given here.

6.2.1 The Hofmann Degradation Reaction

This is a good general method that is useful for the synthesis of alkyl and aryl amines. In the Hofmann Reaction an acid amide (or imide) reacts with bromine to form a bromamide, which can then react with alkali to form an isocyanate. This is hydrolysed in water and excess alkali to form the amine, which can be separated by distillation.

Example one – methylamine hydrochloride from acetamide

Acetamide (20 g) and bromine (54 g) are mixed in a 1 litre round-bottomed reaction flask, which is cooled in ice water. Potassium hydroxide (20 g in 200 ml of water) is added to the reaction flask slowly with the reaction temperature being kept below 10 °C. During the addition the colour of the reacting solution turns from red-brown to yellow. The reaction should be complete after addition of the potassium hydroxide, and the reaction temperature should be left to cool down to a few degrees Celsius. This reaction mixture is now placed in a dropping funnel and fed slowly into a solution of potassium hydroxide (60 g in 100 ml of water) held in another 1 litre rounded-bottomed flask. During the addition the reaction temperature should not be allowed to exceed 60 °C. After the addition is completed the reaction flask is heated on a water bath at 65–70 °C for about half an hour until the yellow colour disappears.

   After this stage the reaction flask should be fitted with a distillation column and the methylamine, together with any ammonia, can be distilled off at low temperature and captured in dilute hydrochloric acid. Evaporation of the hydrochloric-acid wash to dryness yields crude methylamine hydrochloride, which can be recrystallized from ethanol and dried in a vacuum desiccator. The obtained methylamine hydrochloride should be in the form of colourless, crystalline plates with a melting point of 200 °C.

$$CH_3-C=ONH_2 + Br_2 \rightarrow CH_3-C=ONHBr + HBr$$

$$CH_3-C=ONHBr \rightarrow CH_3-N=C=O + HBr$$

$$H_3-N=C=O + H_2O \rightarrow CH_3-NH_2 + CO_2 \xrightarrow{HCl} CH_3-NH_2 \cdot HCl$$

(This synthesis is adapted from [1].)

Example two – anthranilic acid from phthalimide

Potassium hydroxide (28 g) is dissolved in water (280 g) in a 1 litre reaction flask cooled in an ice-water bath. Bromine (16 g) is added slowly to the cooled solution with constant stirring. After the addition is completed, phthalimide (28 g) is added very slowly to prevent any strong exothermic behaviour. After this addition is completed, potassium hydroxide (12 g in 120 ml of water) is added and the reaction mixture is warmed to 80 °C on a water bath. The reaction solution is then acidified with acetic acid (use pH paper to check the value) and saturated copper(II) acetate solution (20 g in the minimum of water) is added. After stirring, the reaction mixture is allowed to stand for 1 h, during which time copper anthranilate can be seen to precipitate out as a pale green solid. This precipitate can be filtered off on a Buchner funnel and washed with cold water to remove any excess copper acetate. The filtrate should then be suspended in the minimum of warm water and sulphurated with hydrogen sulphide gas, which results in copper(II) sulphide being precipitated from solution. The copper(II) sulphide can be filtered off and washed with a little warm water, and the filtered solution and washings can be evaporated on a water bath to crystallize out the anthranilic acid. The yield should be about 85% of theoretical, and the anthranilic acid should be obtained as colourless crystals with a melting point of 145 °C. (This synthesis is adapted from [2].)

6.2.2 From diakylated 4-nitrosoanilines

This is a good method for producing secondary alkyl amines, although in some cases the necessary starting nitrosoaniline may be difficult to obtain.

Example three – dimethylamine from 4-nitroso-N,N-dimethylaniline

4-Nitroso-*N*,*N*-dimethylaniline hydrochloride (20 g in 100 ml of water) is added slowly to a 1 litre reaction flask containing potassium hydroxide (30 g in 500 ml of water). The flask should then be fitted with a distillation arm with a condenser and a dropping funnel containing hydrochloric acid (75 ml of 2 M HCl). The hydrochloric acid should be added slowly to prevent excessive heating and once the addition is complete the flask can be heated to distil the reaction solution (evaporation will start during addition of the hydrochloric acid as the reaction is very exothermic) and oily droplets can be seen distilling over. Once the droplets have stopped flowing the contents of the distillation receiver can be evaporated under reduced pressure to dryness. This will yield crude dimethyl amine as the hydrochloride salt. The yield should be around 75% of theory if the distillation is carried out carefully.

The other reaction product 4-nitrosophenol can be obtained by acidification of the reaction flask and solvent extraction with diethyl ether.

$$ON-C_6H_4-N(CH_3)_2 + H_2O \rightarrow ON-C_6H_4-OH + NH(CH_3)_2$$

(This synthesis is adapted from [3].)

6.2.3 By alkylation of primary and secondary amines

This method can be used to prepare tertiary amines from primary amines by using alkylating agents such as dimethyl sulphate, diethyl sulphate or alkyl halides. In acidic solution dimethyl and diethyl sulphates have a tendancy to form quanternary salts, which must be taken into alkaline solution to liberate the free-amine bases.

Example four – N,N,N′,N′-tetramethylethylene diamine from ethylene diamine

Ethylene diamine (50 g), together with toluene (1 litre), is added to a 2 litre round-bottomed flask fitted with a dropping funnel and overhead reflux condenser. The reaction solution is warmed to about 40 °C and dimethyl sulphate (290 g) is added very slowly via the dropping funnel while the reaction temperature is kept below 60 °C. The reaction with dimethyl sulphate is very exothermic and it should be added to the reaction drop by drop. Extreme care should be taken handling dimethyl sulphate as it is an extremely hazardous chemical with a high toxicity. After the reaction is complete the reaction mixture can be heated to 75 °C and held there for 30 min to ensure the reaction is complete. An excess of dimethyl sulphate is used to ensure that two hydrogen atoms on opposite amine groups are substituted with methyl groups. The reaction mixture should then be cooled and quenched slowly into sodium carbonate solution (10%, 1500 ml). After quenching, sufficient sodium hydroxide solution (30%) should be added to adjust the solution pH to about 10 and the resulting solution should be throughly mixed.

The upper organic layer should be retained and transferred to a distillation flask fitted with a glass-packed distillation column. The distillation first removes any water present as a toluene water azeotrope, and then the toluene should distil over at 110 °C. The product distils at 120–122 °C and provided the distillation is carried out carefully yields of over 80% (*c.* 80 g) can be obtained. (A second distillation may be required to yield pure product.)

$$H_2N-CH-CH-NH_2 + 4(CH_3)_2SO_4 \rightarrow [(CH_3)_2HN-CH_2-CH_2-NH(CH_3)_2]^{2+}$$
$$+ 2[CH_3SO_4]^- + 2CH_3OSO_3H$$
$$\xrightarrow[+NaOH/H_2O]{} (CH_3)_2N-CH_2-CH_2-N(CH_3)_2$$
$$+ 4CH_3OH + 4Na_2SO_4$$

(This synthesis is adapted from [4].)

6.2.4 By the reduction of nitriles, oximes and amides

Amines can be obtained by reduction of nitriles, oximes and amides with sodium and alcohol. Although the main reaction products are primary amines, some secondary amines are usually also formed in the reaction. Some nitriles can be reduced with lithium aluminium hydride, although the reaction also works with sodium borohydride.

Example five – n-*pentylamine from* n-*butyl cyanide*

n-Butyl cyanide (33 g in 76 ml of absolute ethanol) is added slowly to a prepared mixture of sodium metal (55 g) in anhydrous toluene (200 ml) in a 1 litre round-bottomed flask fitted with a reflux condenser. The sodium metal to be dissolved in the toluene should be divided into five small pieces and added slowly to the toluene with cooling in an ice bath. The reaction is exothermic and the mixture should be heated or cooled as required to maintain a reflux period of about 2 h. After the reflux period is complete the reaction mixture should be allowed to cool and then, once cool, a solution of methanol (40 g) in water (20 g) should be added to neutralize any left-over sodium. The flask should then be set up for steam distillation and the reaction mixture steam distilled for about 2 h.

After steam distillation, the distilled fraction should be acidified with 1 M hydrochloric acid (40 ml) and the toluene layer collected and then evaporated to dryness with a vacuum rotary evaporator. After removal of the solvents the crude product left in the evaporation flask should be the amine hydrochloride. The free amine can be recovered by adding sodium-hydroxide solution (40 g in 200 ml of water). The amine layer should settle out at this stage and can be separated off and dried with sodium-hydroxide pellets (5 g, anhydrous, freshly dried). Distillation of the dried product at 102–105 °C gives about 25 g of the pure pentylamine (about 80% of theory). (This synthesis is adapted from [5].)

6.2.5 By rearrangement of N-alkyl anilines

Example six – *mesidine from methyl*-meta-*xylidine*

When the hydrochloride of an α-alkyl aryl amine is heated to around 300 °C the α-alkyl group migrates from the nitrogen to the *ortho* or *para* position on the aromatic nucleus.

$$C_6H_5-NH-CH_3 \rightarrow CH_3-C_6H_4-NH_2$$

In many instances it is not necessary to isolate the secondary amine, and the reaction may be carried out in an autoclave with the primary amine hydrochloride and an alkyl alcohol. *Meta*-xylidine (100 g) and methanol (40 g) are mixed and heated in a sealed autoclave at 300 °C for 24 h. The autoclave contents are then allowed to cool and the reaction products are dissolved in sufficient water to prepare a solution. Sodium hydroxide is added to adjust the pH to above 8 and the aqueous alkaline solution is extracted with diethyl ether (500 ml) in a separating funnel. The ether extract is retained and solid potassium carbonate (25 g) is added to remove any water present. The dried solution is then filtered, the ether removed by vacuum distillation (or on a water bath if the facility is not available). The produced mesidine can then be further purified by distillation at 229–230 °C. The mesidine is an oily colourless liquid and can be obtained in a yield of 70% (77 g). (This synthesis is adapted from [2].)

6.2.6 *From hexamethylene tetramine (the Delépine Reaction)*

Hexamethylene tetramine is a useful alternative to sodium azide for the synthesis of primary amines. In the first stage of the reaction an alkyl halide is added to hexamethylene tetramine to form an *N*-alkylated hexamethylene tetramine, which can then be decomposed with either hydrochloric acid/ethanol to give a primary alkyl amine or with formic acid to give a tertiary dimethylamine. The reaction with formic acid is called the Eschweiler–Clark Methylation.

Example seven – 1-naphthaldehyde from 1-(chloromethyl)naphthalene

1-(Chloromethyl)naphthalene (53 g) and hexamethylene tetramine (84 g), water (125 ml) and acetic acid (125 ml) are added to a 500 ml round-bottomed flask fitted with a reflux condenser. The mixture is then heated and allowed to reflux for 15 min. After the first 15 min of reflux, hydrochloric acid (100 ml, concentrated) should be added to the refluxing reaction mixture, which should then be left to reflux for a further 100 min and then allowed to cool. Once cooled, diethyl ether (150 ml) should be added to the reaction flask and the contents transferred to a separating funnel. The ether layer should be retained in the funnel and washed three times with water (50 ml) and then with aqeuous sodium carbonate (50 ml of 10% solution). The etherial solution should then be dried by mixing with anhydrous sodium sulphate (15 g) and then filtered into a round-bottomed flask suitable for connection to a vacuum rotary evaporator.

   Removal of the ether under reduced pressure should give a crude residue of 1-naphthaldehyde, which can then be recrystallized from dichloromethane. The yield of recrystallized product should be 35 g or 75% of theory. (This synthesis is adapted from [5].)

6.2.7 *From the reduction of nitro compounds*

Although this method can be used for the synthesis of some alkyl amines it is generally used only for aryl amines. The reaction with alkyl amines tends to give low yields and many unwanted by-products. The reduction can be performed with an electropositive metal and acid (iron, zinc or tin in hydrochloric acid all work well).

Example eight – aniline from nitrobenzene

Nitrobenzene (80 g) and tin (150 g, granulated) are added to a 5 litre flask fitted with a reflux condenser and a side port with a stopper. Hydrochloric acid (400 ml, concentrated) is then added to the flask initially in 10 ml lots and then in 20 ml lots once the initial reaction has subsided. Care must be taken not to let the reaction become too vigorous as this will result in a yield loss via the condenser.

After the addition of the hydrochloric acid is complete, the flask should be heated on a boiling water bath for 30 min to ensure the reaction is complete. After this stage, water (200 ml) should be added to the reaction mixture followed by sodium-hydroxide solution (300 g in 400 ml of water). The addition of sodium hydroxide should be done slowly to prevent excess evolution of heat, as before. During addition of the sodium-hydroxide solution a white precipitate of tin hydroxide will be formed, but this should redissolve. If there is still a white precipitate in the reaction flask after addition of the sodium-hydroxide solution, additional sodium-hydroxide solution (100 ml) should be added. At this stage aniline can be seen as dark oily droplets. The reflux condenser should be removed from the reaction flask at this stage and the flask equipped with a distillation condenser. Injection of steam into the reaction solution via the flask side arm will result in the aniline distilling over, together with some of the aqueous mixture.

The steam distillation should be performed until the distillate in the condenser is clear or milky with no trace of brown. The distillate should be a milky liquid on top of which an oily layer of aniline is floating. Sodium hydroxide should then be added to the distillate to saturate the aqueous solution (about 25 g per 100 ml of aqueous solution should be enough) and then the whole distillate extracted with diethyl ether (100 ml should be sufficient). The ether layer should be retained and dehydrated by mixing with solid potassium-hydroxide pellets.

The mixture should then be filtered and the ether removed on a vacuum rotary evaporator with a water bath. The crude aniline left behind in the distillation flask should then be further distilled at about 190 °C by using an oil bath as a heat source and an air condenser. The product should be pure aniline (*c.* 50 g, about 90% of theory), which may be collected as a clear colourless liquid and stored under nitrogen away from light.

$$C_6H_5-NO_2 + 3Sn + 13HCl \rightarrow 2(C_6H_5-NH_2.HCl) + 3SnCl_4 + 4H_2O$$

(This method is adapted from [1].)

6.2.8 *From diazonium salts*

Diazonium salts are best produced by the action of cold hydrochloric acid and aqueous sodium nitrite on primary aryl amines. The resulting solution can be used in further reactions without the need to isolate the diazonium salt. Some primary aliphatic amines may also give diazonium salts, but generally the reaction product is an alcohol. Aliphatic nitrites formed as intermediates in this reaction can be explosive (e.g. isopropyl nitrite and amyl nitrite) and generally the reaction between nitrous acid and primary aliphatic amines is best avoided. Diazonium salts can be used to prepare azo dyes (example ten gives a method for the laboratory-scale preparation of methyl orange).

Example nine – the formation of monochlorobenzene from aniline via phenyl diazonium chloride and the Sandmeyer Reaction

Aniline (10 g) is added to dilute hydrochloric acid (a solution of 25 ml of concentrated hydrochloric acid and 25 ml of water) in a glass round-bottomed flask. The mixture is stirred and then left to cool in an ice bath placed around the flask. After cooling for 1 h, sodium nitrite (9 g in 25 ml of water) is added very slowly to the contents of the flask via a dropping funnel with constant stirring. During the addition of the sodium-nitrite solution the reaction temperature should not exceed 5 °C.

The addition of the sodium-nitrite solution should be continued until free nitrous acid can be detected in the reaction mixture (a drop of the reaction solution withdrawn from the flask on the end of a glass rod can be tested with starch–iodide paper, a blue colour indicating the presence of nitrous acid). The resulting solution of phenyl diazonium chloride should be stored in a stoppered flask in the ice bath until required at a later stage of the experiment.

$$HCl + NaNO_2 \rightarrow HNO_2 + HCl$$
$$C_6H_5-NH_2 + HNO_2 + HCl \rightarrow [C_6H_5-N_2]^+[Cl]^- + 2H_2O$$

Copper(II) sulphate ($CuSO_4.5H_2O$, 12.5 g) and sodium chloride (6 g) are dissolved in 25 ml of water in a preweighed round-bottomed flask. The resulting solution is heated to boiling point, and then hydrochloric acid (50 g concentrated) and copper filings (6.5 g) are added. The flask is then heated until the contents become colourless. Hydrochloric acid (concentrated) is then added to the flask to bring the total solution weight to 102 g. This solution of copper(I) chloride is then used in the next part of the experiment.

The solution of phenyl diazonium chloride is now added slowly to the freshly prepared copper(I) chloride solution (see above) in a round-bottomed flask fitted with a reflux condenser. A yellow precipitate should be formed immediately. The contents of the flask are then heated and the reaction mixture refluxed for 5 min until clear. The monochlorobenzene formed from this reaction can be recovered by steam distillation and then separated by washing in diethyl ether (100 ml) and dried with anhydrous calcium chloride (15 g). Filtration of the etherial solution and evaporation of the solvent on a vacuum rotary evaporator yields about 8.5 g of monochlorobenzene, which is about 70% of theory.

$$[C_6H_5-N_2]^+[Cl]^- + CuCl \rightarrow C_6H_5-Cl + N_2 + CuCl$$

(This method is adapted from [1].)

Example ten – the preparation of an azo dye (Methyl Orange)

Sulphanilic acid (10 g, anhydrous) and sodium carbonate (3.1 g) are dissolved in water (120 ml) in a round-bottomed flask. Sodium nitrite (4 g) is then added to this solution, which is then cooled in an ice bath for 1 h. After 1 h, hydrochloric acid (5.5 g of concentrated acid diluted to 50 ml with water) is added to the cooled mixture with stirring so that the reaction temperature does not rise above 5 °C. After the addition is complete, dimethylaniline (7 g) is added to hydrochloric acid (5.5 g diluted to 50 ml with water as before) and this mixture is added to the cooled solution by using a dropping funnel and constant stirring. Again the reaction temperature should be kept below 5 °C. Once this addition is complete the stirred solution should be allowed to cool for 30 min, after which time sodium-hydroxide solution (concentrated) should be added slowly to the reaction mixture until the pH of the reaction solution is between 8 and 9. Care should be taken to avoid adding excess sodium hydroxide. The produced Methyl Orange should be in the form of an orange layer on top of the reaction solution at this stage. Addition of sodium chloride (50 g) to the reaction solution will assist the formation of the Methyl-Orange layer. The Methyl Orange can easily be separated off at this stage by using a separating funnel and then recrystallized from water. The yield is 18 g or about 95% of theoretical.

The sodium salt of Methyl Orange produced can be used as an indicator, and is yellow in aqueous or alkaline solution but turns red in the presence of acids.

$$[H_2N-C_6H_4-SO_3]^-[Na]^+ + HNO_2 \rightarrow [OH]^- \, ^+[N_2-C_6H_4-SO_3]^-[Na]^+$$

$$\downarrow +C_6H_5-N(CH_3)_2$$

(This method is adapted from [2].)

6.2.9 Quaternary ammonium salts

Quaternary ammonium salts (or 'quats' as they are sometimes called) are widely used as phase-transfer catalysts to enable reactions to take place between immiscible solvents. Two commonly used quats are tetrabutyl ammonium bromide (tbab) and tetramethyl ammonium hydroxide (tmah).

Example eleven – the preparation of tetramethyl ammonium hydroxide

Tetramethyl ammonium chloride (100 g) is dissolved in ethanol (200 ml, absolute). To this solution is added a filtered solution of potassium hydroxide (51.1 g) in water (200 ml). Upon mixing, a white precipitate of potassium chloride is formed.

The mixture is allowed to stand for 2 h, under nitrogen, and is then filtered. The white precipitate in the filter funnel should be washed after filtration with methyl alcohol (10 ml). The filtered solution should then be reduced with a vacuum rotary evaporator with the temperature of the water bath being maintained at 35 °C. This operation removes some of the ethanol and all of the methyl alcohol. After the flow of distillate has stopped, water (100 ml) should be added to the reduced filtered solution and it should be left to stand for 2 h, again under nitrogen, to allow crystallization of the tetramethyl ammonium hydroxide, which can been seen as white crystals. The solution should then be allowed to stand overnight in a fridge at 2 °C for further crystallization to take place. Filtration of the crystals from solution should yield about 70 g of tetramethyl ammonium hydroxide pentahydrate (40% of theory). (This method is adapted from [2].)

6.2.10 The preparation of **ortho**-*phenylene diamine*

The reduction of aromatic nitro compounds to aromatic diamines can be performed with iron or tin in hydrochloric acid (for *meta*-phenylene diamine and *para*-phenylene diamine). *Ortho*-phenylene diamine (or 1,2-phenylene diamine) is best prepared in a slightly different way: the reduction of 1,2-dinitrobenzene in ethanolic sodium hydroxide with zinc dust works best. A strongly alkaline solution is most successful because the *N*-hydroxy derivatives formed in neutral or weakly alkaline solutions may also rearrange to form ring-substituted hydroxy compounds, e.g. *N*-hydroxylaniline forms 4-hydroxy aniline. The *ortho*-phenylene diamine made in this example may be used in examples fifteen and eighteen for the synthesis of heterocycles.

Example twelve – the synthesis of 1,2-phenylene diamine

Ortho-nitroaniline (25 g) is added to a 2 litre round-bottomed flask fitted with a reflux condenser, together with ethanol (200 ml) and sodium hydroxide solution (32 g dissolved in the minimum of water).

The reaction mixture should be heated slowly, during which time zinc powder (30 g) is added to the reaction in small lots (1–2 g at a time). After the addition of all the zinc powder, the reaction mixture should be refluxed for 20 min or until a colourless solution is obtained. This solution should be filtered as rapidly as possible, mixed with water (500 ml) and then allowed to cool. Once cooled, the aqueous solution should be extracted with ether (2 × 200 ml) and the ether layers combined and evaporated to about a quarter of the original volume in a vacuum rotary evaporator. The resulting solution can be left for the *ortho*-phenylene diamine to crystallize out overnight or, alternatively, if anhydrous hydrogen chloride gas is bubbled into the reaction mixture the dihydrochloride salt should precipitate out from solution and this can be filtered off and dried. The yield of *ortho*-phenylene diamine should be around 13 g or 74% of theory. (This method is adapted from [5].)

6.3 Amino-acid protection

The most widely used method of peptide synthesis involves the coupling of resin-bound Fmoc-protected amino acids. The protection reaction is performed under Schotten–Baumann conditions and the product is easily isolated as a crystalline product. However, for amino acids with two amino groups the diprotection (i.e. of both amino groups), although easy from a synthetic perspective, can limit the use of the diprotected product in peptide synthesis. A better strategy is to use orthogonal protection with one Fmoc group and one BOC group, as this allows selective deprotection at the different sites and also increases the solubility (di-Fmoc-protected amino acids are only sparingly soluble and can be used for peptide synthesis only in low dilution). The following synthetic procedure illustrates the formation of bis-(Fmoc)-L-histidine and its subsequent conversion into the orthogonally protected Fmoc-His(BOC)-OH with a BOC group on the imidazole nitrogen.

Example thirteen – the preparation of bis-(Fmoc)-L-histidine

L-Histidine (4.56 g) is added to water (120 ml) in a 500 ml beaker and the pH of the stirred solution is adjusted to 8.2 by using sodium carbonate solution (10%). The solution is then cooled in an ice bath for 30 min.

After cooling, a solution of Fmoc-ONSu (25.3 g) in dioxane (300 ml) is then added slowly to the histidine solution via a dropping funnel with sodium carbonate solution being added if required to maintain the pH at 8.2.

After the addition is complete the resulting solution should be left to stir for 3 h with no external heating. After this time, water (1500 ml) is added to the reaction mixture and the resulting solution is extracted three times with ether (250 ml) and then with ethyl acetate (150 ml) and then is finally acidified with hydrochloric acid to pH5. The precipitated solid can be filtered off, washed with water (25 ml) and vacuum dried over phosphorous pentoxide to yield 16 g of anhydrous product. (This method is adpated from [6].)

Example fourteen – the preparation of N_α-Fmoc-N-BOC-L-histidine

Bis-(Fmoc)-L-histidine (10 g, from example thirteen) is suspended in dimethylformamide (100 ml), and diisopropylethylamine (5.38 g, Hünig's Base) is added with constant stirring. Di-BOC (9.08 g) is then added to the reaction mixture as a molten liquid and the reaction mixture is left to stir overnight to reach completion.

The next day the reaction mixture is poured into a round-bottomed flask connected to a vacuum rotary evaporator, and evaporated to a sticky syrup at 30 °C. The contents of the flask are then dissolved in the minimum of ethyl acetate required to form a solution, transferred to a separating funnel. Note that the distillation flask should be washed out with an additional 25 ml of ethyl acetate and then washed twice with citric acid solution (10% aqueous, 150 ml in each wash) and then four times with water (50 ml each wash).

The organic layer should be mixed with anhydrous sodium sulphate and left for 10 min before filtering and then evaporating the clear solution down to a syrup again for a second time with a vacuum rotary evaporator at 30 °C, as before. The syrupy oil obtained after the solvent stripping should be dissolved in ethyl acetate (75 ml should be sufficient) and then added slowly, with stirring, to petroleum ether (300 ml). This should cause precipitation of the product, which is then filtered off.

The crude product from this stage of the experiment should be redissolved in ethyl acetate (5 ml) and diethyl ether (75 ml), and the resulting solution added slowly to petroleum ether (300 ml). The precipitate that forms is collected by filtration and dried in a vaccum desiccator. After drying, the final product weight should be 5.6 g of N_α-Fmoc-N-BOC-L-histidine with a melting point of not less than 95 °C. In the event of a low melting point the final purification stage should be repeated for a second time. (This method is adapted from [6].)

6.4 The synthesis of heterocyclic amines

In this section, synthetic methods will be given for the small-scale preparation of indole, carbazole, 4-nitropyridine-N-oxide, quinoxaline and indigotin dye.

6.4.1 The synthesis of indole and carbazole

Indole may be prepared from the reaction between 1,2-phenylene diamine and formic acid. The first stage of the reaction is the formation of the N-acyl derivative, which cyclizes in the presence of excess formic acid. If glyoxal is used in this synthesis the reaction product is quinoxaline and the method for this is given here in example eighteen. The synthesis of 1,2,3,4-tetrahydrocarbazole given in example sixteen is an example of the Fischer-Indole Synthesis – see also Appendix 3.

Example fifteen – the synthesis of indole

Formic acid (20 ml) and 1,2-phenylene diamine (5.2 g) are added to a round-bottomed flask (100 ml) fitted with a reflux condenser and are then refluxed for 1 h to allow the reaction to reach completion. After the reaction period, cold water (20 ml) is added to the reaction mixture and sufficient sodium carbonate (dry powder) is added to the flask to adjust the pH to 12. The reaction product is filtered off at this stage and can be washed with ice-cold water (200 ml). The crude product can be recrystallized from cold water as white prisms and the reaction yield should be around 5 g or 86% of theory.

$$\text{(benzene ring with two NH}_2\text{ groups)} + \text{HCO}_2\text{H} \rightarrow \text{(indole, N-H)}$$

(This method is adapted from [7].)

Example sixteen – the synthesis of 1,2,3,4-tetrahydrocarbazole

Cylcohexanone (10 ml), acetic acid (50 ml, glacial) and phenyl hydrazine (10 ml) are added to a 500 ml round-bottomed flask fitted with a reflux condenser and are heated to boiling point and allowed to reflux for 20 min. The hot product from the reflux should then be poured into a glass beaker in a fume hood and the reaction flask should be washed out with acetic acid (30 ml), and the washings added to the beaker. The mixture should then be allowed to stand for 2 h to cool. After cooling, water (200 ml) should be added to the reaction mixture (do not add the water too early or a sticky residue will precipitate out instead of crystals). The product can be filtered off at this stage and washed on the filter paper with ethanol/water (50 ml, 75% ethanol, 25% water). The crude product can be recrystallized from petroleum ether and vacuum dried over phosphorus pentoxide. The reaction yield should be 15 g or 90% of theory.

$$\text{C}_6\text{H}_5\text{–NH–NH}_2 + \text{C}_6\text{H}_{10}\text{=O} \rightarrow \text{C}_6\text{H}_5\text{–NH–N=C}_6\text{H}_{10}$$

$$\rightarrow \text{(1,2,3,4-tetrahydrocarbazole, N-H)}$$

(This method is adapted from [7].)

6.4.2 The nitration of pyridine N-oxide

As discussed in Chapter 3, pyridine is extremely resistant to electrophilic attack and the nitration of pyridine with mixed acid requires activation at high temperatures and does not give a clean product. However, formation of pyridine *N*-oxide enables nitration to proceed smoothly, giving predominantly the 4-substituted product.

Example seventeen – the preparation of 4-nitropyridine N-*oxide*

Pyridine *N*-oxide (10 g) is dissolved in a mixture of nitric acid (10 ml, fuming) and sulphuric acid (30 ml, concentrated) in a 250 ml round-bottomed flask fitted with a reflux condenser and a thermometer (the bulb of which should be below the level of the liquid in the flask). The reaction mixture should then be heated at 130 °C for about 200 min. After the reaction time, the hot mixture should be poured onto crushed ice (250 ml) using a splash guard to protect the operator. The resulting solution should then be stirred, and neutralized to pH7 with ammonium-hydroxide solution (concentrated). A pale yellow product should crystallize out at this stage; this can be filtered off and dried at 100 °C in a vacuum oven. A second crop of product can also be obtained from the filtrated solution, which can be washed twice with dichloromethane (30 ml) and separated (dichloromethane forms the lower level in a separating funnel with the aqueous layer on top). The dichloromethane solution should then be evaporated and the residue combined with the dried product isolated earlier. The blended mixture should then be washed with ether (100 ml) on a filter funnel and dried as before to yield 10 g of product, which is about 66% of theory. (This method is adapted from [5].)

6.4.3 The synthesis of quinoxaline

This is a variation of the earlier synthesis with 1,2-phenylene diamine with formic acid. Here the formic acid is replaced with glyoxal and both nitrogens are retained in the product. The reaction may be performed with just these two reagents but a better yield is obtained with the addition of sodium hydrogen sulphite (note that reagent-grade sodium hydrogen sulphite is a mixture of sodium bisulphite and sodium metabisulphite).

Example eighteen – the synthesis of quinoxaline

Glyoxal (18 ml of 30% aqueous solution) is added to a 500 ml beaker containing sodium metabisulphite (20.9 g) and water (100 ml).

   The contents of the beaker are then heated to 70 °C on a water bath with stirring. A second solution of *ortho*-phenylene diamine (10.8 g) in water (150 ml) is now prepared in a 1 litre beaker, which is also heated to 70 °C on a water bath. Once both solutions have reached 70 °C they may be mixed: the glyoxal/bisulphite solution should be added slowly to the *ortho*-phenylene diamine solution with stirring. The resulting combined solution should then be removed from the water bath and allowed to cool to room temperature. On attaining room temperature, sodium bicarbonate (45 g in 100 ml of water) is added to the cooled solution with stirring. The pH of the solution should now be greater than 7. If not, additional sodium-carbonate solution should be added. The aqueous reaction solution should now be triply extracted with ether (3 × 50 ml) and the combined etherial extracts dried with anhydrous magnesium sulphate, filtered and the solvent removed with a rotary vacuum evaporator. The crude product is a yellow oil. This may be purified by vacuum distillation at 12 mmHg and 109 °C to yield 10 g of the pure product (75% of theory), which is a low-melting solid (*c.* 29 °C).

(This method is adapted from [1].)

6.4.4 The synthesis of indigotin (indigo)

Indigo is a widely used textile dye and will be reviewed in Chapter 8.

The preparation of a blue/purple dye that eluded synthetic chemists for centuries can be performed in the laboratory with easily available reagents. The reaction proceeds via an aldol intermediate, which undergoes an intramolecular reaction to form indigotin.

Example nineteen – the synthesis of indigotin

Ortho-nitrobenzaldehyde (10 g) is dissolved in a mixture of acetone (50 ml) and water (50 ml) in a stoppered conical flask containing a magnetic stirring bar and this flask is then placed on a magnetic stirrer plate. Sodium-hydroxide solution (50 ml, 1 M) is added dropwise to this solution with constant stirring. The reaction mixture should become warm during the addition of the sodium-hydroxide solution and the solution colour should change from green to brown. After the addition of the sodium-hydroxide solution the reaction mixture is allowed to cool and is then filtered.

The crude indigotin crystals collected in the filter funnel are then washed with ethanol (3 × 100 ml) and ether (2 × 200 ml) and dried in a vacuum desiccator. The crude indigo crystals may then be recrystallized from redistilled aniline to yield 4–5 g of pure indigotin as dark blue crystals (about 50% of theory). (This method is adapted from [7].)

6.5 Analytical methods for amines

The majority of amine analysis performed nowadays is by instrumental techniques such as infra-red (IR), ultra-violet (UV) and nuclear magnetic resonance (NMR) spectroscopy. However, classical (wet) methods of analysis can still be used for amines and may help in the elucidation and interpretation of spectroscopic results.

6.5.1 Classical methods of analysis

If the amine under test is a solid then a good first step in its analysis is to determine its melting point, which can be compared with values in published tables [8]. However, some amines do not give clearly defined melting points and may decompose on heating, so it is not possible to identify all amines by this method alone.

As virtually all amines possess basic properties, further information about them can be ascertained by the ease with which they dissolve in water and produce salts. Aliphatic amines are more strongly basic than aromatic amines, and so their aqueous solutions are more strongly alkaline (and this can be measured with litmus). The lower aliphatic amines are generally more soluble than the higher aliphatic amines or aromatic amines. The solubilities of amines in water can be improved by the addition of hydrochloric acid, which produces the corresponding ammonium salt. Nearly all ammonium-chloride salts are soluble, with the exception of the α- and β-naphthyl amines, which are only slightly soluble in cold hydrochloric acid.

A saturated aqueous solution of sodium picrate (2,4,6-trinitrophenol) can be added to this solution of the ammonium-hydrochloride salt and the corresponding picrate salt is precipitated. Determination of the melting point of the ammonium picrate salt, which is different to the melting point of the free amine, allows for further compound identification. Tables of the melting points of picrate salts are given in reference [3] and a few examples are shown in Table 6.1.

Another quick and simple test for amines is the chloranil test. Amines dissolved in dioxane give coloured products on mixing with chloranil (2,3,5,6-tetrachloro-*p*-benzoquinone). Aliphatic amines generally give red colours but other amines give a variety of colours, and these are given in Table 6.2. A few drops of a test solution

Table 6.1. *Melting points of amines and ammonium picrate salts*

Amine	Melting point (°C)	Picrate salt melting point (°C)
Ethylamine	−81	168
Benzylamine	10	199
Aniline	−6	165
Ortho-phenylene diamine	103	208
Piperidine	−13	152
Piperazine	104	280
Triethylamine	−115	173
Dimethylaniline	2	162
Pyridine	−42	167
Quinoline	−15	203

Table 6.2. *The colours of chloranil amine adducts*

Amine	Colour of adduct produced with chloranil
n-Butylamine	Red
Benzylamine	Red
Aniline	Violet
N-methylaniline	Blue-green
N,N-dimethylaniline	Blue
Para-nitroaniline	Orange
Diphenylamine	Blue-green
Sulphanilic acid	No colour

of a microspatula of the amine dissolved in 1 ml of dioxan should be added to a concentrated solution of chloranil in dioxan. The colouration should form instantly, and tables of produced colours have been published [1].

The two most widely used classical-analytical techniques for amines are the Hinsberg Test [9] and the Liebermann Nitroso Test [10].

The basis of the Hinsberg Test is that mono-substituted sulphonamides of primary, secondary and tertiary amines have different solubilities in sodium-hydroxide solution. Benzene sulphonyl chloride or *para*-toluene sulphonyl chloride are used for the derivatization, which is performed cold in a shaken, stoppered flask.

$$R-NH_2 + Cl-SO_2-C_6H_5 \rightarrow R-NH(SO_2-C_6H_5) + HCl$$

$$RR'-NH + Cl-SO_2-C_6H_5 \rightarrow RR'N-SO_2-C_6H_5 + HCl$$

$$RR'R''N + Cl-SO_2-C_6H_5 \rightarrow \text{No reaction}$$

The product from the reaction of a primary amine is soluble in sodium hydroxide. The product from the reaction of a secondary amine is insoluble in acidic and alkaline solutions, and tertiary amines (which do not react with sulphonyl chlorides) can be extracted from the reaction mixture by acidification. Unfortunately some primary amines do form disubstituted products; this is a limitation of the Hinsberg Test. However, these products can be converted into alkali-soluble mono-derivatives by refluxing with ethanolic sodium ethoxide.

Example twenty – the Hinsberg Test

The amine to be tested (0.5 g) is added to a 50 ml conical flask with sodium-hydroxide solution (20 ml, 5 M) and benzenesulphonyl chloride (1 g). The flask should then be stoppered and shaken for 5 min to allow the reaction to take place. After the reaction, a visual examination will show the presence of any insoluble material, which may be:

(i) the original amine if it was a tertiary amine, or
(ii) an alkali-insoluble sulphonamide from a secondary amine, or
(iii) a disubstituted derivative of a primary amine, or
(iv) a sodium salt of a mono-sulphonamide derivative $Na[R-N-SO_2-C_6H_5)]$.

Any precipitate should be filtered off at this stage. If the product is (i), (ii) or (iii) it will be soluble in ether. If it is (iv) it will be soluble in distilled water.

The filtered precipitate should be placed in a separating funnel with ether (15 ml). If it dissolves in the ether then it is (i), (ii) or (iii). If it does not dissolve, then water (15 ml) should be added and the funnel shaken. The aqueous (lower) phase should then be run off and acidified with hydrochloric acid. If a precipiate forms then it is the sulphonamide that was formed from a primary amine. If no precipitate is formed then the amine under test is not a primary amine.

If no precipitate is formed with hydrochloric acid then the etherial solution in the separating funnel must be extracted twice with hydrochloric acid (5 ml), and the acid washes combined and made alkaline with sodium-hydroxide solution. If the original amine was a tertiary amine it will be liberated as the free base at this stage and can be isolated by steam distillation. If nothing is isolated at this stage then the test amine (which must be a secondary amine or possibly a disubstituted primary amine) must still be in etherial solution as a sulphonamide derivative. The ether solution in the separating funnel should be washed with water (10 ml) and then the ether removed by evaporation on a water bath and the product recrystallized from ethanol/water. The purified product can then be refluxed with ethanolic sodium ethoxide. Identification may be possible by determination of the melting point but, if not, if the test amine was a primary amine that formed a disubstituted product then after reflux it will be soluble in sodium-hydroxide solution. However, if it was a secondary amine then it will be unaffected and will remain insoluble in both acidic and basic aqueous solutions.

Unfortunately the Hinsberg Test is not completely accurate, as higher primary amines such as heptylamine form insoluble sulphonamide derivatives, and so can be confused with the sulphonamide derivatives formed from secondary amines. However, the Hinsberg Test is relatively easy to perform and has the advantage that the original amine under test can be recovered by refluxing in sulphuric acid (70%) or hydrochloric acid (25%).

The Liebermann Nitroso Test involves the reaction of amines with nitrous acid. Primary alkyl amines react to form alcohols (e.g. *n*-butylamine gives *n*-butanol) with the evolution of nitrogen gas, which may be observed in the test solution.

$$C_4H_9-NH_2 + HNO_2 \rightarrow C_4H_9-OH + N_2\uparrow + H_2O$$

Primary aromatic amines react with nitrous acid to form diazo compounds, which can be mixed with β-napthol in sodium-hydroxide solution to give a bright red azo dye. The test is very simple and can be performed in less than 10 min. Less than 1 g of the amine under test is added to dilute hydrochloric acid in a test tube. A few drops of sodium-nitrite solution are added to the test tube, which is then stoppered and shaken for 1 min.

The resulting solution is poured into an aqueous solution of sodium hydroxide to which β-naphthol has been added. The formation of a bright red dye or red-brown precipitate is indicative of the presence of a primary aromatic amine. In general, any aromatic alcohol can be used in this reaction except for catechol (which decomposes) and the colour of the azo dye obtained, which may be orange or violet, is dependent upon the alcohol used, e.g. aniline gives the following azo dye with β-naphthol.

$$C_6H_5-NH_2 + HCl/NaNO_2 \rightarrow [C_6H_5-N_2]^+[Cl]^-$$

Secondary aliphatic and aromatic amines react in a different manner with nitrous acid and form *N*-nitroso compounds, which can be detected by the Liebermann Nitroso Test.

$$R_2N-H + HNO_2 \rightarrow R_2N-NO + H_2O$$

Example twentyone – the Liebermann Nitroso Test

Dissolve the amine (1 g) under test in ethanol (8 ml) in a test tube. Cool the test tube in an ice bath for 10 min and then add to it, in the following order, hydrochloric acid (1 ml) followed by sodium nitrite (1 g dissolved in the minimum quantity of water). Mix the content of the test tube well and allow to stand in the ice bath for 10 min until an oily yellow layer or yellow crystals appear. At this stage, transfer the contents of the test tube into a separating funnel and add ether (15 ml). The yellow oil or crystals should dissolve in the ether layer. The ether layer should then be washed with 10% sodium-hydroxide solution (20 ml) and water (20 ml) to ensure that it is free of nitrous acid. The etherial solution should then be evaporated on a water bath to yield the pure *N*-nitroso derivative, which can be detected in the Liebermann Test as follows.

The *N*-nitroso derivative (0.1 g), as formed in the above synthesis, should be added to phenol (0.1 g) and sulphuric acid (0.5 ml) in a test tube and warmed gently. A blue colour should form after 5 min.

The reaction solution should then be diluted with water, which should turn the test solution red. Finally, if sufficient sodium-hydroxide solution is added to the reaction mixture the blue colour should be restored.

The colours formed are due to the liberation of nitrous acid and its subsequent reaction with phenol. This is the reason why it is so important to wash the etherial solution of the *N*-nitroso amine well to remove any traces of free nitrous acid that could react with phenol.

Tertiary amines generally do not react with nitrous acid, with the exception of tertiary alkyl anilines, which form green-coloured *para*-nitroso derivatives (see below) that do not produce any colouration when subjected to the Liebermann Test.

$$(CH_3)_2N{-}C_6H_5 + HNO_2 \rightarrow \underset{CH_3}{\overset{CH_3}{N}}{-}\!\!\bigcirc\!\!{-}N{=}O + H_2O$$

6.5.2 Spectroscopic methods of analysis

Although classical methods of analysis are still used in the laboratory for the determination of amines, the most widely used methods of analysis are spectroscopic methods such as infra-red (IR), ultra-violet/visible (UV/visible) and nuclear magnetic resonance (NMR). Gas chromatography coupled with mass spectroscopic detection (known as GC/MS) is also widely used and most modern GC/MS instrumentation comes with a software library that can identify the common charged fragments and can offer suggestions as to their possible identity. The cleavage

Table 6.3. *Infra-red amine stretching and bending frequencies*

Group	Band (cm^{-1})	Notes
Stretching mode (v)		
>N–H	3500–3300	Primary amines may show two bands here
=N–H	3500–3300	
–NH$_3^+$	3130–3030	Broad bands
>N$^+$H$_2$	2700–2250	Broad bands
Bending mode (δ)		
–NH$_2$	1650–1560	Medium intensity
>N–H	1580–1490	Weak intensity
–NH$_3^+$	1600 and 1500	Strong intensity
		Secondary amines
		display only the 1600 band

of alkyl groups to produce alkenes is a common fragmentation pattern in amines during mass spectroscopy.

$$R'-(CH_2)_n-N^+-CH_2-R'' \rightarrow R'-(CH_2)_n-N^+=CH_2 \rightarrow H-N^+=CH_2$$
$$\quad\quad\quad | \quad\quad\quad\quad\quad\quad\quad | \quad\quad\quad\quad\quad\quad | $$
$$\quad\quad\quad R \quad\quad\quad\quad\quad\quad\quad R \quad\quad\quad\quad\quad R $$
$$\rightarrow [CH_2=NH_2]^+$$
$$(m/e = 30)$$

The ion series $m/e = 30, 44, 58, 72, 86, 100, \ldots$, is commonly seen for the fragmentation of charged amines in mass-spectroscopy experiments.

Amines have characteristic absorption bands that can be detected in infra-red spectroscopic experiments. These are given in Table 6.3. The positions of the absorption bands may be shifted to lower frequencies as a result of hydrogen bonding. For example, v_{N-H} may shift from 3500–3300 cm^{-1} to 3400–3100 cm^{-1} and become significantly broadened. The main problem with infra-red spectroscopy is that the main amine absorption bands overlap with the absorption bands due to hydroxyl groups (3600–3200 cm^{-1}). Although the absorption bands due to water are broad and amine absorption bands tends to be narrower and better defined, it is still easy to confuse v_{O-H} with v_{N-H} especially if the amine under test also contains a hydroxy group (as is the case with amino acids).

Ultra-violet/visible spectroscopy is associated with the measurement of electronic energy-level transitions in molecules. The wavelength of absorption is related to the extent of the energy level separation by the equation

$$E \text{ (kJ mol}^{-1}) = 119\,720/\lambda_{max}(\text{nm}),$$

where $\lambda_{max} = $ the wavelength of maximum absorption.

Unfortunately, the wavelengths associated with the excitation of electrons in carbon–carbon single bonds ($\sigma \rightarrow \sigma^*$) and the excitation of the nitrogen lone pair ($n \rightarrow \sigma^*$) are below 200 nm and are not easily attainable (they have to be measured in a vacuum, as air must be excluded). Also, they are not greatly affected by the molecular structure, so measurement of λ_{max} does not help in understanding the molecular properties of amines. The $n \rightarrow \pi^*$ transition in the $>N^+(CH_3)_2$ group is measurable at 229 nm but is of such low energy that it does not give much information about the molecular structure.

A better option is to measure the $\pi \rightarrow \pi^*$ and $n \rightarrow \pi^*$ transitions in molecules with conjugated double bonds. Many amine derivatives such as azo and azine dyes and phthalocyanines contain chromophores and are strongly coloured. It is here that ultra-violet/visible spectroscopy finds real application in the study of the electronic properties of amines. Also the substitution of an amino group into a conjugated system can affect the absorption wavelength. For example, benzene absorbs at 184 nm, 203.5 nm and 254 nm. This absorption shift to longer wavelength is called a red shift or bathochromic shift. The band present at 203.5 nm is called the K-band and is very susceptible to ring substituents. For example, substitution of one hydrogen by an amino group results in a change in the wavelength of absorption from 203.5 nm to 230 nm. The B-band at 254 nm is similarly affected, and shifts to 280 nm. The B-band in benzene shows fine structure because of associated molecular vibrations. This is termed a $0 \rightarrow 0$ transition and corresponds to a transition from the ground-state vibrational energy of the electronic ground state to the ground-state vibrational energy of the electronic excited state. The substitution of amino or azo groups onto the benzene ring reduces the level of molecular symmetry and the fine structure of the K-band is increased.

When two substituents are present in a benzene ring the extent of the wavelength shift is dependent on the substitution pattern. If, for example, nitro and amino groups are present then in hexene solution it can be seen that when the substituents are *meta* or *para* to each other the complementary effect is greater then for the *ortho*-substituted product.

Amine	K-band	B-band
Ortho-nitroaniline	229 nm	275 nm
Meta-nitroaniline	235 nm	373 nm
Para-nitroaniline	229 nm	375 nm

It is possible to calculate the values of the K- and B-bands of substituted benzenes from tables of absorption data by a simple arithmetic addition developed by Woodward and Fieser [11]. For example, benzoic acid has a K-band at 250 nm. An *ortho-* or *meta*-amino group will cause a shift of 13 nm, and a *para*-amino group will cause a shift of 58 nm. Therefore the K-band of *ortho-* and *meta*-amino benzoic acid will be at 263 nm and the K-band of *para*-amino benzoic acid will be at 308 nm.

The position of ultra-violet absorption bands is also affected by steric factors that might act to destroy or constrain the molecular planarity and decrease the extent of molecular overlap. For example, 3,5-dimethyl-4-nitroaniline has a 10 nm red shift from the 375 nm found in 4-nitroaniline because the presence of the methyl groups affects the coplanarity. Steric shifts in absorption maxima may be either positive or negative but the extent of the absorption of radiation is reduced and the peak intensity is diminished with increasing steric strain away from planarity.

The lone pair of electrons on the nitrogen group of amines is also able to change the position of the absorption maxima in conjugated systems because of the ability of the electrons to interact with the solvent. This can be illustrated clearly by using the example of aniline. In ethanol, the K-band is at 230 nm, but when measured in dilute hydrochloric acid it shifts to 203 nm. In the free base the nitrogen lone pair is able to donate its electron density to the aromatic ring and so it creates a red shift. In hydrochloric-acid solution an ammonium salt is formed that is not able to conjugate with the aromatic ring and the molecule has a K-band value similar to that for unsubstituted benzene. It is possible to compare the intensity of the UV absorption bands in different solvents by measuring the molar extinction coefficient, ε. The units of ε are 1000 cm^2 mol^{-1}, but by convention these are not usually expressed. We have

$$\log_{10}(I_0/I) = \varepsilon l c,$$

where l = path length of the test solution (cm), c = concentration (mol l^{-1}), and I and I_0 = intensities of the transmitted and incident light respectively.

The molar extinction coefficient is also given by

$$\varepsilon = 0.87 \times 10^{20} \, Pa,$$

where P = the probability of transition and a = the target area of the chromophore. (This explains why azo and azine dyes, where the conjugation extends over the whole molecule, are so strongly coloured.)

This effect of changing a solvent on the UV spectra of a heterocycle such as pyridine is very pronounced.

In hexane, an n→π* band is observed at 270 nm with ε = 450. In ethanol this band disappears completely, but the π→π* band shifts from 251 nm to 257 nm and the molar extinction coefficient increases from ε = 2000 to ε = 2700. This type of behaviour is typical for nitrogen heterocycles where protonation of the nitrogen atom reduces or eliminates the n→π band and increases the intensity of the π→π* band.

In general the UV/visible spectra of six-membered heterocycles (e.g. pyridine) and fused six-membered heterocycles (e.g. quinoline) resemble their carbocylic

Table 6.4. *The UV spectra of nitrogen-containing heterocycles in hexane (h),*
chloroform (c) and cyclohexane (ch)

Molecule	λ_{max}(nm)	ε	λ_{max}(nm)	ε	λ_{max}(nm)	ε
Benzene (h)	204	7400	254	200	—	—
Pyridine (h)	195	7500	251	2000	270	450
Pyridazine (h)	—	—	246	1400	340	315
Pyrazine (h)	—	—	260	5600	328	1040
Naphthalene (c)	220	100 000	275	5600	312	250
Quinoline (c)	226	270	270	3880	313	2360
Isoquinoline (c)	217	37 000	266	4030	317	3100
Pyrrole (c)	210	5100	—	—	—	—
Imidazole (c)	206	3500	—	—	—	—
Indole (ch)	219	25 000	261	6300	280	5620
					(also 288	4900)

counterparts but have additional bands because of the $n \rightarrow \pi^*$ transition. The values of λ_{max} and ε for some nitrogen heterocycles are given in Table 6.4.

The UV spectra of five-membered nitrogen heterocycles such as pyrrole and imidazole show only a single UV absorption band with no fine structure and no detectable $n \rightarrow \pi^*$ absorption. Even azoles do not show an $n \rightarrow \pi^*$ absorption, which is unusual as they contain 'pyridine-type' nitrogens.

Macrocyclic nitrogen-containing hetereocycles such as porphyrins, corrins, corroles and phthalocyanines are all strong UV/visible absorbers and display a strong band at 400 nm called the Soret Band ($\varepsilon \approx 100\,000$) and other bands, typically four, in the 400–700 nm region called Q-bands. The spectral properties of these macrocylic heterocycles have been discussed in Chapter 5.

Although infra-red and UV/visible spectroscopy can supply useful information in the study of amines, the most widely used tool for the structural elucidation of amines is probably nuclear magnetic resonance (NMR) spectroscopy. For the study of amines the most commonly used techniques are ^{1}H, ^{13}C, ^{15}N and ^{14}N NMR (the ^{14}N nucleus is quadrupolar). The frequency (v) at which a nucleus will resonate in a magnetic field is given by

$$v = \gamma H / 2\pi,$$

where H is the local field experienced by the nucleus and γ is the gyromagnetic ratio. However, because the nucleus may be shielded by electrons the local field H may not correspond to the applied field H_0. The extent of this shielding effect may be represented by σ, the shielding parameter, where

$$H = H_0(1 - \sigma).$$

Table 6.5. *^1H chemical shifts of protons in amines*

Protons	δ (p.p.m.)
CH_3-N	2.3
CH_3-N^+	3.3
CH_3-N-Ar	3.0
$-CH_2-N$	2.5
$-CH-N$ $\mid$	2.8
RNH_2	1–5
R_2NH	1–5
$ArNH_2$	3–6
$Ar-NHR$	3–6
$C=N-OH$	9–12
$-CH=C-N$ $\mid$	4–5
$-C=CH-N$	5.7–8

Therefore

$$v = \gamma H_0 (1 - \sigma)/2\pi.$$

For protons (^1H NMR) the resonant frequencies are conventionally measured in hertz although measurement in milligauss would be equally valid. The resonance positions of protons in ^1H NMR are measured relative to the resonance of the protons in tetramethylsilane $((CH_3)_4Si)$, and the displacement from the resonance of the tetramethylsilane (TMS) protons is called the chemical shift, denoted as δ. The value of v_{TMS} is assigned as 0 p.p.m., and

$$\delta \text{ (in p.p.m.)} = (v_{sample} - v_{TMS})/\text{operating frequency in MHz}.$$

Because the extent of shielding is directly related to the resonant position of a proton in a ^1H NMR spectrum, measurement of the chemical shift, δ, provides information about the chemical environment of protons in a molecule. For example, the protons in a methyl group attached to a nitrogen centre, CH_3-N, resonate at 2.3 p.p.m., but the protons on a methylene group attached to a nitrogen atom, $-CH_2-N$, resonate at 2.5 p.p.m. Some examples of the chemical shifts of protons in amines are given in Table 6.5.

The move towards higher σ is called a paramagnetic shift. Electron-withdrawing groups that reduce the electron density around an atom cause a deshielding effect and enable resonance to occur at a lower field position (a higher δ value). For an aromatic amine like aniline, the resonance values of the protons *ortho*, *meta* and

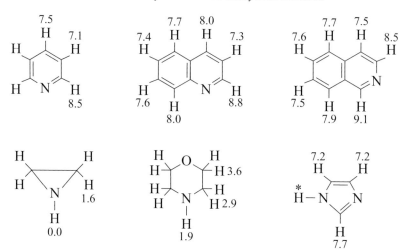

Figure 6.1. Proton NMR signal values (p.p.m.) in nitrogen-containing heterocycles. The N—H marked ∗ gives a broad signal that may occur between 7 and 12 p.p.m., the exact position depending upon the sample concentration and solvent used.

para to the amino group are shifted from the resonance positions in benzene ($\delta =$ 7.27 p.p.m.) by the following amounts.

Substituent (C_6H_5—X)	*Ortho*	*Meta*	*Para*
—NH_2	−0.6	−0.25	−0.63
—NH_3^+	0.4	0.2	0.2
—$NHCH_3$	−0.8	−0.3	−0.6
—$N(CH_3)_2$	−0.6	−0.1	−0.62

The resonant positions of the protons in some nitrogen-containing heterocycles are shown in Figure 6.1. Integration of the peak area of the ^1H NMR spectrum gives the total number of protons present, so it is easy to establish if there are equivalent protons in the molecule under test as is the case in pyridine, imidazole, aziridine and morpholine, as shown.

The peaks in NMR spectra may be split into several lines because of spin–spin coupling. The origin of this splitting may be caused by coupling with other adjacent protons or with other nuclei such as ^{13}C or ^{15}N. Measurement of the extent and character of the line splitting can provide useful information about the other atoms that are in the vicinity of the test proton.

For a proton coupling with n_a, n_b, n_c, . . . , sets of other equivalent protons, the multiplicity of the resonance will be $(n_a + 1)(n_b + 1)(n_c + 1)$.

Coupling constants between adjacent protons in heterocyclic systems are of the order of 1–10 Hz, coupling between protons and ^{13}C is of the order of 100–300 Hz, and ^1H to ^{15}N coupling constants are in the region of 30–100 Hz (between a half and a third of the value of ^1H to ^{13}C coupling).

Table 6.6. *Selected examples of ^{13}C chemical shifts*

Functionality	Chemical shift (p.p.m., TMS = 0)
CH_3-	0–60
$-CH_2-$	5–70
$-\overset{\mid}{\underset{\mid}{C}}-H$	22–78
$-C\equiv C-$	75–105
$-C=C-$ (aromatic)	105–155
$-CO_2H$	160–230
$>C=N-$ (azomethine)	145–162
$>C=N-$ (heteroaromatic)	142–160
$-C\equiv N$	110–125
$CH_3-N<$	15–45
$-CH_2-N<$	40–58
$>CH-N<$	50–68
$-\overset{\mid}{\underset{\mid}{C}}-N<$	62–75

The large extent of coupling between carbon and hydrogen in ^{13}C NMR leads to very complex spectra, which is why ^{13}C spectra are usually obtained with the whole of the proton spectrum being simultaneously irradiated. This technique is called proton decoupling. In proton-decoupled ^{13}C NMR each carbon atom exhibits one signal in the recorded spectrum. Unfortunately the low natural abundance of ^{13}C (1.108%) requires pulsed or Fourier Transform NMR techniques to collect spectra, but the ^{13}C nucleus has the advantage of yielding sharp signals and displays a wide range of chemical shifts depending on its chemical environment. As in proton NMR, tetramethylsilane is widely used as a reference standard in ^{13}C NMR, but unlike proton NMR the signals in ^{13}C NMR are usually not integrated as the effects of proton decoupling remove the accuracy and senstivity of the instrumental calibration. Integrated ^{13}C NMR spectra can be obtained by using inverse-gated decoupling techniques. Some typical chemical shifts for ^{13}C NMR are given in Table 6.6.

If it is impossible to interpret the spectral information despite having access to combined ^1H and ^{13}C data then two-dimensional (2D) NMR techniques can be used; these show the atom connectivity in the molecule under examination.

There are several 2D NMR techniques available. They are known by their acronyms including COSY (correlated spectroscopy) and NOSEY (nuclear Overhauser enhancement spectroscopy) [12].

Although it might be thought that nitrogen NMR would offer chemists studying amines an unparalleled understanding of the bonding and precise chemical

Table 6.7. *Chemical shifts in nitrogen NMR*
$(CH_3NO_2 = 0$ *p.p.m.*$)$

Molecule	δ (p.p.m.)
Aziridine	-280 to -420
Azetidine	-300 to -320
Ammonium $[NR_4]^+$	-270 to -380
Aliphatic amine	-280 to -375
Aromatic amine	-280 to -375
Imine	15 to -60
Nitrile	-110 to -150
Hydrazides	-230 to -350
Azoles	-100 to -280
NF_3	-20 to -30
Diazonium $[Ar-N^1=N^2]^+$	$N^1 = 0$ to -80
	$N^2 = -110$ to -170
Pyridine	15 to -51
Pyrrole	-181 to -231
Metal ammine ($M-NH_2$)	-300 to -500
Metal amine ($M-NR_2$)	-280 to -420
Metalloporphyrin ($N-M$)	-200 to 270
($N-H$)	-250 to -350
Coordinated dinitrogen	300 to -150
Nitrido	380 to 100

environment of the nitrogen atoms in this class of compound, in practice there are many problems to be overcome before measurement of nitrogen NMR can be facilitated. Nitrogen has two NMR active nuclei, which are ^{15}N and ^{14}N. However, ^{15}N yields sharp spectral lines but is rather insensitive (and in low natural abundance, 0.36%), whereas ^{14}N, which is quadrupolar and in 99.64% natural abundance, is a medium-sensitivity nucleus giving very broad signals that are often unresolvable even on a high-resolution spectrometer.

Fortunately, ^{14}N and ^{15}N measurements are interchangeable because the primary isotope effects on the chemical shifts of nitrogen are tiny. In fact in some tables of spectral data the nitrogen nucleus is not identified.

There are many reference standards used for nitrogen NMR work, but the most commonly used seem to be pure nitromethane (for liquids, where $\delta = 0$ p.p.m.) and ammonium chloride (for solid state work, where $\delta = -341$ p.p.m. relative to nitromethane). Some examples of chemical shifts in nitrogen NMR are given in Table 6.7.

The lone pair of electrons on the nitrogen atom is highly susceptible to the effects of solvents, pH, temperature and the formation of hydrogen bonds. For example, the indole nitrogen resonates at $\delta = -251$ p.p.m. in dioxan, but at $\delta = -246$ p.p.m.

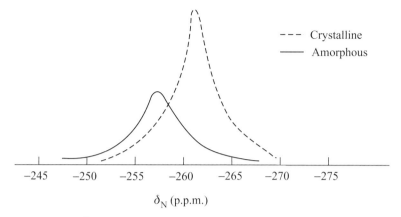

Figure 6.2. The ^{15}N NMR spectrum of Nylon 6, showing the crystalline and amorphous forms.

in methanol. Pyridine as a pure liquid has a nitrogen resonance of $\delta = -62$ p.p.m. but in water this shifts to $\delta = -85$ p.p.m.

The low isotopic abundance of ^{15}N may be compensated for by using ^{15}N-enriched samples for high-resolution NMR determination, and this field does seem to be more widely researched than ^{14}N NMR. There are two recent reviews of this area of ^{15}N spectrocopy [13] and [14], both of which contain detailed tables of chemical shift assignments. Reference [14] is of particular interest as it describes the analysis of amines by ^{15}N NMR.

Nitrogen NMR has been used for the analysis of synthetic polyamides, and for nylon it has been determined that the ^{15}N signal moves downfield with increasing chain length [15]. A solid-state ^{15}N NMR spectrum of Nylon 6 is shown in Figure 6.2, and shows how the amorphous and crystalline forms of Nylon 6 can be determined [15]. Nitrogen NMR has also been extensively used to study the process of nitrogen fixation and ammonia synthesis in root nodules. It is hoped that these studies will enable a better understanding of this biochemical process and the design of bioreactors that can facilitate this process on a commercial scale to supply the growing needs of mankind for fixed nitrogen. Recent work on this area [16] at the University of Roskilde has identified the formation of an ammonium pool in bacteroid cyctoplasm and the time-dependent formation of glutamine amide and aspargine amide in plant cytoplasm. The bacteriod cyctoplasm appears to be appreciably basic, between pH 7.2 and pH 8.9 [16]. A good recent review of the use of NMR techniques in structural elucidation (which also includes an explanation of the position of the USA FDA and European Department of the Quality of Medicines (EDQM) on the use of NMR data in drug master files) has been written by Holzgrabe *et al.* [17] and is highly recommended by the present author.

References

[1] P. A. Claret, *Experimental Chemistry Part II, Small-Scale Organic Preparations* (London: Pitman and Sons, 1961).

[2] J. J. Sudborough and T. Campbell-James, *Practical Organic Chemistry*, 2nd edition (Glasgow: Blackie and Sons, 1949).

[3] F. G. Mann and B. C. Saunders, *Practical Organic Chemistry*, 4th edition (Singapore: Longman Scientific and Technical, 1994).

[4] R. W. Alder, P. S. Bowman, W. R. P. Steele and D. R. Winterman, *J. Chem. Soc. Chem. Commun.* (1968), 723.

[5] A. I. Vogel, A. R. Tatchell, B. S. Furniss, A. J. Hannaford and P. W. G. Smith, *Vogel's Textbook of Practical Organic Chemistry*, 5th edition (Harlow, UK: Pearson Higher Education, 1990).

[6] E. Atherton and R. C. Sheppard, *Solid Phase Peptide Synthesis: A Practical Approach* (Oxford: Oxford University Press, 1989).

[7] A. O. Fitton and R. K. Smalley, *Practical Heterocyclic Chemistry* (London: Academic Press, 1968).

[8] D. R. Lide (ed.), *CRC Handbook of Chemistry and Physics 1999–2000*, 79th edition (Florida: CRC Press, 1998).

[9] O. Hinsberg and J. Kessler, *Ber.*, **38** (1905), 906.

[10] C. T. Leibermann and P. Jacobsen, *Ann.*, **211** (1882), 49.

[11] L. F. Fieser, M. Fieser and S. Rajagophan, *J. Org. Chem.*, **13** (1948), 800.

[12] W. S. Brey (ed.), *Pulse Methods in 1D and 2D Liquid NMR* (San Diego: Academic Press, 1988).

[13] R. K. Harris (ed.), *Encyclopedia of Nuclear Magnetic Resonance*, Volume 5 (Chichester, UK: J. Wiley, 1996).

[14] W. Schlif and L. Stefanick in *New Advances in Analytical Chemistry*, ed. A. Ur-Rahman (Amsterdam: Harwood, 2000).

[15] D. G. Powell and L. J. Mathias, *J. Am. Chem. Soc.*, **112** (1990), 669.

[16] A. M. Schaff, H. Egsgaard, P. E. Hansen and L. Rosendahl, *Plant Physiol.*, **131** (2003), 367.

[17] U. Holzgrabe, I. Wawer and B. Diehl, *NMR Spectroscopy and Drug Development and Analysis* (Weinheim: Wiley VCH, 1999).

7

Amine protection, amine oxides and amino acids

7.1 Amine protection

Both amines and amino groups are powerful nucleophiles as a consequence of the electron lone pair on their nitrogen atoms that is readily donated to electrophilic reagents. The problem of how to achieve selectivity in bi- or poly-functional molecules was not solved until the 1930s, when Berman and Zervas used the carboxybenzoyl group (usually abbreviated to CBZ or just Z) to protect the amine groups in amino acids, thereby allowing reaction to take place at the carboxylic acid function [1]. Until the mid 1960s this was the only commercially available protecting agent available in large quantities, usually as the chloride CBZ-Cl or Z-Cl. The introduction of CBZ-Cl was a major breakthrough for chemists as it allowed amino acids to be coupled cleanly for the first time to produce peptides. In the mid to late 1960s two new protecting agents were discovered by different research groups. Carpino and Han discovered the protecting agent N-(9-fluorenylmethoxycarbonyl)succinimide (Fmoc-ONSu) [2] and Tarbell and co-workers discovered di-tertiary-butyl dicarbonate (Di-BOC) [3]. Although the use of tertiary butoxy carbonyl group (BOC) had been known since the 1950s, its use was severely limited by the lack of a suitable stable precursor.

At the time of writing (late 2003), although over 100 protecting agents were known, these three protecting agents, Di-BOC, CBZ-Cl and Fmoc-ONSu, dominated the worldwide market and were the only protecting agents with significant commercial use.

In 2001, the world markets for these three protecting agents were 800 tonnes per year for Z-Cl, 300 tonnes per year for Di-BOC, and 40 tonnes per year for Fmoc-ONSu [4]. By late 2003, the market for Fmoc-ONSu had increased because of its increasing application in peptide-based life-science molecules. Significant price reductions in the cost of the Di-BOC and Fmoc-ONSu have also made these reagents attractive for the synthesis of novel pharmaceuticals for the management of HIV/AIDS that are affordable to developing countries.

All three protecting agents convert amines into amides and the overall reaction upon protection is as follows.

$$R_2-NH + R' \text{ (or Ar)}-CO-X \rightarrow R_2N-CO-R' \text{ (or Ar)}$$

The amine group replaces a hydrogen atom with a carbonyl group.

Amides are actually resonance hybrids, and the carbon–nitrogen bond has partial double-bond character with donation of the nitrogen lone pair to the electrophilic carbonyl group. The extent of the donation of the nitrogen lone pair reduces the nucleophilic properties of the amine group accordingly and so passivates its reactivity with electrophiles. In addition, bulky alkyl or aryl substituents on the protecting group provide steric blocking to incoming reagents.

$$
\begin{array}{ccc}
R-C=O & & R-C-O^- \\
| & \rightleftharpoons & || \\
R'-N-R'' & & R'-N^+-R''
\end{array}
$$

The first reported use of protecting agents was the work of E. Fischer in the 1870s. Helfreich reviewed this work, and extended it by using acetate and acetonide groups in carbohydrate chemistry in the 1920s [5].

The criteria for an ideal protecting group are as follows.

(a) The protecting-group precursor must react selectively in good yield with the target site.
(b) The protecting group must be stable under the subsequent reaction conditions and must not contain reactive sites.
(c) The protecting group must be easy to remove in good yield by non-aggressive reagents after completion of the reaction step.
(d) The protecting-group precursor should form a stable, easily isolatable derivative with the target compound.
(e) The protecting-group precursor should be easy to handle and store.
(f) The protecting-group percursor should not be so expensive that its use in commercial synthesis is precluded.
(g) The protecting group should not alter the stereochemistry of a chiral substrate molecule on either protection or deprotection, and should also protect the substrate from racemization or epimerization.

7.1.1 Carboxybenzoyl chloride (CBZ-Cl)

The use of CBZ-chloride to convert amines into *N*-benzoyl amides was first reported in 1932 by Bergmann and Zervas [1]. Amines protected by CBZ groups are stable in alkaline solution but are readily cleaved in the presence of acids. The introduction of CBZ-chloride opened up the study of synthetic peptides, although the protection and deprotection cycles required to manufacture peptides generally work best with

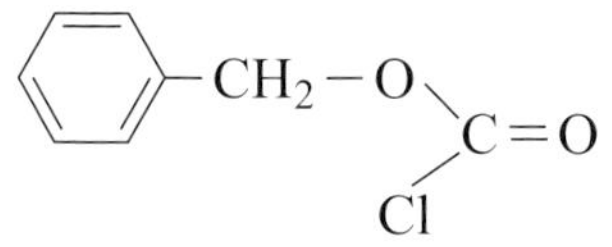

CAS = [501-53-1]
EINECS = 2079250
Aspect = colourless, oily liquid
Melting point = 0 °C
Boiling point = 85 °C at 7 mmHg

CBZ-Cl

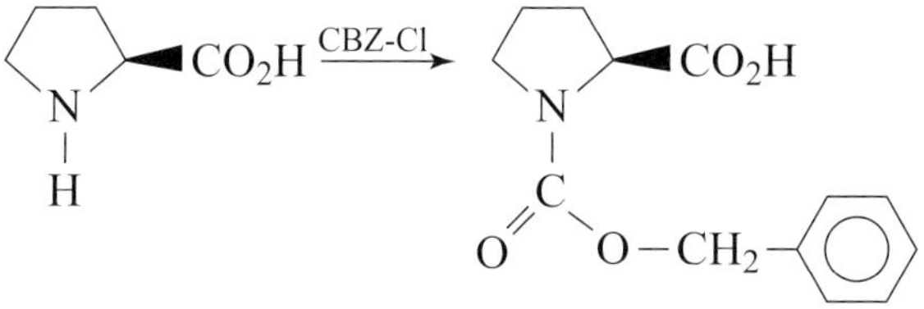

Figure 7.1. The structure and properties of CBZ-Cl, and its reaction with proline.

short-chain peptides. However, the problem with CBZ-chloride is that it is difficult to handle as it is very susceptible to hydrolysis by water and atmospheric moisture. CBZ-protected amines and amino acids can be prepared only in alkaline conditions. A further difficulty with CBZ is that amino acids that contain two or more amino groups require orthogonal protection with each amine unit requiring a different protecting group to allow selective protection and deprotection.

CBZ-Cl (see Figure 7.1) can be prepared by the reaction between benzyl alcohol and phosgene either in toluene solution or in a solvent-free system. Commercially available CBZ-Cl often contains up to 5% impurities that can include benzyl chloride, benzyl alcohol, toluene and hydrochloric acid. Residual toluene does not usually interfere with reactions between CBZ-Cl and amines (in fact, CBZ-Cl is often used in toluene solution), but the other impurities are more problematic.

If pure CBZ-Cl is required it is best to flush commercial CBZ-Cl with dry air to remove any dissolved carbon dioxide and hydrogen chloride. The purged CBZ-Cl should then be mixed with anhydrous sodium sulphate to remove any water, filtered and vacuum distilled at below 85 °C.

$$C_6H_5-CH_2-OH + Cl_2C{=}O \rightarrow C_6H_5-CH_2-O-CO-Cl + HCl$$

Primary and secondary amines are protected by reaction with CBZ-Cl to form the corresponding benzyl carbamates. The reaction may be done in aqueous sodium-hydroxide solution, or under Schotten–Baumann conditions with aqueous sodium carbonate and an organic solvent, or under anhydrous conditions such as with triethylamine in dichloromethane. For bi-functional amines, the amine group is often preferentially protected over functional groups such as alcohols, thiols and phenols. CBZ-protected amino acids are generally crystalline and are soluble in common solvents. The use of low temperatures (down to -20 °C) for the protection reaction is advantageous in that it suppresses racemization at the chiral centre.

The protection of proline with CBZ-Cl works well in 2M sodium-hydroxide solution. The yield at 5–10 °C is over 90%, and the reaction is complete after 1 h. A slight stoichiometric excess of 1.1 mol of CBZ-Cl should be used relative to proline.

Some amino acids can be protected by refluxing with CBZ in ethyl acetate as a mono-solvent with no additional base, although yields are usually around 10% lower than under Schotten–Baumann conditions.

Electron-deficient amines such as anilines can be protected with CBZ by using magnesium oxide in acetone as a solvent system, and pyrrole nitrogens can be protected by prior deprotonation to form a potassium salt.

Once formed, CBZ-protected amines are stable to a range of weakly acidic and basic conditions but subsequent reactions are best carried out under basic conditions.

CBZ-protected amines can be cleaved by three main methods: acidolysis, catalytic hydrogenation and reduction with dissolved metals. The reaction with dissolved metals usually refers to sodium metal in liquid ammonia, but this is rather a blunt instrument and will simultaneously deprotect any other protected amine groups present that have been protected with other protecting agents and so is unhelpful for synthesis with orthogonally protected molecules with two or three differently protected amine groups.

Deprotection with acidolysis can be performed with protic acids or Lewis Acids and allows selective deprotection. For example, CBZ–amine groups are generally unaffected by 4M HCl in dioxane and by sulphuric acid in dichloromethane. Preferred reagents for deprotection by acidolysis include hydrogen bromide in acetic-acid solution (which is widely used) and also triflic acid. Trifluoroacetic acid is widely used to cleave BOC–amine groups and will also cleave CBZ–amines, but the reaction kinetics are very slow unless a catalytic quantity of thioanisole is added to the reaction mixture.

Catalytic deprotection with Pd/H_2 works well, and either palladium black or palladium on carbon can be used as the reaction catalyst. However, transfer catalytic hydrogenation with 10% Pd–C and cyclohexadiene is probably the most widely used method of deprotection in a glacial-acetic-acid solvent.

The deprotection products are the parent amine plus carbon dioxide and either toluene (if catalytic hydrogenation is used) or benzyl bromide (if hydrogen bromide/ acetic acid is used). The acid stability of CBZ–amine groups can be enhanced by substitution of electron-withdrawing groups into the phenyl group in the *ortho* position, and the use of 2-chloro and 2-bromo CBZ derivatives has been described [6].

Apart from a slight tendency to racemize during peptide synthesis, CBZ is an excellent protecting group and has revolutionized peptide synthesis since its introduction. However, there is a major drawback to the use of CBZ: this is the lack of an easy-to-handle precursor. CBZ-Cl is highly toxic, is a potent lachrymator and is also a suspected carcinogen. It reacts with atmospheric moisture and must be handled in a fume hood.

However, CBZ-Cl is widely used as a protecting agent because of its widespread availability and low cost. Its use is not restricted to amines and it can be used to protect other functional groups such as thiols, alcohols, phenols and even amides. The protection of alcohols and phenols can be achieved by using CBZ-Cl in the presence of a tertiary amine base (such as triethylamine) in dichloromethane at 0 °C to −20 °C. For thiol protection, sodium bicarbonate is used instead of an amine base as this encourages the selective protection of thiols even in the presence of amines, as in the case of cysteine. For amides, CBZ-Cl can be used with triethylamine but a better strategy is to use butyl lithium to deprotonate the amide group first before adding the carboxybenzoyl chloride.

Other benzyloxycarbonylating reagents such as dibenzyl carbonate, *N*-CBZ-imidazole and CBZ-ONSu are known, but these all have to be prepared first from CBZ-Cl. The quest to develop an easy-to-handle protecting agent that could be manfactured in commercial quantities and used with CBZ in orthogonal protection continued until the late 1960s when two new protecting agents, Di-BOC and Fmoc-ONSu, were discovered.

7.1.2 Di-tertiary-butyl dicarbonate (Di-BOC)

Di-BOC is a white crystalline solid at ambient temperatures, but is usually encountered in the form of a fused melt as it becomes molten at 22–24 °C. It is chemically stable and resistant to hydrolysis under alkaline conditions. The oxazolones formed from the reaction of BOC-protected amino acids do not racemize during peptide synthesis and preserve the stereochemistry of the original amino acid. The ease of use of Di-BOC coupled with its retention of molecular stereochemistry have led to it becoming the main source of the BOC protecting group used in amino acid and amine chemistry.

The original synthesis for Di-BOC published by Tarbell *et al.* [3] still forms the basis for the commercial manufacture of Di-BOC. In this synthesis, potassium tertiary butylate is reacted with carbon dioxide to form potassium tertiary butyl carbonate. This carbonate is then treated with phosgene in toluene to form the

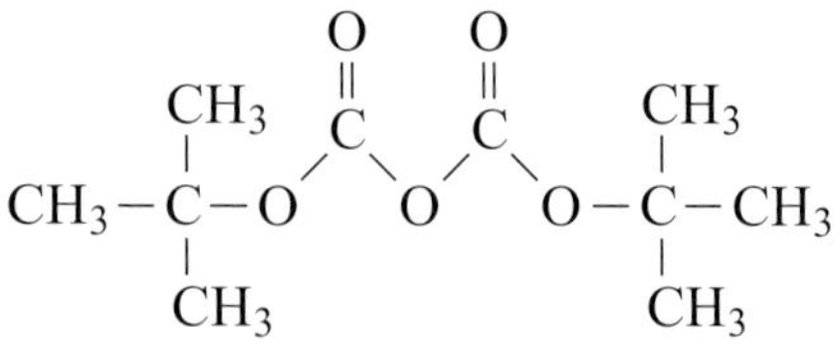

$$CH_3-\underset{\underset{CH_3}{|}}{\overset{\overset{CH_3}{|}}{C}}-O \overset{\overset{O}{||}}{\underset{}{C}} \diagdown_{O} \diagup \overset{\overset{O}{||}}{\underset{}{C}} \diagdown_{O}-\underset{\underset{CH_3}{|}}{\overset{\overset{CH_3}{|}}{C}}-CH_3$$

Di-tert-butyl dicarbonate
CAS = [24424-99-5]
EINECS = 246240l
Aspect = low-melting solid
Melting point = 22–24 °C

Figure 7.2. The structure and properties of Di-BOC.

unstable di-tertiary-butyl-tricarbonate, which disproportionates in the presence of a basic catalyst to form Di-BOC.

$$\text{t-BuOK} + CO_2 \rightarrow \text{t-BuOCO}_2\text{K}$$

$$\text{t-BuOCO}_2\text{K} + Cl_2C{=}O \rightarrow (\text{t-BuOCO}_2)_2\text{CO} + 2KCl$$

$$(\text{t-BuOCO}_2)_2\text{CO} + \text{amine catalyst} \rightarrow (\text{t-BuCO}_2)_2\text{O} + CO_2$$

Some recent modifications of this synthesis [7] have replaced phosgene with thionyl chloride or phosphorus oxychloride. In these instances the intermediate takes the form $(CH_3)_3CO{-}CO{-}OXO_2{-}CO{-}O(CH_3)_3$, which disproportionates to yield Di-BOC and XO, where X = P–Cl or S=O.

The amine catalyst used in the final stage can be 1,4-diazabicyclo-(2,2,2)-octane. This was first used by Tarbell *et al.*, but other amines such as *N*-methyl piperidine, *N*-ethyl piperidine, tris-(2,12-methoxyethoxyethyl)amine and triethylbenzylammonium chloride have also been reported as being effective catalysts [7]. The structure and properties of Di-BOC are given in Figure 7.2.

Di-BOC reacts smoothly and rapidly with amines in organic solvents, mixed solvents such as aqueous dimethylformamide, and alkaline aqueous solutions to form pure derivatives in high yields. The tertiary-butoxy carbonylation of alkyl amines usually occurs at room temperature within 20–30 min. This reaction can occur up to 2.5 times faster in the presence of trace amounts of hydroxylamine than with Di-BOC alone. Hydroxylamine acts as a catalyst, and its addition is thought to promote the *in situ* formation of BOC-ONH$_2$, which reacts at a faster rate with alkyl amines than Di-BOC alone [8]. The protection of aryl amines is not so easy to accomplish and usually requires heating; for example, the reaction between aniline and Di-BOC requires heating under reflux in tetrahydrofuran. The BOC-NH-Ar group in aryl systems is usually *ortho* directing and allows electrophilic substitution into the aromatic ring.

After the desired reaction, BOC deprotection can be accomplished by using mild acidic conditions such as trifluoroacetic acid (BOC-aniline can be deprotected in excess of 99% yield in the presence of excess trifluoroacetic acid for 1 h at 20 °C). A scavenger is usually added to the reaction mixture during deprotection to remove any t-butyl cations, thereby preventing secondary methylation or butylation of the target molecule. Some typical scavengers for butyl cations are anisole, thiophenol, cresol and dimethyl sulphide.

Protection

$$t\text{-Bu}-O-CO-O-CO_2-t\text{-Bu} + H_2N-R \rightarrow t\text{-Bu}-O-CO-NHR + CO_2 + t\text{-BuOH}$$

Deprotection

$$t\text{-Bu}-O-CO-NHR \, (+ \, CF_3COOH) \rightarrow H_2N-R + CO_2 + CH_3C(CH_2)CH_3$$

The t-butyl group cleaves as a butyl cation, which can react with hydroxyl groups in aqueous solutions to form t-butanol as well as butene.

Di-BOC reacts with alkyl diamines in dioxane to form a mono-protected derivative with only 3% of the di-protected derivative.

$$H_2N-(CH_2)_n-NH_2 + \text{Di-BOC} \rightarrow H_2N-(CH_2)_n-NH-BOC$$

In cases where an alkyl diamine has one secondary amine and one primary amine (e.g. monomethylethylene diamine, $CH_3-NH-CH_2-CH_2-NH_2$) it is possible to protect the secondary amine selectively, but the primary amine must first be treated with benzaldehyde to form a Schiff Base.

$$H_2N-(CH_2)_n-NHR + C_6H_5CHO \rightarrow C_6H_5-CH{=}N-(CH_2)_n-NHR$$
$$\downarrow + \text{Di-BOC}$$
$$C_6H_5-CH{=}N-(CH_2)_n-NR(BOC)$$
$$\downarrow + KHSO_4$$
$$H_2N-(CH_2)_n-NR(BOC)$$

For heterocyclic amines such as indole, reaction with Di-BOC in dichloromethane results in the ring nitrogen being protected and allows the synthesis of α-substituted derivatives by lithiation. Tryptophan reacts with Di-BOC to form a doubly BOC-protected derivative that may be selectively cleaved with hydrochloric acid in dioxane (the BOC group on the alkylamino side chain is cleaved preferentially) [9].

When alkyl amides are treated with an excess of Di-BOC it is possible to obtain a di-protected derivative $R-CO-N(BOC)_2$, which can be aminated with a primary amine to form di-(tertiarybutylcarboxy) amine $HN(BOC)_2$.

$$R-CO-NH_2 + 2\text{Di-BOC} \rightarrow R-CO-N(BOC)_2$$
$$R-CO-N(BOC)_2 + R'-NH_2 \rightarrow R-CO-NH-R' + HN(BOC)_2$$

Hydrazines are acylated by Di-BOC in isopropanol or diacylated by Di-BOC in benzene or dioxane. BOC-hydrazide, formed by the mono-protection of hydrazine with the BOC protecting group, is a useful synthetic reagent and can be used in the synthesis of complex hydrazides where hydrazide cannot be used because a mixture of mono- and bi-substituted products would be obtained [10]. BOC-hydrazide can be used in the Japp–Klingemann Reaction to produce hydrazones and osazones and also in the synthesis of heterocyclic compounds with adjacent ring nitrogen atoms (e.g. pyridazine) or with one ring nitrogen N-bonded to a BOC-protected amine (e.g. N-(BOC)-aminopiperidine).

7.1.3 N-(9-fluorenylmethoxycarbonyl)succinimide (Fmoc-ONSu)

The development of the 9-fluorenylmethoxycarbonyl (Fmoc) group in the 1960s by Carpino and Han [2] was in response to the increasing need to use amino acids for the solution-phase synthesis of peptides and life-science molecules in acidic solutions. The Fmoc group is exceptionally stable to acidic conditions and is tolerant to sulphuric and hydrochloric acids and also to hydrogen bromide in acetic acid or trifluoroacetic-acid solution [11].

Two advantages of the acid stability of the Fmoc group are that it allows protected amino acids to be either converted to acid chlorides with thionyl chloride or converted to t-butyl esters with butene in sulphuric acid. Resonance stabilization of the cyclopentadienide ion creates an acidic hydrogen at ring position 9, which confers acid stability to the Fmoc group. A major problem with the Fmoc group prior to Carpino and Han's synthesis of Fmoc-ONSu was that the only available precursor to Fmoc was the chloride Fmoc-Cl, which is corrosive and also can be fatal by inhalation. By comparison, Fmoc-ONSu is a safe, easy-to-handle white crystalline solid at ambient temperatures. Other advantages of Fmoc-ONSu over Fmoc-Cl·are that it generally produces cleaner reaction products that do not require recrystallization, and also that optical rotation is better retained with Fmoc-ONSu.

The two main methods of synthesis of Fmoc-ONSu reported in the chemical literature are (i) from 9-fluorenylmethyl alcohol with a carbonylation agent such as phosgene, triphosgene or dimethyl carbonate, and (ii) directly from Fmoc-Cl (although this method is rarely used commercially). The structure of Fmoc-ONSu is shown in Figure 7.3.

The protection of amino groups with Fmoc-ONSu is accomplished under Schotten–Baumann conditions with phase-transfer conditions for water-soluble amines or under anhydrous conditions for other amines. Under phase-transfer conditions, solvent mixtures such as dioxane/water and dimethylformamide/water are used with the addition of sodium bicarbonate to maintain pH control during the

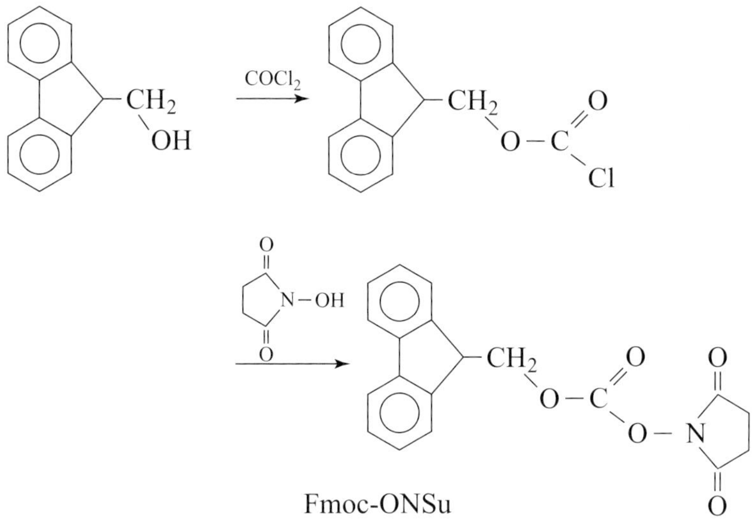

Fmoc-ONSu

Figure 7.3. The synthesis and structure of Fmoc-ONSu.

reaction. When the reaction is done under anhydrous conditions a small quantity of a tertiary amine (such as trimethyl amine or tetramethylethylene diamine) is added to the reaction solution to maintain pH control.

The protection reaction usually takes place at, or below, room temperature in good yield with retention of chirality. After the reaction, washing with water removes the succinimide by-product that crystallized out from the organic layer. In instances where the Fmoc-protected amine has a low solubility in the reaction solvent, the best method of recovery is to strip off the reaction solvent under vacuum at low temperature and to replace it with a second solvent such as ethylacetate/ether or dichloromethane. This recovery method is especially useful when di-protected Fmoc amines are required, as they often have poor solubilities in conventional solvents. Deprotection of Fmoc-protected amines is usually accomplished with 30% piperidine in dimethylformamide, although the Fmoc group is generally labile in the presence of any secondary amine in basic conditions. The mechanism of deprotection is by proton abstraction to form the dibenzocyclopentadienide anion, which on cleavage from the parent amine gives dibenzofulvene. The reaction obeys E1cb kinetics and is fairly fast when piperidine is chosen as the secondary amine to initiate cleavage. Other amines such as morpholine and diisopropylethylamine have also been reported to be effective in the cleavage of the Fmoc group, but generally the reaction times are slower [11]. Also because piperidine forms an isolatable adduct with dibenzofulvene it remains the first-choice reagent for deprotection of Fmoc-amines. The deprotection of an Fmoc-protected amine is shown schematically in Figure 7.4. Fmoc-ONSu is also widely used as a reagent for the analysis of

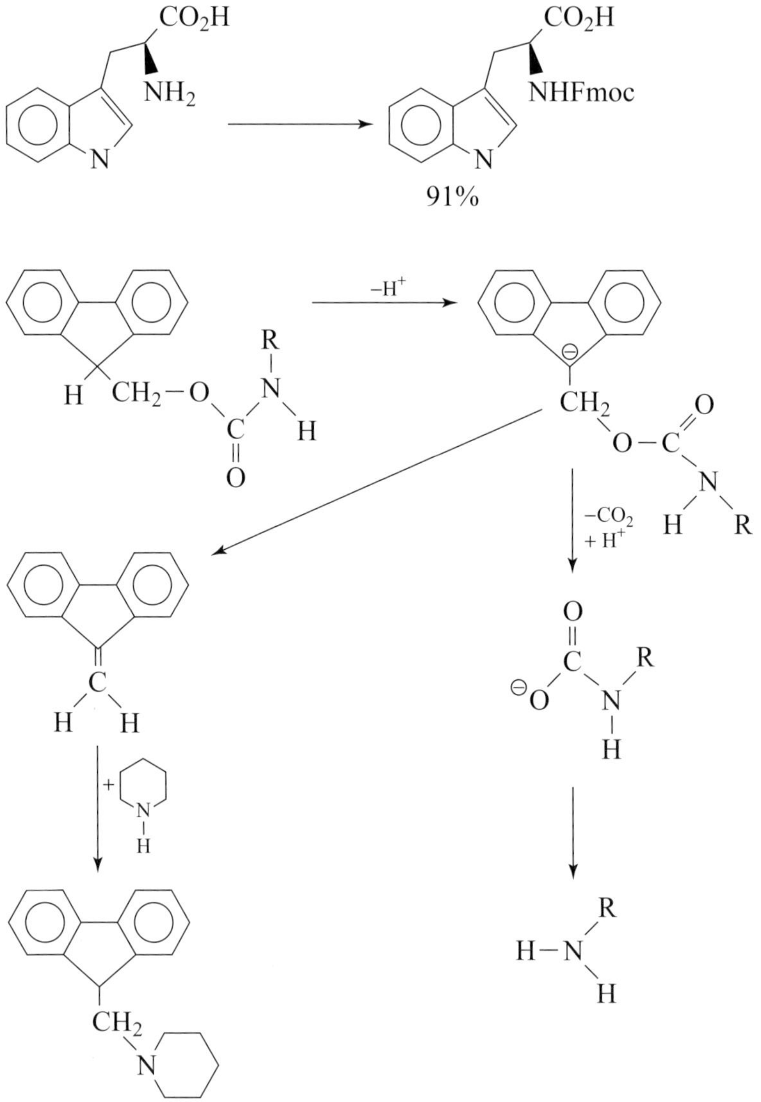

Figure 7.4. The protection and deprotection of an amine group with Fmoc-ONSu.

amine mixtures by high-performance liquid chromatography (HPLC). The problem with many amines is that they do not possess aromatic or heterocyclic rings and so are not amenable to UV detection (refractive index detection can be used but is limited to single-solvent, isocratic systems and is not amenable to gradient-elution HPLC). Pre-column derivatization with Fmoc-ONSu in water/dioxane followed by acidification with hydrochloric acid and extraction into ethyl acetate generally gives

Table 7.1. *A comparison of CBZ-Cl, Di-BOC and Fmoc-ONSu*

Property	CBZ-Cl	Di-BOC	Fmoc-ONSu
Physical characteristics			
CAS number	[501-53-1]	[24424-99-5]	[82911-6-1]
Molecular weight	170.60	218.45	337.34
State at ambient	Liquid	Low-melting solid	Solid
Melting point	$-65\,°C$	22 to 24 °C	150 to 153 °C
Solubility in			
Organic solvents	Good	Good	Good
Mixed solvents	Good	Good	Good
Aqueous	Good in alkaline bases	Good in alkaline bases	Poor, needs phase-transfer catalyst
Stability of protected amines and amino acids to			
Acidic conditions	Stable in dil. acid	Unstable	Stable
Basic conditions	Stable	Stable	Unstable
Heat	Good	Good	Good
Catalytic hydrogenation	Readily cleaved	Stable under certain conditions	Stable under most conditions
Retention of chirality	Good	Excellent	Excellent
Reaction by-products			
On protection	Gas	Gas or liquid	Solid (H_2O-soluble)
On deprotection	Gas or liquid	Gas or liquid	Solid (hexane-soluble)

yields of over 95% and allows UV detection to be used for the analysis of amine mixtures [11]. A comparison of the three protecting agents CBZ-Cl, Di-BOC and Fmoc-ONSu is given in Table 7.1.

7.2 Amine oxides

The term amine oxide is generally accepted as the trivial name for the oxides of tertiary amines. The amine may be aliphatic or aromatic and have mixed alkyl/aryl side chains attached to it, but must not have any nitrogen–hydrogen bonds, because primary and secondary amines react to form hydroxylamines and also nitroamines (for primary amines only).

$$R–NH_2 + H_2O_2 \rightarrow R–NH–OH + H_2O$$

$$R–NH–OH + 2H_2O_2 \rightarrow R–NO_2 + 3H_2O$$

Amine oxides are usually shown with a dative covalent bond between the amine nitrogen atom and the oxygen atom with formal positive and negative charges. The dipole moment of the N—O bond is fairly high at 4.3 D, and this gives the amine oxides their special character.

Aliphatic or aryl amine oxides

$$R^1-\overset{\overset{\displaystyle R^2}{|}}{\underset{\underset{\displaystyle R^3}{|}}{N^+}}\!\!\rightarrow O^-$$

R^1, R^2 and R^3 may be alkyl or aryl groups and may be the same, or different, e.g. cetyldimethyl dimethyl amine oxide.

Aromatic amine oxides

The aromatic ring may be pyridine, quinoline or another heteroaromatic ring.

Aliphatic amine oxides are widely used in surfactant formulations and aromatic amine oxides are used in pharmaceuticals and some agrochemicals, but their major commercial application is in the formulation of detergent blends for cleaning and personal-care products [12]. Small amounts of amine oxides in detergent blends can dramatically improve the quality and quantity of foam produced in water and also they have low irritability to skin and eyes.

The properties of amine oxides vary depending upon the nature of the side chains attached to the central nitrogen atom. However, all amine oxides have large dipole moments. Aliphatic amine oxides are generally hydroscopic solids that are readily soluble in water. Aromatic amine oxides are generally less hydroscopic and have better solubilities in non-polar solvents. The differences between aliphatic amine oxides and aromatic amine oxides can be explained in terms of their structures. Aliphatic amine oxides such as tributylamine oxide have tetrahedrally configured nitrogen atoms, whereas aromatic amine oxides such as pyridine oxide have a trigonal nitrogen that directs the oxygen atom into the same plane as the aromatic π-orbitals, which can interact with non-bonded electron density on the oxygen. This interaction leads to resonance stabilization and the special characteristics of aromatic amine oxides.

The amine oxides usually exist as unionized hydrates in aqueous solution and it is difficult to isolate them in a pure anhydrous form, although aromatic amine oxides generally have a lower affinity for water. Below pH3 in aqueous solution, amine oxides exist in a cationic form and are capable of protonation reactions ($R_3NO + HB \rightarrow [R_3NOH]^+[\mathbf{B}]^-$). Amine oxides are weaker bases than the parent amines and do not exhibit such a wide range of pK_a values (see Table 7.2). This is explained by

Table 7.2. *The pK$_a$ values of some protonated amine oxides and their parent amines*

Amine	Amine pK$_a$	Amine-oxide pK$_a$
Trimethylamine	9.74	4.65
Triethylamine	10.76	5.13
N,N-diethylphenylamine	6.56	4.53

the fact that the formation of a dative covalent bond with oxygen is more important in defining the properties of amine oxides than the effect of substituents that are only adjacent to the basic centre.

The non-ionic/cationic behaviour of amine oxides defines their properties and also their chemical reactions. For example, their critical micelle concentrations (CMC) are stongly affected by solution pH and temperature, and the number of monomers per micelle is also strongly affected by these parameters. On changing the solution from water to 0.2 M hydrochloric acid the CMC for a fatty alkyl amine oxide approximately halves and the number of molecules per micelle roughly doubles. Increasing the solution temperature from ambient to 50 °C has a less-pronounced effect but generally lowers the CMC and number of monomers per micelle by around 80% [13]. Typically for a fatty amine oxide such as dimethyldodecylamine oxide, a micelle at pH7 and ambient temperature contains 76 monomers and has a CMC of 0.5 g l^{-1}.

Amine oxides have reported applications as phase-transfer catalysts, in pharmaceuticals and agrochemicals, as fabric softeners and as fuel additives for petroleum to prevent icing in cold weather.

7.2.1 Preparative methods for amine oxides

The simplest method of synthesis for aliphatic tertiary amine oxides is direct oxidation, which works well with hydrogen peroxide, Caro's Acid and ozone. Somewhat surprisingly, the oxidation reaction does not work with potassium permanganate and if the amine used contains α-hydrogens an enamine is produced; this may hydrolyse to form an aldehyde together with a secondary amine. Typically, hydrogen peroxide is used at 60–80 °C and yields of around 90% are common.

$$R_3N + [O] \rightarrow R_3N^+\!-\!O^-$$
(When [O] is supplied from H_2O_2, O_3 or H_2SO_5.)
$$R'\!-\!CH_2\!-\!NR''_2 + [O] \rightarrow R'\!-\!CHO + R''_2NH$$
(When [O] is from $KMnO_4$, MnO_2 or $Hg(OAc)_2$.)

Generally the oxidation of tertiary amines is performed in water but occasionally alcoholic solvents are used. Hydrocarbon solvents can be used for the reaction with ozone, especially if the reaction is to be performed a low temperature; for example, chloroform can be used at $-50\,°C$.

With hydrogen peroxide, the first stage of the reaction is the formation of a trialkylammonium peroxide complex, $R_3N.H_2O_2$, which can often be isolated. This complex is then attacked by $[OH]^+$ and subsequently decomposes to form the corresponding N-oxide. Caro's Acid works in a similar manner, as shown in the reaction scheme below.

$$R_3N + HO-OSO_3H \rightarrow [HSO_4]^- [R_3N-OH]^+ \rightarrow R_3N-O + H_2SO_4$$

Another method of synthesis is from the alkylation of hydroxylamines. However, this depends on finding a suitable hydroxylamine precursor.

$$RNH-OH + 2R'-Cl \rightarrow RR'_2N-O + 2HCl$$

(The oxidation of mono-*N*-substituted hydroxylamines gives nitroso compounds, $RNH-OH + [O] \rightarrow R-N=O + H_2O$.) This reaction is not used commercially for the synthesis of amine oxides but works well in the laboratory.

Because of the lower basicity of aromatic amine oxides, they do not react well with hydrogen peroxide; peracids such as perbenzoic acid or monopermaleic acid must be used instead. A good method of synthesis is to use glacial acetic acid as a solvent for the amine and to add hydrogen peroxide to the heated solution at 80–95 °C [14].

Most commercial amine oxides are made from the reaction of tertiary alkyl amines with hydrogen peroxide in water or isopropyl alcohol as a solvent, or from the reaction of aromatic amines with peracetic acid formed *in situ*. The manufacturing process can be carried out in glass (or glass-lined) reactors, and for alkylamine oxides it can be performed in stainless steel (type 316 works well). Care must be taken that the reactors are not overfilled as amine oxides can yield foams, so sufficient space (typically 30%) must be left at the top of the reactor. A small quantity (0.05%) of ethylenediaminetetraacetic acid (EDTA) is usually added to the reaction mixture to chelate any trace metals present that might catalyse the decomposition of hydrogen peroxide. When the reaction is completed any unreacted hydrogen peroxide can be destroyed by the addition of sodium bisulphite, although this can produce a coloured product. Commercial amine oxides usually contain only about 50% amine oxide, with the balance being water and about 2% unreacted amine. Many manufacturers of amine oxides produce them for captive applications, for example in the preparation of surfactant blends for cosmetic and toiletry applications.

However, before this operation is performed any free hydrogen peroxide or peracids present must be neutralized (some commercial amine oxides contain free peracids and peroxides up to 0.1%). If a low-colour amine oxide is required then the use of sodium bisulphite should be avoided; another effective method of removing any excess peroxide, such as reacting with manganese dioxide at 50–60 °C followed by filtration, should be used instead. If pure, anhydrous, amine oxides are required for study or research applications then the residual water content has to be removed by vacuum stripping. Another method of producing anhydrous amine oxides is to start from a commercially available amine hydrate that can be azetropically distilled with dimethylformamide to remove the water present.

7.2.2 Reactions of amine oxides

Although amine oxides are mainly manufactured for their properties in detergent formulations and personal-care products, they do exhibit a range of unique chemical properties that are of interest to the synthetic chemist.

(a) The Cope Elimination [15]

Aliphatic amine oxides with a β-hydrogen or hydrogens form olefins and dialkyl-hydroxylamines when heated over 90 °C in a vacuum. This intermolecular reaction proceeds through a five-membered cyclic ring and is known as the Cope Elimination.

$$CH_3-\underset{\underset{O}{|}}{\overset{\overset{CH_3}{|}}{N}}-CH_2-CH_2-R \rightarrow \quad H\overset{\overset{\overset{R}{\diagdown}}{CH-CH_2}}{\diagdown}\underset{O}{\overset{}{}}\overset{\diagup}{N}\overset{\diagup CH_3}{\diagdown CH_3} \rightarrow CH_2{=}CHR + (CH_3)_2N-OH$$

When two possibilities exist for elimination the alkyl group with the greatest number of β-hydogens is preferentially eliminated. The presence of electron-withdrawing groups on the β-carbon also favours elimination. Because the mechanism of the Cope Elimination requires the formation of a five-membered intermediate, some amine oxides may not give this reaction if steric considerations rule out the formation of the intermediate, e.g. *N*-methylpiperidine oxide. Because the five-membered intermediate is planar, the Cope Elimation favours *cis* elimination, which allows for the steric control of the eliminated olefins. With the Cope Elimination it is possible to produce unsymmetrical olefins of the type $RR'C{=}CH_2$ from amine oxides of the type $RR'-CH-CH_2-N(CH_3)_2O$. At lower temperatures (below 90 °C) some amine oxides may deoxygenate back to the parent amine rather than giving the Cope Elimination. In such instances the reaction temperature should be increased slightly to achieve the required elimination reaction.

(b) The Meissenheimer Rearrangement [16]

For aliphatic amine oxides without a β-hydrogen the Cope Elimination will not
work, and instead they rearrange on heating to between 100 and 200 °C to form trisubstituted hydroxylamines. This is the Meissenheimer Rearrangement. It works
best when the amine has an allylic or benzyl system attached to the nitrogen and
it is catalysed by the presence of strong bases. The rate of rearrangement is not
dependent on the concentration of the base but is generally found to increase as the
basicity of the parent amine decreases. The exact reaction mechanism is unclear
and may involve a cyclic intramolecular species or a free-radical reaction. Overall
the reaction is as shown below.

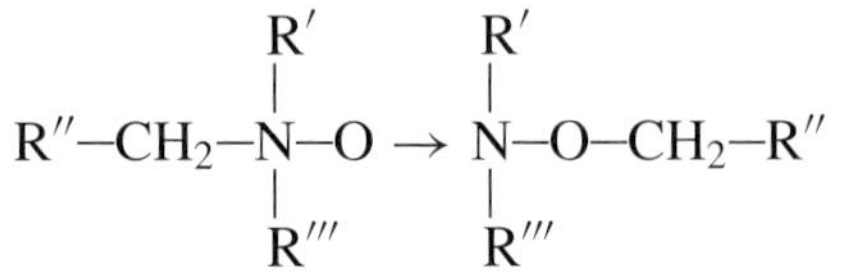

$$
\underset{\underset{R'''}{|}}{\overset{\overset{R'}{|}}{R''-CH_2-N-O}} \rightarrow \underset{\underset{R'''}{|}}{\overset{\overset{R'}{|}}{N-O-CH_2-R''}}
$$

(c) Substitution reactions

Amine oxides are able to be protonated at low pH to form cationic species.

$$R_3N-O + [H]^+ \rightarrow [R_3N-O-H]^+$$

Aromatic amine-oxide systems can stabilize this positive charge by resonance and
may then react with nucleophilic agents such as cyanide or methoxide ions. The
stabilization of protonated amine oxides is greatest for benzylic systems, which can
eliminate a dialkyl hydroxylamine to form a substituted benzylic fragment.

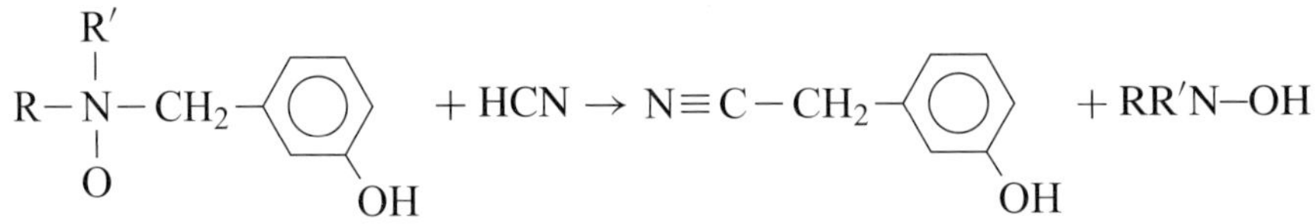

$$
\underset{\underset{O}{|}}{\overset{\overset{R'}{|}}{R-N-CH_2}}\text{—}\langle\bigcirc\rangle\text{—OH} + HCN \rightarrow N\equiv C-CH_2\text{—}\langle\bigcirc\rangle\text{—OH} + RR'N-OH
$$

Aromatic amine oxides such as pyridine *N*-oxide may undergo both electrophilic
and nucleophilic substitution because the dipolar N—O group can act as an electron donor or acceptor. In total, five resonance structures are possible for pyridine
N-oxide, and these are shown in Figure 7.5.

   The pyridine ring is passive towards electrophilic reagents and does not give the
Friedel–Crafts Reaction, and it is nitrated only with difficulty under conditions of
high pressure and temperature. Pyridine *N*-oxide will nitrate at the 4-position when
reacted with concentrated sulphuric acid and fuming nitric acid at 100 °C. Reaction
at higher temperatures produces a mixture of 2- and 4-substituted isomers. Pyridine
N-oxide is a useful synthetic reagent and is able to oxidize alkyl halides to carbonyls

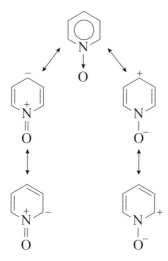

Figure 7.5. The resonance structures of pyridine *N*-oxide.

and α-halo acids to aldehydes and ketones with the loss of a carbon atom.

$$Br-CH_2-CH_2-CH_2-CH_2-CH_2-OAc + py-O/NaHCO_3/toluene \rightarrow$$
$$O=CH-CH_2-CH_2-CH_2-CH_2-OAc$$

$$CH_3-CH_2-CH_2-CHBr-CO_2H + py-O/xylene \rightarrow CH_3-CH_2-CH_2-CHO$$

(Here py = pyridine and py−O = pyridine *N*-oxide.)

(d) Reduction reactions

Aliphatic amine oxides may be easily reduced to amines by sulphurous acid (H_2SO_3) at room temperature, but aromatic amine oxides are resistant to reduction and so reagents such as zinc/hydrochloric acid or iron/acetic acid or catalytic hydrogenation with Raney nickel are required for deoxygenation. Some other reagents reported to be effective for the reduction of amine oxides are samarium diiodide, iron pentacarbonyl and acetic-formic anhydride (CH_3CO_2CHO).

(e) Alkylation

Alkylating agents such as dimethyl and diethyl sulphate and alkyl halides react at the oxygen centre in amine oxides to form alkyloxy quaternary salts. For example, ethyl bromide and triethylamine oxide give triethylethoxy ammonium bromide.

$$(C_2H_5)_3N-O + C_2H_5-Br \rightarrow [(C_2H_5)_3N-O-C_2H_5]^+ [Br]^-$$

(f) Acylation

Aliphatic amine oxides react with acyl halides and acid anhydrides but the reaction products are strongly dependent upon the solvent used and other experimental conditions. The reaction may also be catalysed by trace quantities of metal ions, which can affect the product mixture.

The nitrogen-containing reaction product is usually either an *N,N*-dialkylamide or a complex ammonium salt that may decompose to yield the parent amine plus an ester.

$$
\underset{\underset{R'}{|}}{\overset{\overset{O}{|}}{R-N-CH_3}} + (CH_3-CO)_2O \rightarrow CH_3-CO-NRR' +
$$

$$
[RR'HN-CH_2OCO-CH_3]^+[CH_3CO_2]^-
$$
$$
\downarrow
$$
$$
RR'NCH_3 + CO_2 + CH_3CO_2CH_3
$$

(g) With alkenes

Alkenes can react with specific amine oxides such as *N*-methyl morpholine oxide (NMMO) in the presence of osmium tetraoxide to form 1,2-diols.

$$
\overset{\backslash}{\underset{/}{C}}=\overset{/}{\underset{\backslash}{C}} + OsO_4/NMMO \rightarrow HO-\overset{|}{\underset{|}{C}}-\overset{|}{\underset{|}{C}}-OH
$$

In the presence of a strong base such as lithium diamide (LDA) in tetrahydrofuran (THF), amine oxides are converted into ylids, which can then react with alkenes to form heterocyclic compounds.

$$
(CH_3)_3N-O + LDA/THF \rightarrow
$$

This reaction can be used to prepare cyclic species such as octahydroindolizines and octahydroisoindoles from trimethylamine oxide and *N*-methyl piperidine oxide and cyclohexene, respectively [17].

7.3 Amines and amino acids

Although there are over 300 known amino acids, both natural and synthetic, only about 20 α-amino acids that are found in mammalian proteins and a few others that are found in the bodies of mammals (either bound or in the free state) are important to the chemistry of life. α-Amino acids consist of a carboxylic group and an amino group bound to the same carbon atom. The two other groups on the α-carbon are usually a hydrogen and an R (alkyl) group (a third functional group), which distinguishes one amino acid from another. The nature of the R group defines whether an amino acid is acidic, basic or neutral in character and also whether it is hydrophilic or hydrophobic. β-Amino acids have an amine group on the carbon atom one away from the α-carbon (e.g. β-alanine, which is $H_2N-CH_2-CH_2-CO_2H$); δ-, γ-, ε- and other amino acids are also known. At physiological pH most amino acids exist in the form of doubly charged (zwitterionic) ions with a positively charged amino group and a negatively charged carboxyl group.

$$R-\overset{\displaystyle H}{\underset{\displaystyle NH_2}{C}}-CO_2H \rightleftharpoons R-\overset{\displaystyle H}{\underset{\displaystyle NH_3^+}{C}}-CO_2^-$$

Many α-amino acids exist in two enantiomeric forms, (+) and (−), and nearly all of the amino acids used in animal and plant metabolism are present in the (−) configuration. Exceptions are non-protein amino acids such as (+)-glutamic acid found in bacterial cell walls, (+)-alanine found in the larvae of some insects, and (+)-serine, which is found in earthworms. Amino acids with (+)-configuration are used in 'antisense' pharmaceuticals and can block receptor sites in enzymes and so inhibit normal metabolic functions [18].

$$H \blacktriangleright \overset{\displaystyle CO_2^-}{\underset{\displaystyle R}{C}} \blacktriangleleft NH_3^+ \qquad H_3N^+ \blacktriangleright \overset{\displaystyle CO_2^-}{\underset{\displaystyle R}{C}} \blacktriangleleft H$$

(+)-configuration   (−)-configuration

A condensation reaction between the carboxyl group on one amino acid and the amino group on another amino acid leads to the formation of a single peptide bond and the elimination of water. The resulting dimer formed from two amino acids is called a dipeptide. The peptide bond has partial double-bond character (like amides) and is stabilized by electronic resonance [19].

$$NH_2-CH_2-CO_2H + NH_2-CH_2-CO_2H \rightarrow$$
$$NH_2-CH_2-CO-NH-CH_2-CO_2H + H_2O$$

Two other resonance hybrids are possible but they make little contribution to the stabilization of the peptide bond.

$$-N{=}\underset{|}{C}{-}O^- \underset{[H]^+}{\rightleftharpoons} -NH{-}\underset{|}{C^+}{-}O^-$$

Although amide groups have a pK_a for protonation of less than zero they do have a significant dipole moment of around 3.63 D. The direction of this dipole (which indicates a net movement of electrons towards the arrow head) favours the formation of *trans*-peptide linkages over *cis*-peptide linkages by around 8 kJ mol^{-1}.

Cis

R^1 ＼ ／ R^2
    C — N
O ∥ ＼ H

Trans

R^1 ＼ ／ H
    C — N
O ∥ ＼ R^2

＼N — θ C ∥ O
$\theta = 46.7°$

For the reaction above with two molecules of glycine there is only one possible outcome – glycyl-glycine (gly–gly) – but if two different amino acids are used, for example alanine and glycine, then two reaction products are possible, alanyl-glycine (ala–gly) or glycyl-alanine (gly–ala), depending upon which amino and carboxyl groups react. For this reason protecting agents such as Fmoc are widely used in peptide synthesis to block the reaction at designated amino sites and to control the reaction outcome. For example, the reaction between Fmoc-protected glycine with alanine can yield only Fmoc-gly–ala as the reaction product, which can then either be deprotected to give gly–ala or used in a further peptide-coupling reaction.

$$Fmoc{-}NH{-}CH_2{-}CO_2H + H_2N{-}CH(CH_3){-}CO_2H$$
$$\rightarrow Fmoc{-}NH{-}CH_2{-}CO{-}NH{-}CH(CH_3){-}CO_2H$$

The Fmoc strategy of peptide synthesis has been developed and reviewed by Atherton and Brown [19] and is the most widely used method of peptide synthesis today.

The chemistry of amino acids has been extensively reviewed elsewhere [20] and for this reason is not repeated here. Although it is possible to prepare amino acids in the laboratory by the Strecker Reaction [21] or by amination of α-halo propionic acids [22], most amino acids are produced commercially by bio-fermentation or by extraction from animal or vegetable sources. However, in addition to simple α-amino acids like glycine, there are five amino acids that are essential to mammalian life that contain an additional amine group or groups to the amine group on the α-carbon atom. These are shown below.

Arginine $HN-CH_2-CH_2-CH_2-CH-CO_2^-$
$\quad\quad\quad\quad |\quad\quad\quad\quad\quad\quad\quad\quad |$
$\quad\quad\quad\quad C=NH \quad\quad\quad\quad NH_3^+$
$\quad\quad\quad\quad |$
$\quad\quad\quad\quad NH_2$

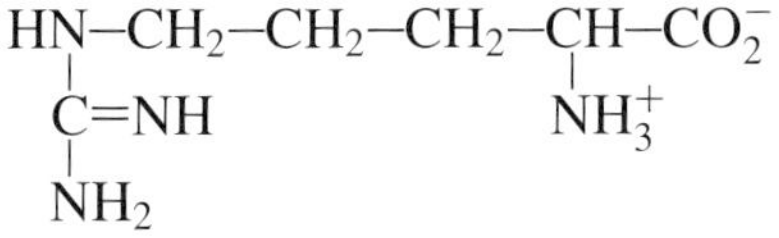

Ornithine $NH_2-CH_2-CH_2-CH_2-CH-CO_2^-$
$\quad\quad\quad\quad\quad\quad\quad\quad\quad\quad\quad\quad\quad |$
$\quad\quad\quad\quad\quad\quad\quad\quad\quad\quad\quad\quad\quad NH_3^+$

Lysine $H_2N-CH_2-CH_2-CH_2-CH_2-CH-CO_2^-$
$\quad\quad\quad\quad\quad\quad\quad\quad\quad\quad\quad\quad\quad\quad\quad\quad |$
$\quad\quad\quad\quad\quad\quad\quad\quad\quad\quad\quad\quad\quad\quad\quad\quad NH_3^+$

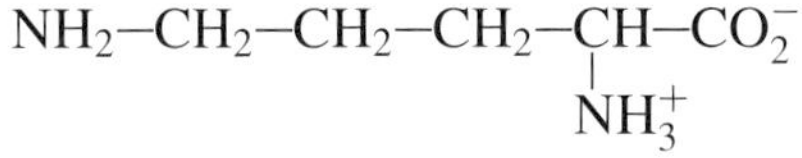

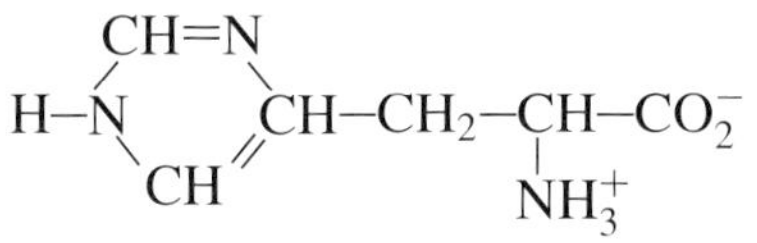

Histidine imidazole ring structure: $CH=N$ ring with $H-N$, CH, $CH-CH_2-CH-CO_2^-$ with NH_3^+

Tryptophan indole ring structure: $=CH_2-CH$ with CO_2^- and NH_3^+, ring nitrogen bonded to H

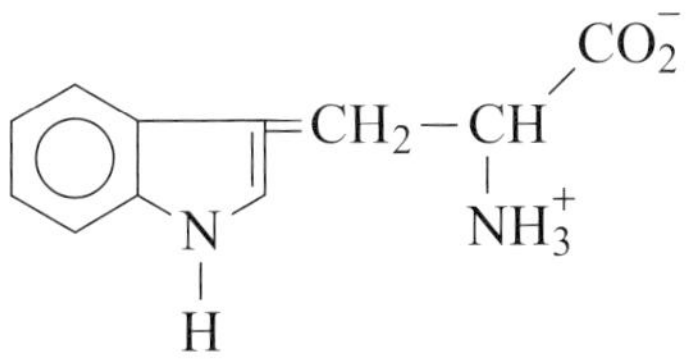

All of these amino acids with a secondary amino group are classified as basic amino acids; however, arginine, ornithine, lysine and histidine are hydrophilic and tryptophan is hydrophobic. Hydrophobic amino acids tend to be located inside protein chains and generally do not ionize or take part in hydrogen bonding. Hydrophilic amino acids can interact with aqueous environments and are usually found on the outside surfaces of proteins or in the reactive centres of enzymes.

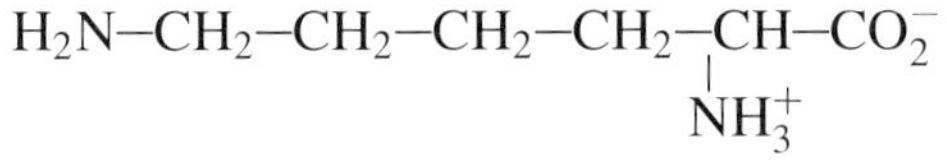

Amino acids with secondary amino groups (like histidine) have three pK_a values. (The reader may have noticed three nitrogen centres in the histidine molecule, but one of the ring nitrogens is already protonated so only one ring-nitrogen centre is able to accept a hydrogen ion under physiological conditions; the same is true for tryptophan where the benzindole nitrogen atom already has a bonded hydrogen atom.) This means that below pH1 the molecule is dipositive with two positively ionized nitrogens and a neutral carboxylic function. At around pH1.5 the carboxylic acid ionizes and the overall charge is +1. At around pH6 the amino group in the imidazole side chain loses its positive charge and the molecule has an overall charge of zero. At around pH9 the amino group on the α-carbon deionizes and the molecular overall charge is -1. Therefore the three pKa values are as follows.

$$pK_a\,1 = 1.82 \quad -CO_2^- + H^+ \rightarrow -CO_2H$$

$$pK_a\,2 = 6.00 \quad NH^+_{imidaz} \rightarrow N_{imidaz} + H^+$$

$$pK_a\,3 = 9.17 \quad NH^+_{3\alpha\text{-carbon}} \rightarrow NH_{2\alpha\text{-carbon}} + H^+$$

Of course, when histidine is present in a peptide chain the only group that titrates is the amino group on the imidazoline side chain, as the carboxyl and the amino groups on the α-carbon are present in covalent peptide linkages. With a pK_a of 6 the imidazole amino group in histidine is a very weak base and at pH7 may protonate to form the conjugate acid or may not, depending upon its environment. In fact the ring-nitrogen site of histidine is the only amino-acid site with a pK_a in the physiological range. A pK_a of 6 is roughly equivalent to a pH of 7. At pH7 about 10% of ring nitrogens in histidine will be protonated and at pH6 about 50% will be protonated. Because of the imidazole amino group, histidine is able to act as an effective buffer under physiological conditions. For example, the sp^2-hybridized ring-nitrogen atom in the imidazole group of histidine is able to form a π-bond to the Fe^{2+} centre in haem. Protonation of the amino group allows the transfer of negative electronic charge to the iron centre and assists in oxygen bonding.

Each molecule of haem actually contains two active histidine fragments: a proximal histidine group bonded to the iron centre through the nitrogen HisF-8, and a distal histidine fragment located on the other side of the bonded oxygen that is able to form a hydrogen bond to the oxygen via the nitrogen HisE-7 [23]. Amino groups in haem are also able to transport carbon dioxide in the blood stream by formation of carbamates.

$$Hb(O_2) + R{-}NH_2 + CO_2 \rightleftharpoons Hb{-}NH{-}CO_2H + O_2$$

Here $R{-}NH_2$ is an α-amino group present in the α- or β-chains of haemoglobin (Hb).

Histidine is able to complex with metals such as zinc and forms metal-bonding pairs with cysteine in biological systems. Histidine fragments are also responsible for the catalytic activity of some enzymes such as chymotrypsin, which can split peptides [24].

The terminal amino group in lysine has a pK_a of 10.5 and it is a strong base that is protonated even in neutral solution. The guanidinium side chain in arginine has a pK_a of over 12 and it is protonated under most conditions.

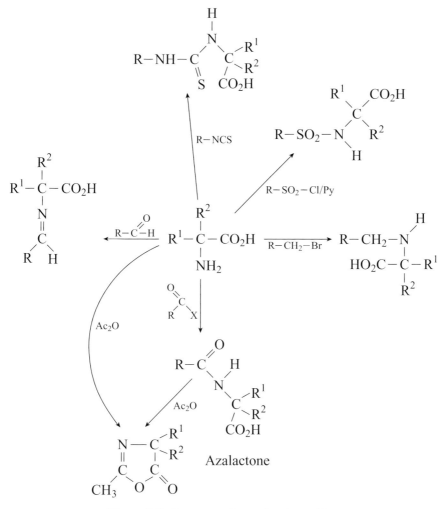

Figure 7.6. Some reactions of amino acids.

The side-chain amino group in ornithine has a pK_a of 10.76, and the indole amino group in tryptophan does not protonate under normal conditions but will accept a proton in the presence of a strong acid such as concentrated sulphuric acid. Unfortunately strong acids can also denature the amino acid so protonation of tryptophan is rare in physiological systems.

The α-amino groups in amino acids display many characteristic reactions of amines and can be alkylated, acetylated and will react with akyl sulphonates and thiocyanates. The alkylation of the amino group with alkyl bromides is difficult to control and a better route to *N*-alkyl amino-acid derivatives is via formation of a Schiff Base followed by reduction. A schematic representation of some reactions of amino acids is given in Figure 7.6.

However, the presence of the secondary amino groups of lysine, arginine, histidine, ornithine and tryptophan enables these amino acids to take part in some unusal chemistry in which other amino acids are unable to participate. The pendant amino groups on lysine and arginine are hydrophilic and stick out from the surface of proteins, forming sites for bonding with phosphate or carboxylate groups [25]. As plant and animal proteins contain on average 5.7% lysine and 5.7% arginine, there will be a relatively high number of pendant amino groups on a protein surface. Unfortunately in any protein most of the pendant amino groups will be protonated, but there will always be a finite fraction of the non-ionized form, available for reaction, that are potent nucleophiles.

A typical prosthetic group that attaches to proteins is pyridoxal phosphate, the active form of vitamin B6, which forms a Schiff Base with the pendant ε-amino group of lysine that is bonded to transaminase protein.

Lysinyl pyridoxamine phosphate

The ring nitrogen of the pyridine ring is strongly electron withdrawing and destabilizes the aldimine. When pyridoxamine phosphate reacts with α-amino acids at the amino group on the α-carbon, the lysinyl fragment is cleaved first to allow reaction and the resulting aldoxime is decarboxylated with the release of carbon dioxide. The reactions of amino acids with pyridoxamine phosphate are described in detail by Bender [26]. Deamination of amino acids is important because the reaction products are often neurotransmitters such as dopamine, adrenaline, histamine and γ-aminobutyric acid (GABA). Deamination also provides a mechanism for the synthesis of the polyamines described in Chapter 8 that are involved in the regulation of DNA metabolism. Pyridoxamine amine is also able to take place in transamination reactions and can generate amino acids from their corresponding 2-oxo-acids (e.g. histidine from imidazolepyruvate, serine from hydroxypyruvate, glycine from glyoxylate, etc.). The amino group of lysine may also undergo a coupling reaction with the glycosidic hydroxyl groups of sugars. This is called the

Maillard Reaction [27]. This reaction is responsible for the browning that occurs during cooking, such as the caramelization of sugar or the roasting of meat or vegetables. The Maillard Reaction was discovered by Louis Camille Maillard and occurs in three main stages.

In the first stage, an amino group reacts with a sugar to form *N*-glycoside. In the second stage of the reaction an immonium ion is formed that isomerizes to form a ketosamine; this is called the Amadori Rearrangement [28]. Once formed the ketosamine may then either dehydrate to form reductones and dehydroreductones (caramel), or form short-chain hydrolytic fission products that can undergo the Strecker Degradation to form Strecker aldehydes. In many cases there is a mixture of products and the mechanism is unclear.

(i) Formation of the *N*-glucosylamine

$$H-\underset{\displaystyle \underset{|}{H-C-OH}}{\overset{\displaystyle \overset{H}{\underset{|}{C}}\overset{\diagdown}{=}O}{\underset{|}{C}}}-OH + H_2N-R \rightarrow \underset{\displaystyle \underset{|}{H-C-OH}}{\overset{\displaystyle \overset{\displaystyle \overset{H}{|}}{H-C-N\overset{\nearrow H}{\diagdown R}}}{\underset{|}{C}}}-OH + H_2O$$

(ii) The Amadori Rearrangement to 1-amino-1-deoxyketose

$$-\underset{\displaystyle \underset{OH}{|}}{\overset{\displaystyle \overset{H}{|}}{C}}-\underset{\displaystyle \underset{NHR}{\|}}{C}-H \;\rightleftharpoons\; -\underset{\displaystyle \underset{O}{\|}}{\overset{\displaystyle \overset{H}{|}}{C}}-\underset{\displaystyle \underset{NHR}{|}}{C}-H \qquad [\text{(iii)} \rightarrow \text{various products}]$$

The Maillard Reaction is thought to be involved in many medical conditions, particularly diabetes, cataract formation and Alzheimer's disease. The Maillard Reaction can lead to protein cross linking, and when two proteins are linked *in vitro* it is known that their functions and properties change. Protein aggregation is certainly responsible for many diabetic complications and it remains a target of medical research to find out how to control and reverse this process.

The guanido group of arginine is strongly basic and is also planar because of resonance stabilization. It can be cleaved off by reaction with arginase *in vivo* and with hydrazine *in vitro* to yield ornithine (but note that the free carboxyl group and the α-amino group require protection).

$$-CH_2-NH-C\overset{\displaystyle \nearrow NH_2}{\underset{\displaystyle \searrow \overset{+}{NH_2}}{}} \xrightarrow{\text{(arginase/H}_2\text{O)}} -CH_2-NH_2 + (NH_2)_2CO$$

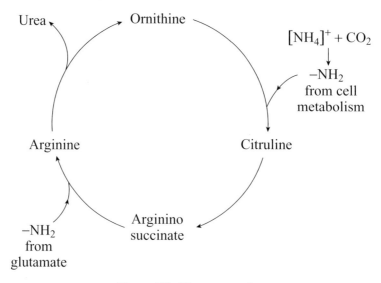

Figure 7.7. The urea cycle.

The guanidine group can also react with 1,2- and 1,3-dicarbonyls to form cyclic adducts, e.g. with cyclohexanedione.

One problem with amino acids is that they are not stored in animals and plants and any excess must be converted into fuel. The carbon-containing fragment can be easily converted into glucose or metabolized via the citric acid cycle. However, the excess nitrogen must also be disposed of as either ammonia (in fishes), uric acid (birds and reptiles) and urea (most vertebrates except those adapted to living in extremely arid conditions). Arginine and ornithine are used in the urea cycle (Figure 7.7) and enable the excretion of urea from living cells.

Amino acids such as arginine, histidine and tryptophan can also be decarboxylated to form biologically important amines such as agmatine (a precursor of spermine and spermidine), histamine (a vasodilator released into tissues as a result of allergic reaction or inflammation) and tryptamine (a precursor of indol-3-yl-acetate). Tryptophan is the least abundant amino acid in plant and animal proteins and is present only at levels of typically 1% to 1.5%. Corn is especially low in tryptophan. However, tryptophan is converted into a larger number of metabolites

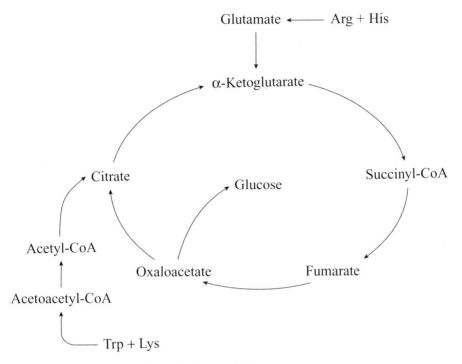

Figure 7.8. The citric-acid cycle.

than any other amino acid [29]. Some examples are shown below.

(i) Tryptophan → 3-hydroxyanthranilate → NAD^+
   ↓
   acetyl-CoA → citric-acid cycle

(ii) Tryptophan → 5-hydroxytryptamine → 5-hydroxyindol-3-yl-acetate

(iii) Tryptophan → tryptamine → indol-3-yl-acetate
   ↳   indol-3-yl-pyruvate ⬏

Tryptophan is present in such tiny quantities in proteins that its degradation is negligible to carbon metabolism, and it is far more important as a precursor to serotonin and melatonin. When tryptophan is metabolized it is converted to acetoacetyl-CoA and then to acetyl-CoA, which can enter the citric-acid cycle (Figure 7.8). The side chains of arginine and histidine (like proline) are degraded to glutamate, which can then either be converted to α-ketoglutarate by transamination with aspartate aminotransferase or oxidation with glutamate dehydrogenase, or used in the biosynthesis of non-essential amino acids as a source of nitrogen by transamination.

$$[NH_4]^+ + \alpha\text{-ketoglutarate} + NADPH + [H]^+ \rightleftharpoons \text{glutamate} + NADP^+ + H_2O$$

$$\text{Glutamate} + [NH_4]^+ + ATP \rightleftharpoons \text{glutamine} + ADP + P_i$$

Here NADPH is nicotinamide adenine dinucleotide phosphate, ATP is adenosine triphosphate, ADP is adenine diphosphate and P_i is inorganic phosphate.

Lysine, like tryptophan, is degraded to acetoacetyl-CoA, which enters the citric-acid cycle.

Overall, the nitrogen content of amino acids is ultimately transferred to urea [30]. This may be represented by three reactions.

(i) α-Amino acid + α-ketoglutarate → glutamate + α-keto acid
(ii) Glutamate → α-ketoglutarate + ammonia
(iii) Ammonia + carbon dioxide → urea

The first reaction is catalysed by the enzymes called aminotransferases. The second reaction is catalysed by glutamate dehydrogenase and the third reaction is catalysed by *N*-acetyl glutamate, with which ammonia and carbon dioxide react to form carbamoyl phosphate, which reacts with ornithine to enter the urea cycle.

Of course once nitrogenous waste has left the bodies of animals and humans the story does not end there. Animal dung has been ploughed back into the soil for centuries as fertilizer for crops and will probably continue to be used in this way in the future as long as there are animals working on farms. Dung is also still used in some parts of the world for fuel.

Before the discovery of guano in the nineteenth century, urine also had a long history of use. The problem facing our ancestors was that there are few naturally occurring sources of ammonia. When urine leaves the bladder it is sterile but on coming into contact with air ammonia is formed in small quantities. For this purpose, the longer urine is exposed to air the better: stale urine has a higher ammonia content than fresh urine (which must be left to mature before it can be used). Male urine was collected by the Romans in major towns in special pots that were left outside buildings at night, and it was used as a bleaching agent for wool in laundries. This is how the Romans used to keep their togas white for special occasions. The Romans also used urine mixed with pigeon droppings and wine as a hair bleach. In Elizabethan England, puppy urine was also widely used as a hair bleach (it was used by Queen Elizabeth I). By the nineteenth century the collection of urine had turned into a major industry in England and about 10 000 tonnes of urine per year where collected nationwide in all major towns and cities and used as a source of ammonia to make synthetic alum from grey shale for use in the textile industry. The development in the late nineteenth century of new dyes that did not require alum as a mordant put an end to this industry and the collection and utilization of urine.

References

[1] M. Bergman and L. Zervas, *Ber.*, **65** (1932), 1192.
[2] L. A. Carpino and G. Y. Han, *J. Am. Chem. Soc.*, **92** (1970), 5748.

[3] D. S. Tarbell, Y. Yamamoto and B. M. Pope, *Proc. Natl Acad. Sci. USA*, **69** (1972), 730.

[4] K. Fischer, *The global market for protecting agents*, CPhI Conference, Paris 2002 (London: Manufacturing Chemist, 2003).

[5] B. Helfreich, *Angew. Chemie*, **41** (1928), 871.

[6] S. A. Lawrence, *Chemical Specialities* (July/August 2000), 60.

[7] S. A. Lawrence, *Chimica Oggi* (May/June 1999), 15.

[8] E. Ponnusamy, U. Fotador, A. Spisni and D. Fiat, *Synthesis* (1986), 48.

[9] H. Franzen and U. Ragnarsson, *Angew. Chemie Int. Ed.*, **23** (1984), 296.

[10] S. A. Lawrence, *Chimica Oggi* (May 2000), 20.

[11] S. A. Lawrence, *Pharmachim.*, **1**: 1/2 (2002), 32.

[12] A. Nowak, *Cosmetic Preparations* (London: Micelle Press, 1991).

[13] W. J. Priestly, *Res. Disc.*, **164** (1977), 75.

[14] M. N. Sheng and J. G. Zajacke, *J. Org. Chem.*, **33** (1968), 588.

[15] A. C. Cope, *Org. Synth. Coll.*, **IV** (1963), 612.

[16] J. Meissenheimer, *Ber.*, **59** (1926), 1848.

[17] J. Chastanet and G. Roussi, *J. Org. Chem.*, **50** (1985), 2910.

[18] M. D. Matteucci and N. Bischofberger, *A. Rep. Med. Chem.*, **26** (1991), 287.

[19] E. Atherton and R. C. Sheppard, *Solid Phase Peptide Synthesis: A Practical Approach* (Oxford: Oxford University Press, 1989).

[20] G. C. Barrett and D. T. Elmore, *Amino Acids and Peptides* (Cambridge: Cambridge University Press, 1998).

[21] A. Strecker, *Ann.*, **75** (1850), 27.

[22] J. M. Orton and R. M. Hill, *Org. Synth. Coll. Vol.*, **1** (1932), 300. (See also B. De, J. F. De Barnardis and R. Prasad, *Synth. Comm.*, **8** (1988), 481.)

[23] M. F. Perutz, A. J. Wilkinson, M. Paoli and G. G. Dodson, *Ann. Rev. Biophys. Biomol. Struct.*, **27** (1998), 1.

[24] R. M. Stroud, *Scient. American*, **231** (1974), 24.

[25] T. Imoto and H. Yamada in *Protein Structure: A Practical Approach*, ed. T. E. Creighton (Oxford: IRL Press, 1989), p. 247.

[26] D. A. Bender, *Nutritional Chemistry of the Vitamins* (Cambridge: Cambridge University Press, 1992).

[27] R. Ikan (ed.), *The Maillard Reaction* (Chichester: J. Wiley, 1996).

[28] M. Amadori, *Atti. Accad. Lincei*, **2** (1925), 337.

[29] H. Sidransky, *Tryptophan: Biochemical and Health Implications* (Florida: CRC Press, 2002).

[30] M. L. Batshaw, *Ann. Neurol.*, **35** (1994), 133.

8

Selected commercial applications of amines

8.1 Amines as corrosion inhibitors

Corrosion is the deterioration and loss of material due to chemical attack. The most common type of corrosion involves the electrochemical process of metal oxidation, which is the removal of electrons from a metal. Electrochemical oxidation can occur in metals at grain boundaries (see Figure 8.1), at points of stress and also at localized impurities, which have higher potentials than the surrounding metal and form the anodes of microgalvanic cells. Corrosion occurs at these anodic sites of higher potential and deposition of the oxidized metals occurs at the cathodic surrounding metal surfaces, which are connected to each other by the presence of an electrolyte solution.

The supply of corrosion inhibitors in the USA alone was a business worth about USD1.1 billion in the year 2000. The annual cost of corrosion worldwide is currently thought to be around USD13 billion, indicating that the market for corrosion inhibitors is still far from saturated.

Corrosion inhibitors find widespread application in the petrochemical industry where they are used to inhibit the corrosion of mild-steel pipelines and on drilling rigs (the cost of stainless steel would be too high so it is more cost effective to use mild steel with corrosion inhibitors), and also in the construction industry where they are used to prevent corrosion in ferroconcrete. Other applications include their uses in paints, in engineering tools and equipment and also in the chemical-manufacturing industry.

To understand the changes that occur when a corrosion inhibitor is added to a metal surface exposed to an electrolyte, it is necessary to understand the factors that determine the potential difference at the metal/electrolyte interface. At this interface an electrical double layer is formed. This is shown schematically in Figure 8.2(a).

Region A is formed by adsorbed anions and their associated water molecules. The loci of these ions make up the inner Helmholtz plane; this is formed when

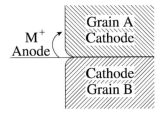

Figure 8.1. A schematic representation of grain-boundary corrosion.

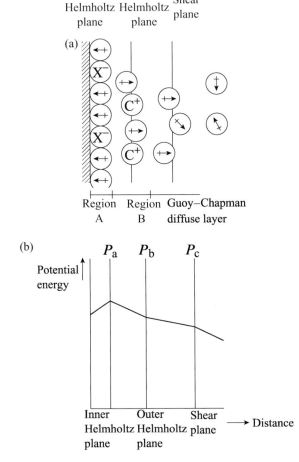

Figure 8.2. The metal/electrolyte interface.

adsorbed anions displace water molecules that were already present on the metal surface. Region B is formed by cationic counter ions which balance the charges of the adsorbed anions. The loci of these cations is called the outer Helmholtz plane. Outside this region is the Guoy–Chapman diffuse layer where the concentration of anions and cations slides back towards that in the bulk electrolyte. The anions on the metal surface are called the potential-determining ions, and their concentration is controlled by the nature of the specific chemical interactions between the electrolyte and the metal surface.

Regions A and B may be regarded as a parallel-plate capacitor across which there is a potential difference; this potential difference cannot be measured directly but can be measured against a standard hydrogen electrode. In Figure 8.2(b) the Stern potential is shown at P_b and the zeta potential is at P_c. When an amine corrosion inhibitor A is introduced to the electrolyte it adsorbs at the metal/electrolyte interface and displaces some of the adsorbed water. This ion exchange can be written as shown below; the rate constant for adsorption of the amine is K_a and the rate constant for the desorption is K_b.

$$\text{Metal}_{(n\text{H}_2\text{O})} + \text{A}_{(\text{solution})} \rightleftharpoons \text{Metal}_{(n\text{A})} + n\text{H}_2\text{O}_{(\text{solution})}$$

The corrosion inhibitor has the effect of ordering the dipoles of the water molecules in the Helmholtz planes and so affecting the dielectic constant of these layers. The dielectric constant in water is about 80; in the outer Helmholtz layer it is 40 and in the inner Helmholtz layer it can be as low as 6. With amine corrosion inhibitors the presence of a nitrogen lone pair assists adsorption, as the lone pair can be donated to the metal surface according to classical Lewis Acid/Base theory. Because amines contain nitrogen they are considered to be hard bases and so adsorb best to metal surfaces that are considered to be hard acids. Iron and steel are particularly well suited to being protected with amine-based corrosion inhibitors in acidic media. Generally amines with higher electron densities on their nitrogen atoms bind more strongly to metal surfaces and are more effective corrosion inhibitors. Cyclic amines are often better corrosion inhibitors than acyclic amines as ring strain leaves the nitrogen lone pair more exposed for donation. Also the presence of electron-donating groups on substituent aromatic rings helps to increase the electron density at the nitrogen atom and improve the corrosion-inhibition properties. Cyclic imines with sp^2-hybridized nitrogen and C–N–C bond angles often perform best of all at corrosion inhibition.

The presence of an adsorbed amine layer on the metal surface creates a zone of high (ohmic) resistance between the metal surface and the electrolyte. This resistance slows down or stops galvanic reactions and the loss of electrons from the the metal surface that leads to corrosion.

Hydrophobic alkyl groups are also important in defining corrosion inhibition as they form a physical barrier preventing non-bonded, mobile water molcules from gaining access to the metal surface. For example, diamines with bridging methylene chains ($\text{H}_2\text{N}-(\text{CH}_2)_n-\text{NH}_2$) work best as corrosion inhibitors when n is in the region of 3 to 8. Longer chain lengths are thought to interfere with the corrosion-inhibition process because instead of both amines groups bonding to the metal surface one group often sticks out into the solution away from the surface and disrupts the favourable arrangement of the water dipoles.

For N-alkyl ammonium the corrosion effectiveness increases with increasing alkyl chain length up to maximum efficiency at about C_{10} to C_{12}.

Two effective corrosion inhibitors are Rose Amine D and dehydroabietyl amine, which are used as pure amines and also as their Mannich Bases (formed with aldehydes such as formaldehyde and acetaldehyde and ketones such as cyclohexanone, acetone and acetophone).

Rose Amine D   Dehydroabietyl amine

The aromatic rings in these two amines are thought to form π-bonds with metal surfaces.

Hexamethylene tetramine is also used as a corrosion inhibitor, but it can react with hydrochloric acid to form chloromethyl ether and bis(chloromethyl) ether, both of which are potent carcinogens, so this restricts its application. Corrosion inhibitors for use at high temperatures often contain acetylenic side chains; one example is the propargyl group, which forms protective polymer films with iron surfaces at high temperatures.

Plots of the fractional surface covered against solution concentration resemble classical Langmuir adsoption isotherms. The fractional surface S covered by adsorption is related to the concentration C of the adsorbed species by the relationship

$$S = KC/(1 + KC)$$

where $K = K_a/K_b$ (as previously defined). Therefore an infinite concentration of a corrosion inhibitor is required to cover completely the total surface area of a metal in contact with an electrolyte. However, it is not necessary to have complete coverage of the total metal/electrolyte surface, although the corrosion rate does generally fall with the concentration of inhibitor in solution. For example, for an $N-C_{10}$ to C_{12} akyl isoquinolinium halide, an application dosage of 1×10^{-4} to 1×10^{-5} mol l^{-1} is sufficient for protecting steel from acidic corrosion. Amine corrosion-inhibiting agents with π-electrons (such as quinolinium and isoquinolinium salts) are able to exert strong ordering effects on the surrounding water molecules, and the presence of a relatively few molecules at the metal surface is sufficient to cause a big change in the dielectric constant of the Helmholtz layer.

Unfortunately, corrosion inhibitors at best delay the corrosion process and will not stop it altogether, because corrosion is an inescapable consequence of the Second Law of Thermodynamics (in an isolated system, spontaneous processes occur in the direction of increasing entropy). In practical terms this means that there must be an economic balance between the cost of corrosion inhibitor used and the cost of replacing the object to be protected, which is why simple alkyl amines are sometimes used for many applications instead of more expensive amines with complicated structures. For a review of the subject of corrosion inhibition see reference [1].

8.2 Amines in pharmaceuticals

There are very few drugs that are purely acyclic compounds. In most instances the acyclic part of the drug structure is not of comparable importance to the functionality attached to it (some notable exceptions are inhalation anaesthetics). Overall, the majority of pharmaceuticals are composed of an acyclic pendant side chain attached to a cyclic or heterocyclic moiety. The side chain often has a terminal carboxylic acid or amine group. Some examples are these side chains and the pharmaceuticals that contain them are shown below.

Tamoxifen and Meclofenoxat $-CH_2-CH_2-N(CH_3)_2$

Promethiazine and Methadone $-CH_2-CH_2-CH_2-N(CH_3)_2$

Amitriptyline and Promazine $CH_3-CH-CH_2-N(CH_3)_2$
$\qquad\qquad\qquad\qquad\qquad\qquad\quad |$

Trimipramine and Dimethothiazine $-CH_2-CH-CH_2-N(CH_3)_2$
$\qquad\qquad\qquad\qquad\qquad\qquad\qquad\qquad |$
$\qquad\qquad\qquad\qquad\qquad\qquad\qquad\quad CH_3$

Chloroquine and Phenglutarimide $-CH_2-CH_2-N(CH_2-CH_3)_2$

The *N,N*-dimethylethylamino group is found as a side chain in 24 marketed drugs covering virtually all therapeutic applications [2]. Some important applications include use in antibiotics, antithelmintics, CNS drugs, anti-fungals and in neoplastic chemotherapy. Some advantages of this and other types of amine-based side chain are as follows.

(a) They are relatively easy to attach to the substrate compound with a readily available precursor.
(b) They are reasonably stereorigid (long alkyl side chains may get in the way of the receptor site and block access).
(c) They are stable under physiological conditions.

(d) They are resistant to self-alkylation.
(e) They do not interact with excipients and other active substances in the finished pharmaceutical dosage form.
(f) They are resistant to metabolism (except at the target receptor site).

One of the keys to developing a successful pharmaceutical is that it should be able to reach the required target receptors in the body. Unfortunately for drug researchers, these receptors are usually hidden inside protective biological membranes, which are specifically designed to keep out invading foreign molecules. Once pharmaceuticals enter the body (either orally, through the skin or by injection) they are trespassing on transport mechanisms that have evolved to serve other functions. For example, many biological membranes are polarized with one side being a positively charged lipid surface and the other a negatively charged lipid surface, in between which is sandwiched a layer of water-filled pores. In order to cross this type of membrane successfully many drugs have been designed to be weak electrolytes and are only partially dissociated in solution. In general the undissociated molecule is lipid soluble and the dissociated molecule is water soluble.

The rate of dissociation of an electrolyte is proportional to its concentration (or in the case of drugs to their activities). If the rate of dissociation of a weak acid HA is $k_1[HA]$, with k_1 as the rate constant and $[HA]$ the concentration of the acid, then at equilibrium we have the following situation.

$$[HA] \rightleftharpoons [H^+] + [A^-]$$
$$k_1[HA] = k_2[H^+][A^-]$$

This can be rewritten as follows.

$$[H^+] = k_1[HA]/k_2[A^-]$$

Now if we replace k_1/k_2 with a single rate constant K_a then the above equation may be simply rearranged.

$$1/[H^+] = [A^-]/K_a[HA]$$

If logarithims are taken on both sides then this equation becomes the following.

$$pH = pK_a + \log([A^-]/[HA])$$

This is the Henderson–Hasselbach Equation and it defines the degree of dissociation of a weak acid and the pH of the solution in which it is dissolved [3]. The equation was first formulated in 1908 by Henderson, who established the relationship between hydrogen ion concentration and the ratio of unionized to ionized acid. In 1917, Hasselbach was able to adapt this equation to include the concepts pH and pK_a, which were introduced only in 1909. The resulting equation is now always referred to as the Henderson–Hasselbach Equation. For bases, the Henderson–Hasselbach Equation can be rewritten for a base **B** where

$$\mathbf{B} + H^+ \rightleftharpoons \mathbf{BH}^+$$

$$pH = pK_a + \log([\mathbf{B}]/[\mathbf{BH}^+])$$

and K_a is k_1/k_2, where k_1 refers to the reaction that supplies hydrogen ions to the system and k_2 is the reaction that removes hydrogen ions from the system.

Drugs with oral-dosage forms enter the blood plasma via the gastrointestinal tract upon being swallowed. The stomach has a pH of around 1 to 2 but the pH of the small intestine is much higher (it is less acidic). In order to become transported to the target site in the body (assuming that this is not a gastric drug) it is first necessary for the drug to pass through cell membranes in the wall of the gastrointestinal tract and to enter the blood plasma, which has a pH of 7.2. With intravenous administration this problem is negated, as the drugs are injected directly into the blood stream or into a body cavity at the point of application, such as the bladder. One of the goals of pharmaceutical research is to produce oral-dosage forms of injectable drugs as this opens up a wider sales market (this issue is particularly acute for injectable antibiotics, where it is often hoped that a small side-chain modification to the drug structure could produce an oral-dosage form).

For an orally administered drug like diazepam with a pK_a of 3.3 at pH1, the log ratio of the unionized base to the ionized base will be given by the $pH-pK_a$. Therefore in stomach acid (pH $= 1.2$) this ratio is $= 1.2 - 3.3 = -2.1$ and in plasma (pH $= 7.2$) the ratio is $7.2 - 3.3 = 3.9$.

Remembering that this is a log ratio, the ratio of ionized to unionized base is as follows:

$$\text{in stomach acid, } \log - 2.1 = 0.008;$$

$$\text{and in blood plasma, } \log 3.9 = 7943.$$

We can say that, in stomach acid, we have

$$100 - \left(\frac{0.008 \times 100}{1 + 0.008} \right)$$

and in blood plasma we have

$$100 - \left(\frac{7943 \times 100}{1 + 7943} \right).$$

Therefore in the stomach the drug diazepam is 99.2% ionized and in blood plasma it is only 0.013% ionized. So there is not very much unionized diazepam available to enter blood plasma from the stomach.

However, in the duodenum the pH is around 6, and we can say that $6 - 3.3 = 2.7$ and $\log 2.7 = 501$, so diazepam is 99.8% unionized. In the jejunum, where the pH is 7.5, the percentage of unionized diazepam becomes ($7.5 - 3.3 = 4.2$, and $\log 4.2 = 15\,848$) and so is 99.99% unionized. Therefore for a drug like diazepam absorption into blood occurs mainly in the small intestine where it is present in its unionized form.

The rate of passage with time of a drug across a cell membrane of constant thickness can be expressed by Fick's Law [4], as follows.

$$T = -PA(C_a - C_b)$$

Here, T = rate of passage per unit time, P = permeability constant of the membrane, A = surface area of the membrane, C_a = concentration of the diffusing material on the side of the membrane with the higher concentration, and C_b = concentration of the diffusing material on the side of the membrane with the lower concentration.

For most drugs, transport across the gastrointestinal tract relies on diffusion and so can be defined by Fick's Law. Exceptions to this include iron, amino acids and drugs related to specific nutrients that cross the intestinal wall via facilitated diffusion making use of the body's own specific transport systems.

From Fick's Law the most important factor is A, the available surface area, which is much greater in the small intestine (duodenum, jejunum and ileum) than elsewhere in the digestive tract (the available surface area of the small intestine in most adult humans is around 250 m^2). The relative alkalinity of the small intestine compared with the stomach favours the absorption of drugs with basic functions. However, it should be noted that the high surface area also allows acidic drugs to enter the blood plasma, although they may also start to do this this earlier in the digestive process through the stomach wall. Therefore the use of pendant amines can be demonstrated to help facilitate transport from the digestive tract into blood plasma compared with other chemical groups.

Once in the blood plasma, the drug molecules can be transported around the body and can enter the cell tissue fluid to gain access to the targeted receptors. However, one exception to this is the passage of drugs into the brain as they first have to cross the blood–brain barrier. The blood–brain barrier is present in all vertebrates and is formed early in foetal development. The tubular blood vessels in the brain are made up of endothelial cells with tight intracellular junctions. These tight junctions eliminate the fenestrae that are found in other blood vessels in the body and allow solutes to pass freely.

The blood–brain barrier is actually composed of two membranes (the lumenal and ablumenal membranes), which are separated by a 300 nm layer of endothelial cytoplasm. In order to penetrate the brain–blood barrier, drug molecules should either be

(a) less than 700 in molecular weight and lipid soluble, or

(b) amenable to catalysed transport with carrier-mediated processes (for example the processes that enable the brain to take up nutrients such as glucose, choline and amino acids, etc.).

Many natural biochemicals that are able to cross the blood–brain barrier contain pendant alkyl amine chains; these include serotonin, adrenaline, nicotine, pyridoxine, morphine, quinine, mescaline and choline. The structures of these naturally occurring biochemicals are related to the structure of phenylethylamine $C_6H_5-CH_2-CH_2-NH_2$, which has an ethylamine attached to an aromatic nucleus and seems to be able to help facilitate transport across the blood–brain barrier. Another widely used class of psychotropic drugs is based on piperidone derivatives, which have been examined in Chapter 2.

The total worldwide market for CNS drugs, based on 1998 figures, is approximately USD40 billion and antidepressants are worth USD13 billion alone (in 2002 the total worldwide market for pharmaceuticals was GBP290 billion) but could be much higher as it is difficult to account for the use of patent-infringing drugs in countries that do not respect intellectual property (this figure does not include the use of illegal drugs or nicotine from tobacco). It is interesting to look at the structures of some successful, market-leading CNS drugs that have annual sales values of over USD200 million because they all contain amine side chains (see below).

Prozac (Fluoxetine)   Seroxat (Paroxetene)

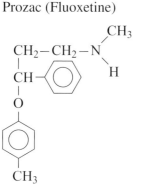

Diazepam (Valium)   Citalopram

Of course transport is irrelevant if the target receptor is not engaged by the pharmaceutical molecule being carried by the blood plasma and there are many receptors within the body and brain that bind to amines.

All receptors are thought to have evolved from five ancient families of receptors that are over 800 million years old. At the time of writing (in 2003) there are so many receptors being discovered in humans and animals that many reviews of receptors end up being out of date within six months of publication. However, despite recent advances, most psychotropic drugs bind to five types of receptor that have been known for decades (seratonin, dopamine, noradrenaline, acetylcholine and γ-aminobutyric acid). The current philosophy on drug action at receptors, proposed by Mautner as early as 1967 [5], is that bioactive chemicals and pharmaceuticals contain two specialized regions, which can be distinguished. One part of the molecule enables the drug to lock onto the receptor site, which has been identified for many drugs to be the amine-containing fragment. The other portion of the molecule acts by inducing a configurational change in the receptor and determines the drug's bioactivity by initiating a pharmacological repsonse.

Therefore it can be seen that, in many instances, the presence of an amine side chain serves two functions in medicine: by assisting pharmaceuticals to be transported to the target site in the body, and then to lock onto the desired receptor. The total quantity of amine-based pharmaceuticals produced worldwide is very small compared with the total ammonia production, because of the high relative cost of pharmaceuticals. For example, many patented drugs cost over GBP1 million per kilogram and even low-cost generic pharmaceuticals produced in India and the Far East still cost over GBP20 per kilogram. However, the potential for adding increased value to basic chemicals by converting them into active pharmaceuticals means that pharmaceutical companies, and the sales revenues they generate, are important to the worldwide economy. In Chapter 1 a figure of 5–7 million tonnes per year was given as the total quantity of ammonia used in the production of amines. With an approximate sales price of ammonia of GBP200 per tonne, this represents a sales revenue of about GBP900 million. This figure is dwarfed by the total global market for pharmaceuticals, which is estimated to be around GBP290 billion with 11% annual growth. Many of these pharmaceuticals that are important in defining our quality of life have humble beginnings and are synthesized from ingredients like air, water and natural gas and would not be available today without the production of ammonia from the Haber–Bosch Process.

8.3 Amines as anti-cancer agents and DNA alkylators

A malignant neoplasm (a new growth) is a proliferation of cells that is no longer under the control of the organism that harbours it. The word cancer is widely used to describe this type of malignant tumour. Cancer is Latin for crab, and it suggests the slow but inevitable spread of neoplasms as they claw their way to other parts of the body and, if not stopped, can result in the death of the patient. Surgery and X-ray treatment are often the primary course of action but chemotherapy can completely cure certain types of cancer (chorioepithelioma and Burkittlymphoma are two examples, and more recently testicular cancer has been successfully treated), and is also effective as an auxiliary treatment used in tandem with surgery and X-rays to help destroy any residual cancer cells that may have been left behind by the primary treatment.

It is in this final chapter of this book that the name of Fritz Haber appears for a second time. Fritz Haber's contribution to ammonia synthesis was discussed in Chapter 1 but, in addition to this work, he also developed the concept of chemical warfare, which was used for the first time in 1915 during the First World War, and he also helped to develop some of the chemicals used for warfare [6]. The first warfare gas used was chlorine but this was replaced by the more effective (and more toxic) sulphur mustard gas in Ypres on 22 April 1917. Exposure to mustard gas causes severe blistering of the skin, serious damage to the eyes and necrosis of the respiratory tracts. Soldiers who were not killed by the primary effects of mustard gas often died from the delayed effects such as vomiting, diarrhoea, loss of appetite and bradycardia, all of which weakened the victim who then became prone to secondary infections in the unsanitary conditions of the trenches. Although the gas was first used by the German army against the Allies in Flanders, its effects were short lived and not decisive in determining the outcome of the war as the Allies quickly learned to take evasive actions including the banning of wearing kilts for the Scottish regiments (to reduce skin exposure) and the use of gas masks. However, on the Eastern Front, the gas caused a greater loss of life because the Russian army did not have the resources to manufacture and distribute gas masks to their soldiers fighting on the front line.

In addition to sulphur mustard gas, a range of nitrogen mustard gases was also developed during the First World War. These had similar effects, but were not as toxic.

Nitrogen mustard

$$R-N \begin{matrix} CH_2-CH_2-Cl \\ CH_2-CH_2-Cl \end{matrix}$$

Sulphur mustard

$$S \begin{matrix} CH_2-CH_2-Cl \\ CH_2-CH_2-Cl \end{matrix}$$

(mustine R=CH_3, normustine R=H)

During the Second World War, a joint American–British effort to determine the effect of mustard gases on cell growth found that their effect was similar to that of X-rays [7]. From this early lead the effects of mustard gases on tumours were studied in mice in 1946, leading to the eventual development of nitrogen-mustard-based anti-tumour agents that were used as neoplastic drugs. The first anti-neoplastic drug developed was mechloroethamine, also known as mustine [8]. When nitrogen-mustard alkylating agents come into contact with aqueous media they form the cyclic ethyleneimmonium ion, which can react with the negatively charged (nucleophilic) chemical centres in tissue proteins. All nitrogen mustards are 2-haloethylamines and the presence of at least two haloethyl groups is essential for cytotoxic activity. The N-alkyl (R) function does not seem to exert any influence on the cytotoxic activity, but it does seem to be important in assisting drug transport in the body to the desired site of action and in limiting side effects.

With free DNA and RNA, akylation occurs predominantly at the N-7 position of the guanine base, although minor alkylation also occurs at guanine O-6 and N-3, at adenine N-1, N-3 and N-7, and at cytosine N-3. The guanine N-7 site is 70 times more likely to be selectively alkylated than the other two sites. Although DNA and RNA are able to repair minor damage, nitrogen mustards are able to cross link two deoxyguanylate nucleotides and so effectively prevent cell replication.

However, the problem with nitrogen mustards is that in addition to alkyating the DNA in cancerous cells, normal healthy cells are also targeted; in addition, the side effects of their application are very severe, and include myelosuppression, skin pigmentation, hair loss, bone pain, vomiting and loss of appetite. Treatments are usually given in clusters with intervals of two or three days to allow the patient to regain some strength to continue the course and for the physician to assess the severity of the toxic effects.

Some other nitrogen mustards that have been developed include melphalan, ifosamine, busulfan, cyclophosphamide, mannomustine, mitoclomine, chlorambucil and thiotepa. They are used for different types of cancer, as shown in Table 8.1.

Table 8.1. *Nitrogen-mustard-based alkylating agents*

Drug name	Therapeutic use	Approximate dose
Mechlorethamine	Hodgkin's Disease Chronic lymphatic leukaemia Polycrythaemia Rubra Vera	250 μg kg^{-1} for 4 days by IV
Uramustine	Malignant lymphomas	50–70 μg kg^{-1} for 3 weeks by mouth
Mannomustine	Hodgkin's Disease Chronic lymphatic leukaemia Plueral tumours	100 mg daily
Chlorambucil	Ovarian and mammary carcinoma	4–8 mg m^{-2} daily
Cyclophosphamide	Burkitt's Tumour Ovarian and lung tumours	30–40 mg kg^{-1} for 3 weeks by IV or 2 mg kg^{-1} by mouth
Melphalan	Multiple myelomata Macroglobulinaemia	1 mg kg^{-1} for 3–7 days
Mitoclomine	Lymphatic leukaemia	100 mg daily by mouth

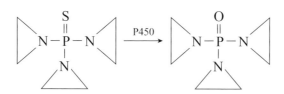

Figure 8.3. Thiotepa is converted to tepa by liver cytochrome P450.

However, the most widely used akylating agent today is thiotepa, which instead of containing two 2-chloroethyl groups contains three ethyleneimine rings. Thiotepa is administered by intravenous injection and is carried by red blood cells to the liver where it is converted to tepa, which is able to alkylate DNA via azidirinidum intermediates (see Figure 8.3).

Although thiotepa is an old drug, it has been 'rediscovered' by the medical community in recent years. The main problem with the use of thiotepa has always been myelosuppression (it attacks bone marrow). However, recent developments in stem-cell technology have led to the technique of removing a fraction of a patient's (or donor's) bone marrow prior to treatment. This can then be replaced after treatment or, in some instances, if a suitable donor can be found the bone marrow can be transplanted from one patient to another [9]. The worldwide production of thiotepa is less than 500 kg (about 100 kg are used per year in the USA) but this figure seems to be growing annually.

It is interesting that aziridine usually undergoes instant polymerization in water unless stabilized with alkali, but thiotepa is stable in blood plasma at pH7.2 for up to 72 hours and is able to cross the blood-brain barrier.

8.4 Amines in the cosmetics and toiletries industries

Unlike anionic and amphoteric surfactants, cationic surfactants are very rarely used for cleaning applications. This is because most surfaces carry negative surface charges onto which the cationic surfactants will become rapidly adsorbed and depleted from solution. Because of this surface adsorption, amine-containing surfactants possess excellent cleaning properties and surface-conditioning properties and are widely used as emollients for skin, shampoos, conditioning agents and dyestuffs for hair, as fabric softeners, as foam baths, as sanitizing agents and as complexing agents. The current worldwide market for cationic and amphoteric surfactants is around USD3.3 million per year (about 2 million tonnes). The main usage breaks down roughly as shown below.

$$
\begin{aligned}
\text{Fabric softeners} &= 23\% \\
\text{Personal-care products} &= 19\% \\
\text{Textile manufacturing} &= 12\% \\
\text{Dishwasher detergents} &= 10\%
\end{aligned}
$$

Other applications for cationic and amphoteric surfactants are as asphalt emulsions for road building, corrosion inhibitors and in organo clays used in paints, drilling muds, resins and surface coatings.

The single most-important family of fatty alkyl quaternary ammonium salts are the cetyl ammonium chlorides. Cetyl trimethylammonium chloride (CTMAC) has excellent water solubility with light to moderate conditioning ability. Mono fatty alkyl ammonium salts can be made either (i) by reaction of a fatty alcohol (e.g. cetyl alcohol) with dimethylamine and then with methyl chloride to form the ammonium salt, or (ii) from the reaction of fatty acids with ammonia followed by reduction and then secondary reaction with either formaldehyde and formic acid or methyl chloride.

$$\text{(i) } R\text{-}CH_2\text{-}OH + H\text{-}N(CH_3)_2 \rightarrow R\text{-}CH_2\text{-}N(CH_3)_2 + H_2O$$
$$\downarrow {\scriptstyle CH_3Cl}$$
$$R\text{-}CH_2\text{-}N(CH_3)_3Cl$$

$$\text{(ii) } R\text{-}CO_2H + NH_3 \qquad \rightarrow R\text{-}C\equiv N + 2H_2O$$
$$\downarrow {\scriptstyle CH_2/catalyst}$$
$$R\text{-}CH_2\text{-}NH_2$$
$$\downarrow {\scriptstyle 2H_2C=O/HCO_2OH}$$
$$R\text{-}CH_2\text{-}N(CH_3)_3Cl$$

Varying the reaction stoichiometry and adding excess fatty alcohols and acids to the reaction mixtures allows di- and trisubstituted quaternary salts to be manufactured. The di-fatty-alkyl quaternaries are stronger conditioning agents but because of the additional alkyl chain they are less soluble in water although they can form stable dispersions. Dicetyl dimethyl ammonium chloride is one example, but the best known is based on hydrogenated tallow and is widely used in fabric softeners. Hydrogenated tallow octyl diammonium chloride has the following structure.

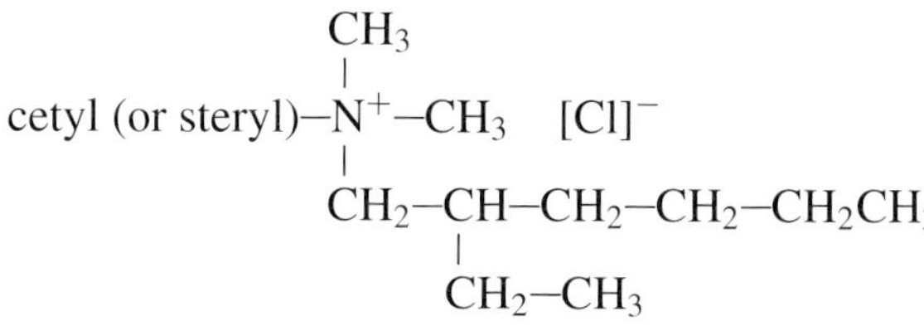

$$
\begin{array}{c}
CH_3 \\
| \\
\text{cetyl (or steryl)}-N^+-CH_3 \quad [Cl]^- \\
| \\
CH_2-CH-CH_2-CH_2-CH_2CH_3 \\
| \\
CH_2-CH_3
\end{array}
$$

The tri-fatty-alkyl quaternary salts are superior conditioners and exhibit the best hair-detangling properties and static control. They are used when intensive hair care is required, for example for split-end control. Although not water soluble, tricetylammonium chloride forms stable solutions in anionic surfactants and so can be successfully formulated into hair-care products [10].

Benzyl quaternary salts formed by reaction of alkyl dimethyl amines and benzyl chloride have good anti-microbial properties. This property is related to the chain length, and products with carbon chain lengths of C_{12} to C_{14} have the most effective anti-microbial activity [11]. Another benzyl quaternary salt, stearyl benzyl ammonium chloride, is widely used in cosmetics [12].

Ester quaternaries are widely used as fabric conditioners. The ester group aids their biodegradability and they are not considered to be environmentally harmful. In the early 1990s, dihydrogenated tallow dimethylammonium chlorides were widely used as fabric softeners, but there was widespread concern in Europe when these amines were found in lake sediments and were found to be non-biodegradable, and so their use was phased out [13]. The ester linkage in the molecular structure is shown below.

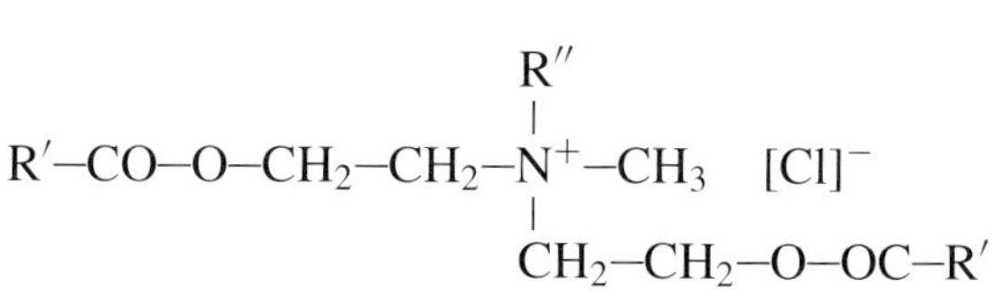

$$
\begin{array}{c}
R'' \\
| \\
R'-CO-O-CH_2-CH_2-N^+-CH_3 \quad [Cl]^- \\
| \\
CH_2-CH_2-O-OC-R'
\end{array}
$$

However, in order to prevent hydrolysis, ester quaternaries have to be kept at slightly acid conditions. This limits their applications in cosmetics, which have near-neutral pHs so that they do not cause adverse effects when in contact with skin.

Ethoxylated quaternary amines ('ethoquats') are also used as conditioners for hair and textiles and because of the presence of hydrophilic, ethoxylate chains they are soluble in anionic surfactants.

$$3CH_3Cl + NH_2-(O-CH_2-CH_2)_n-OH \rightarrow (CH_3)_3N-(O-CH_2-CH_2)_n-OH$$

Another class of amine-based surfactants that are widely used as fabric conditioners are imidazole derivatives. The imidazoline ring contains two nitrogen atoms and was first synthesized in 1888 by Hofmann from an acid and a diamine with the elimination of water (see Chapter 4).

Commercially manufactured imidazoles are prepared from fatty acid ethyl esters and di- and polyamines such as ethylene diamine, diethylene triamine and triethylene tetramine.

However, imidazoles are very susceptible to hydrolysis, so the quaternary ammonium salt is usually employed because the alkylation of the second nitrogen leads to ring stabilization, improved stability to hydrolysis and increased water solubility.

Dimethyl sulphate is preferred as the methylation reagent as the rate of methylation is between 10 and 100 times faster than with methyl chloride [14]. A synthetic method for the methylation of amines with dimethyl sulphate is given in Chapter 6 (Example four).

In addition to amine-based cationic surfactants, many amphoteric surfactants can also contain amine groups. Amphoteric surfactants are widely used in combination with anionic surfactants as they form 'milder' formulations that are less irritating to skin. The structure of a typical amphoteric surfactant is shown below.

$$CH_3-(CH_2)_{11}-CH_2-\overset{\overset{\displaystyle CH_3}{|}}{\underset{\underset{\displaystyle CH_3}{|}}{N^+}}-CH_2-CO_2^-$$

Because they have two functional groups, amphoteric surfactants behave as cationic surfactants at low pH and as anionics at high pH. At intermediate pH they carry both negative and positive charges. Some amphoteric surfactants called betaines retain both positive and negative charges over a wide pH range.

Alkyl betaines can be prepared by a condensation reaction between a fatty dimethyl amine and sodium chloroacetate. In addition to modifying the skin irritability of an anionic surfactant, the presence of small quantities of betaines in shampoo formulations provides thickening and also improves the foam structure. In conditioning shampoos betaines can enhance the effect of quaternary ammonium salts used to detangle hair. Commonly used betaines include lauryl and stearyl betaines. Betaines take their name from the trimethylglycine found in sugar beets. However, alkylamido betaines are more widely used in cosmetic and toiletry formulations than alkyl betaines because they are less irritating than alkyl betaines and also because they are cheaper. Alkylamido betaines are usually manufactured by condensing fatty acids with dimethylaminopropylamine (DMAPA), followed by a secondary reaction with sodium chloroacetate. The general structure of amidopropyl betaines is shown below.

$$R-CO-NH-CH_2-CH_2-CH_2-\overset{\overset{\displaystyle CH_3}{|}}{\underset{\underset{\displaystyle CH_3}{|}}{N^+}}-CH_2-CO-O^-$$

The most commonly used amido betaine is cocamidopropyl betaine (CAPB), where R = lauryl in the above structure. It has good detergency and foaming properties, and has good compatability with skin. CAPB is widely used in shampoos, foam baths and even dishwashing liquids. If the carboxyl group is replaced with a sulphonate group, the resulting surfactant is called a sulphobetaine. Sulphobetaines are used in soaps and prevent limescale dispersion in hard water.

If the nitrogen group is replaced by a phosphorous group very little difference is observed in the properties of the resulting betaine; however, if a change is made to the alkyl chain by adding or removing a methylene group more pronounced changes in the properties of the betaine will be observed. For example, the number of monomers that make up a micelle can change dramatically when adding or removing $-CH_2-$, but is hardly affected when a nitrogen is replaced with a phosphorus. This indicates that it is the alkyl chain that is important in influencing the properties of the betaines. Betaines find wide application in Europe, but in North America glycinates and propionates are more common although they form a less-stable foam in hard water.

If the group $R-CO-NH-CH_2-CH_2- = Y$, then the structures of glycinates and propionate surfactants can be represented as follows.

$$
\begin{array}{ll}
\text{Acylamphoglycinate} & Y-N-CH_2-COONa \\
& \quad\quad\quad\; | \\
& \quad\quad CH_2-CH_2-OH \\[4pt]
\text{Acylamphopropionate} & Y-N-CH_2-CH_2-COONa \\
& \quad\quad\quad\; | \\
& \quad\quad CH_2-CH_2-OH
\end{array}
$$

The acylamphopropionate surfactants have an additional methylene group. If the terminal hydroxy group of acylamphoglycinates is replaced by the $-CH_2-CO_2Na$ group from sodium chloroacetate, then they are converted into acylamphocarboxy-glycinates. Similarly, acylamphopropionates react with acrylic acid followed by sodium hydroxide which converts the OH group to $-CH_2-CH_2-CO_2Na$ to form acylamphocarboxy propionates.

The most important products in these surfactant groups are the cocoderivatives such as cocoamphoglycinate, cocoamphocarboxyglycinate, cocoamphopropionate and cocoamphocarboxypropionate.

When these classes of surfactants were first developed it was widely thought that they contained imidazole rings, but it is now known that the imidazole ring hydrolyses and that all these surfactants are purely acyclic.

The final type of surfactant used in cosmetic and toiletry applications is the amine oxides. Unlike betaines, the amine oxides do not display cationic behaviour and are either anionic or non-ionic depending on the pH. Amine oxides are manufactured by the action of hydrogen peroxide on tertiary alkyl amines. They are usually colourless liquids and are excellent foamers. However, amine oxides are also good hair conditioners and so are widely used with anionic surfactants in conditioner-shampoos.

There are two conventions used when naming amine oxides. For example, the amine oxide below is called dodecyldimethylamine-*N*-oxide (i.e. nitrogen substituents in alphabetical order) or dimethyldodecylamine-*N*-oxide (i.e. substituents in order of increasing size).

$$
\begin{array}{c}
CH_3 \\
| \\
C_{12}H_{25}-N \rightarrow O \\
| \\
CH_3
\end{array}
$$

Amine oxides have found application in pharmaceuticals and agrochemicals and also are extensively used in detergents and cosmetics. They were first used in the 1960s in household detergents for dishwashing at about 2.5 weight % as a replacement for fatty alkanolamides, because they gave better foam and produced a smooth rather than oily feeling on the skin. Amine oxides were also then used in shampoos up to around 7 weight %. Typically, dimethylcocamine-N-oxide and dimethyldodecylamine-N-oxide are used. Amine oxides with higher alkyl chains are used in industrial applications as phase-transfer catalysts and in the production of acrylonitrile polymers and polyesters [15].

Other uses of amine-N-oxides are as corrosion inhibitors, as spinning bath additives for rayon and polyamides and as antistatic agents for textiles [16].

Apart from their use as surfactants, amine derivatives are also widely used as hair dyes. Hair keratin is a protein and contains about 25 amino acids in the form of a copolymer. Keratin differs from many other proteins in that it contains a high elemental percentage (about 20%) of sulphur. There is also an excess of carboxylic groups over amino groups in the copolymer side chains; this has the net effect of making the surface of the hair slightly negatively charged.

This property is exploited in the formulation of hair dyes, which are of a cationic character. Hair dyes can be grouped into two categories: firstly temporary and semi-permanent dyes, which can be removed by washing with shampoo; and permanent dyes, which enter the hair shaft and last for the lifetime of the hair until it breaks or drops out of the scalp. With a temporary dye a soluble colour is used. This does not penetrate the hair shaft and is kept in place with a setting solution. Usually the dye is mixed with a cationic surfactant to help it substant onto the hair. Semi-permanent dyes are based on cationic dyestuffs that have a natural affinity for hair and are formulated into slightly alkaline solutions with surfactants such as cocoamidopropyl betaine and cetyl trimethyl ammonium chloride. Triethanolamine is usually added to bring the pH to around 8.5.

With permanent dyes the colouration is formed during the dyeing process (i.e. it has not been preformed) and three products are used – a dye intermediate, an oxidizing agent and a coupler. In the dyeing process the dye intermediate penetrates the hair shaft and then is oxidized by hydrogen peroxide to form a receptor for a coupling agent, which produces the hair colouration. The high-molecular-weight dyestuffs produced in the hair shafts after reaction with the coupler are unable to leave the hair shaft and are fixed in place, so providing permanent coloration.

The dye intermediates used are usually aromatic with electron-donating groups in 1,2 or 1,4 positions. The most commonly used molecules for this application are 1,4-phenylenediamine, 4-toluylene diamine and 4-aminophenol. Reaction with hydrogen peroxide then proceeds to form quinonimines.

$$\text{(structure with NH}_2 \text{ groups)} + H_2O_2 \rightarrow \text{(structure with NH groups)} + 2H_2O$$

$$\text{(structure with NH}_2 \text{ and OH groups)} + H_2O_2 \rightarrow \text{(structure with NH and O groups)} + 2H_2O$$

   Coupling can also take place outside the hair shaft but hopefully any dye formed outside the hair will be washed away when the hair is rinsed. The coupling takes place in alkaline media and pH control is usually effected by using ammonia or triethylamine. Generally this reaction is not sensitive to overprocessing but it is sensitive to underprocessing, which is why strange colours are sometimes obtained if insufficient reaction times are used.

   Surfactants are used in the hair-dye solutions to ensure an even distribution of the dye precursor and coupler on the hair and to provide good wetting properties. Ammonium oleate is often added to dye intermediates formulated in shampoos to ensure an even hair coverage during the first stage of the dyeing process [17].

8.5 The applications of amines in colour reprography

Chemicals exhibit colour when they absorb electromagnetic radiation in the visible-wavelength range (400–700 nm), resulting in excitation of an electron from its ground-state orbital into a higher-state orbital. The transitions of interest for organic chemicals are $n \rightarrow \pi$ and $\pi \rightarrow \pi^*$, where π denotes the pi-orbitals and n denotes the non-bonding orbitals on, for example, an oxygen or nitrogen atom. The excitation energy, E, is related to the absorption wavelength, λ, by the equation below, where h is Planck's constant, c is the velocity of light and k is a constant.

$$\lambda E = hc = k$$

For organic molecules with conjugated double bonds, increasing the conjugation lowers the excitation energy and increases the absorption wavelength. The addition of an electron-withdrawing group (e.g. a carbonyl group or an azo group) and an electron-donating group (e.g. an amine) at opposite ends of a conjugated system act to increase the degree of delocalization, which can assist the formation of colouration in organic molecules to produce a specific shade. The fact that amine

CI Pigment Yellow 13

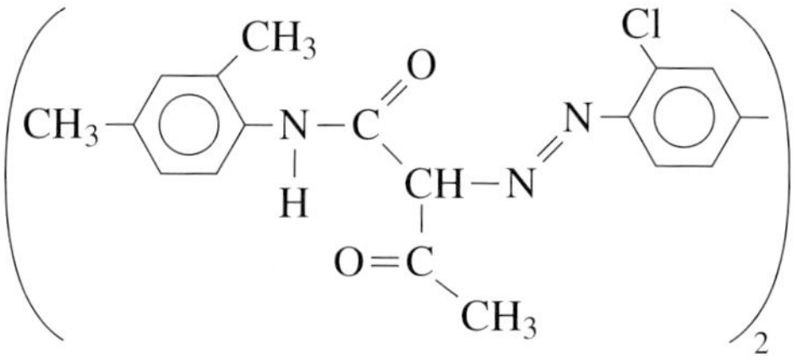

CI Pigment Blue 15.3

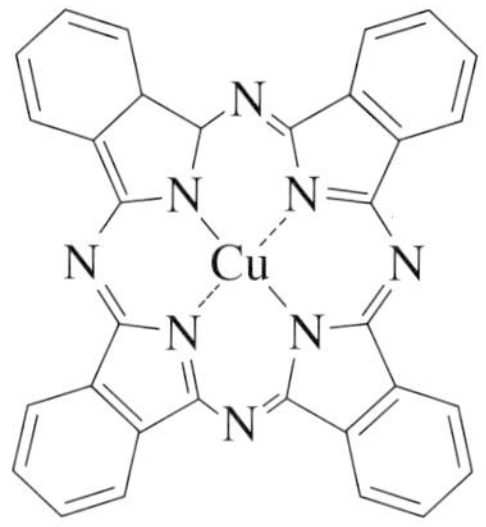

CI Pigment Red 184

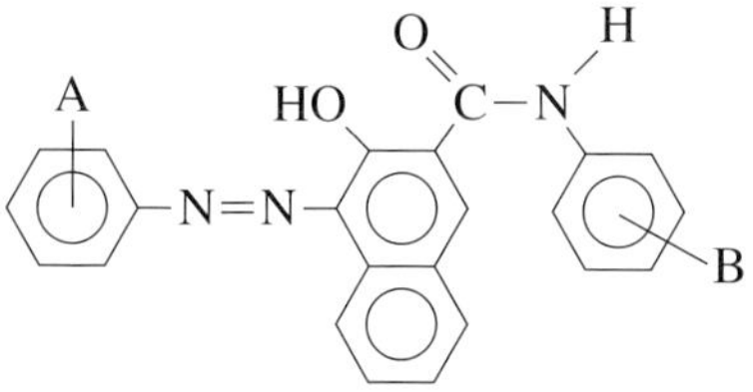

Figure 8.4. The structures of some colour dyes used in ink-jet printing (A and B are substituent groups).

and azo groups are both useful in modifying colouration explains their widespread use in reprography. Because the human eye does not respond to continous visible radiation and has only three receptors for violet to blue (400–500 nm), blue to green (450–610 nm) and green to red (500–700 nm), three optical signals are all that are required to provide colour perception. Hence the object of reprography is to provide dyestuffs in the three additive primaries (blue, green and red) by using the three subtractive primary dyes (cyan, magenta and yellow) to produce colour images. The first coloured dyes for use on paper were merely textile dyes and they were not very successful as they were of low purity and did not aggregate well on acidic paper surfaces. These initial problems were overcome by purifying the dyes and also by replacing the sulphonic acid groups with carboxylic groups, thereby making the dyes less water soluble. Some typical dyes used for ink-jet colour printing are shown in Figure 8.4.

Black colours are usually produced by using carbon black, which is superior to earlier black dyes such as the copper complex CI Reactive Black 31 and CI Food Black 2. Although phthalocyanine was actually discovered in 1907 [18] and copper phthalocyanine was discovered in 1927 [19], the development of phthalocyanine dyes was initiated by a fortuitous accident at an ICI subsidiary in Scotland (Scottish Dyes Ltd) in 1928 during the manufacture of phthalimide from the action of ammonia on molten phthalic anhydride in iron reactors, when a blue impurity was noticed in some batches. The potential of metal–phthalocyanine complexes as dyestuffs was immediately recognized and the first copper phthalocyanine blue dyes were marketed in the 1930s as Monastral Fast Blue BS and Monastral Fast G. Copper phthalocyanines are thermally and chemically stable and can be sublimed at 580 °C and also dissolved in sulphuric acid without reaction.

Halogen substitution in the phthalocyanine ring produces blue-green colourations, and a green dye called Pigment Green 7 is obtained from 15 chlorine substitutions into the ring.

Phthalocyanines can be prepared from the reaction between phthalodinitrile and metals, or from phthalic anhydride, urea and metal salts, or from *ortho*-cyanobenzamide with metals or metal salts. The reactions may be carried out in a solvent but also work well in solventless systems. Phthalocyanines are used for the dyeing of textiles, paper and leather and can be formulated into printing inks, paints and plastic/resin based surface coatings.

Amines also have application in colour photography. The photographic process consists of three actions that may be combined in some instances (such as instant photography) but that nevertheless form three distinct chemical processes. These are:

  (i) exposure or irradiation,
 (ii) development (or amplification) of the image, and
(iii) fixation of the image.

The process of irradiation in conventional photography (i.e. not digital) is almost exclusively the exposure of silver halides to a light source, which causes the silver and halide to dissociate leaving behind an opaque film of colloidal silver.

$$AgX + light \rightarrow Ag\bullet + X\bullet$$
$$(2Ag\bullet \rightarrow Ag-Ag \text{ and } 2Br\bullet \rightarrow Br_2)$$

However, silver halides such as silver bromide do not absorb visible radiation but instead absorb and are dissociated by UV light. In order to produce a visible image, most photographic films contain chemical colour developers and image couplers to convert the silver emulsion into images that can be recognized by receptors in the human eye.

In colour photography, three dyes are used to produce red, green and blue colours. However, in the film negative the colour must be complementary to the colour required in the final print (for example, a magenta dye in the film negative will produce a green reflection from the finished print). A typical colour film consists of nine or ten layers in total, which are given in the correct order below.

OBJECT

↓

CAMERA LENS

↓

[Film overcoat]
[UV filter]
[Blue-sensitized AgX + yellow coupler]
[Blue light filter]
[Green-sensitized AgX + magenta coupler]
[Interlayer]
[Red-sensitized AgX + cyan coupler]
[Light filter to reduce scatter]
[Cellulose acetate or other transparent support]
[Magnetic backing (if APS film)]

After irradiation of the film, two processes take place in the film-development operation. Firstly, a substituted paraphenylenediamine is added, which reacts with any silver-atom clusters to form an oxidized intermediate with a quaternary cationic nitrogen. This intermediate then reacts with the colour coupler to form a leuco dye, which undergoes an intramolecular rearrangement to form a stable colour dye. The leuco dye is actually colourless but is still referred to as a dye although it is actually a 'pro-dye'. Schematically this process may be represented as follows.

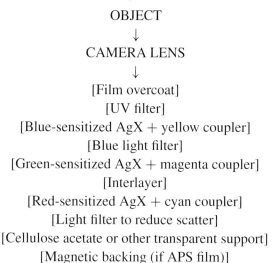

   Cyan couplers are typically substituted naphthalenes, yellow couplers are keto-carboxamides, and magenta couplers are based on the pyrazolone ring and contain amine functions. The structure of a typical magenta coupler based on a substituted pyrazolotriazole is shown below.

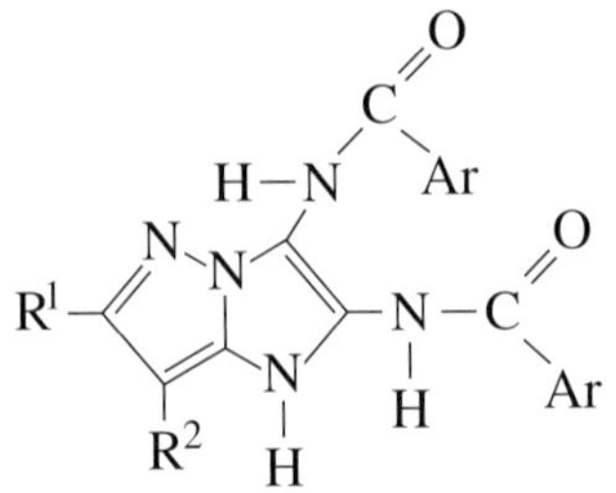

The chemistry of colour reprography has been reviewed in reference [19].

8.6 The applications of hydrazines

Approximately 100 000 tonnes of hydrazine and hydrazine derivatives are produced every year for use in defence applications, primarily as a propellant fuel for missiles and rockets. This figure accounts for less than 10% of the current total worldwide production of hydrazine and is falling from a Cold War high of about 25% of the total worldwide hydrazine production. The important uses for hydrazine and hydrazine derivatives are as follows.

Agrochemicals	$= 40\%$	(or about 500 000 tonnes)
Blowing agents	$= 33\%$	(or about 400 000 tonnes)
Water treatment	$= 15\%$	(or about 190 000 tonnes)
Rocket fuel	$= 8\%$	(or about 100 000 tonnes)
Other (including pharma)	$= 4\%$	(or about 50 000 tonnes)

   Hydrazine is particularly well suited to application in rocket fuels as it is readily decomposed to nitrogen gas by catalysts or oxidizing agents. The ejection of the gases from the decomposition of hydrazine provides the propulsion system for rockets and missiles. The main applications of hydrazine in rocket propulsion are as two hydrazine derivatives, monomethylhydrazine (MMH) and unsymmetrical dimethyl hydrazine (UDMH). They are used as hypergolic fuels, where the fuel and oxidant ignite spontaneously. Hydrazine can be used as a monopropellant and works well with an iridium/alumina catalyst, but for most applications dual propellants are preferred and MMH and UDMH are used with an oxidant such as nitric acid or (more commonly) dinitrogen tetroxide, N_2O_4 [20]. NASA uses MMH and dinitrogen tetroxide in the Space Shuttle, whereas UDMH is preferred by the European Space Agency and by the Russian and Chinese space programs.

In the past NASA used a 50/50 mixture of MMH and USMH called Aerozine in its Titan rockets, but has now switched to MMH-based systems. Hypergolic fuels do not require cryogenic handling; they are liquids and can be pumped into the reactor chamber of a missile or rocket. Unfortunately they are extremely toxic and require special handling. The effectiveness of a hypergolic fuel is usually expressed as its specific impulse, which is the change in momentum per unit mass consumed (i.e. better fuels produce a higher specific impulse), and is given by the following expression.

$$I = u/g$$

Here I = specific impulse, u = total impulse/mass of expelled propellant and g = 9.8 m s^{-2}.

Hydrazine as a monofuel produces I = 230–240 s; MMH/N_2O_4, UDMH/N_2O_4 and Aerozine/N_2O_4 are all very similar and produce I = 290–315 s.

Hydrazine is also used in the production of blowing agents, the most important of which is azodicarbonamide. This decomposes in air at 195–200 °C and in plastics at 160–200 °C. The gas yield from the decompostion of azodicarbonamide is 220 ml g^{-1}, and is mainly carbon monoxide (32% vol.) and nitrogen (65% vol.). The solid residue from the decomposition of azodicarbonamide is 41% urazole and biurea (which reacts to form urazole).

$$H_2N–CO–N=N–CO–NH_2$$
Azodicarbonamide

↓

$$H_2N–CO–NH–NH–CO–NH_2 \rightarrow$$
Biurea

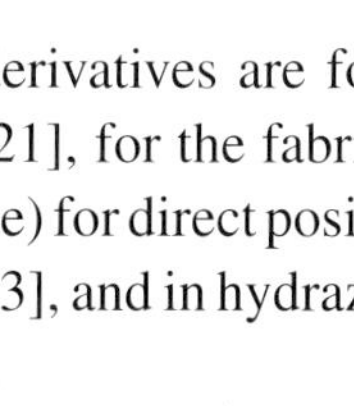

Urazole

Some other applications of hydrazine and its derivatives are for electrodeless plating and the reprocessing of nuclear fuel rods [21], for the fabrication of 'fast' radiographic films (hydrazinium chloride and oxalate) for direct positive instant processing [22], in the fabrication of semiconductors [23], and in hydrazine/hydroxide-based fuel cells (the reaction is shown below) [24].

$$N_2H_4 + 4OH^- \rightarrow N_2 + 4H_2O + 4e^- \quad [E = 1.56 \text{ V}]$$

Hydrazine was used in the past as an additive to water in boiler systems, but has now been replaced by carbohydrazide (morpholine is also used for this application). The addition of hydrazine or hydrazides prevents metal corrosion at low pH and is even effective in the presence of oxygen [25].

However, about 40% of the world's total production of hydrazine and hydrazides is used in the manufacture of agrochemicals. The total worldwide production of herbicides is around 450 000 tonnes per year, of which about 27% contain a heterocyclic nitrogen atom. The three most important hydrazine-containing intermediates for the production of agrochemicals are semicarbazide, thiosemicarbazide and aminoguanine, and the most important hydrazine-containing herbicides are Metribuzin (a grass killer used in soybean farms) and Spike (a selective weedkiller).

Metribuzin   Spike

Maleic hydrazide, which is manufactured from hydrazine and maleic acid (see below), is widely used as a plant and crop growth retardant.

$$HO_2C-CH{=}CH-CO_2H + H_2N-NH_2 \rightarrow$$

However, the use of hydrazines in agrochemicals is small compared with the total useage of amines in agrochemicals. This will be examined in the next section.

8.7 The applications of amines in agrochemicals

Chemical analysis of dried green-plant matter taken from many plant species gives the following averaged elemental contents.

Element	g per kg of plant material
Oxygen	440
Carbon	420
Hydrogen	60
Nitrogen	30
Potassium	20
Phosphorus	4
Others (total)	26

It should be noted that as 80% to 90% of a plant's total weight is water, this elemental balance refers only to the dried matter, which is a fifth to a tenth of the plant's actual weight. Therefore in 100 g of green plant there will be on average 45 mg of nitrogen. In a plant nitrogen may be absorbed by the roots in the form of $[NH_4]^+$ and $[NO_3]^-$ and by the leaves as NH_3 and nitrogen oxides. Given that the metabolic pathways exist in plants for the uptake of nitrogen it is not surprising that so many agrochemicals either contain nitrogen or are amines or amine derivatives.

The worldwide market for agrochemicals is worth around USD33 billion per year (the UK annual market is only about USD1.5 billion), and agrochemicals can be divided into four main classifications depending upon their application. These are herbicides, fungicides, pesticides (primarily insecticides, rodenticides and molluscicides), and other applications such as growth retardents, adjuvants, bird repellants, etc., which are grouped together. Some agrochemicals are pure amines, such as the herbicides based on pyridine, triazine or phenylene diamine, but often an amide-based or urea-based agrochemical is used that is metabolized to an amine. Studies of several agrochemicals often reveal that a common amine is formed during their metabolism. For example, the acylaminofungicides Benalaxyl, Metalaxyl and Furalaxyl (shown in Figure 8.5), all metabolize to 2,6-dimethyl aniline.

This type of metabolic degradation is common to many agrochemicals. When the end product is an amine it is thought that the resulting bioactivity results from the formation of glucose-ester conjugates. For example, Metribuzin (which was discussed in the previous section) is thought to form a conjugate of the type shown below.

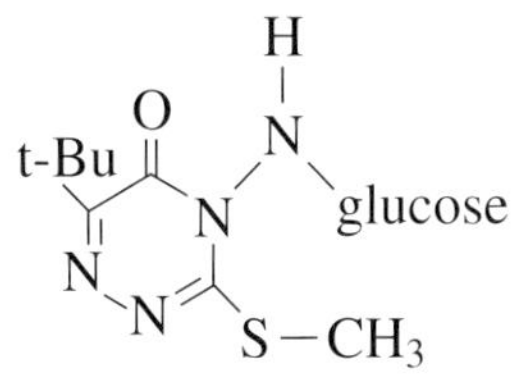

Maleic hydrazide is also thought to form a glucose conjugate, but this may be formed at either a nitrogen or oxygen centre.

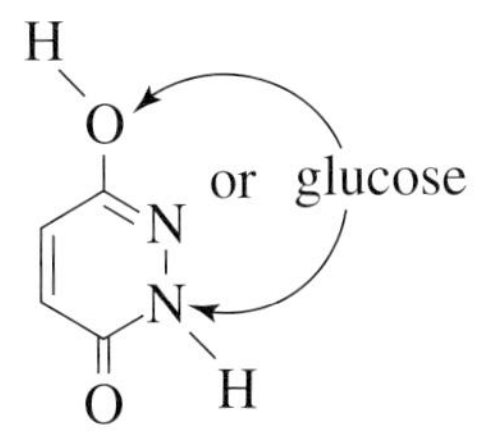

Guanidino- and guanine-containing fragments are also common in agrochemicals and the molecular fragment shown below is found in many agrochemicals and also

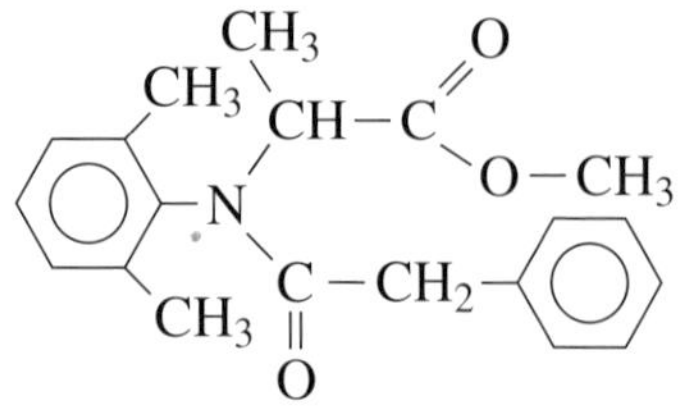

Benalaxyl

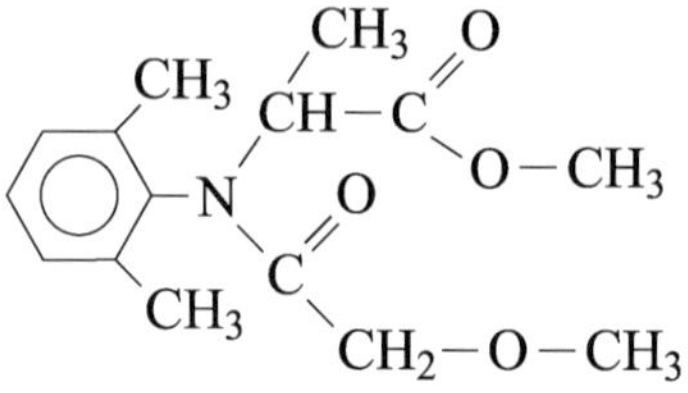

Metalaxyl

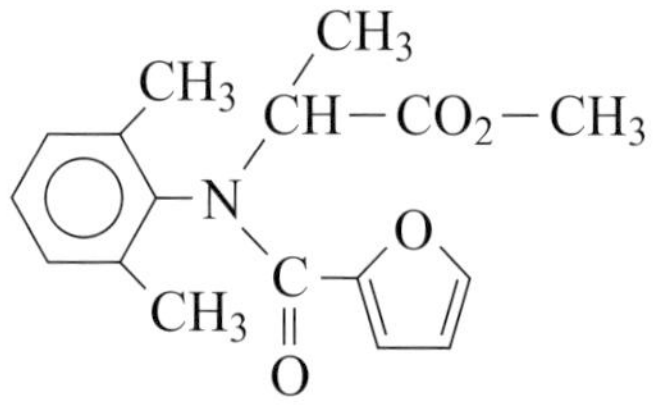

Furalaxyl

Figure 8.5. The structures of Benalaxyl, Metalaxyl and Furalaxyl.

in their metabolized by-products.

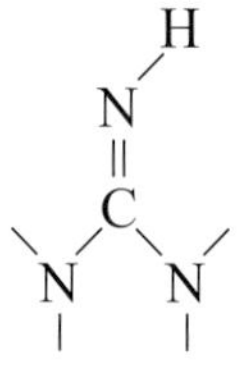

The specific actions of some fungicides, herbicides and pesticides will now be examined.

8.7.1 Fungicides

Agrochemical fungicides containing amines fall into approximately four main types:

(a) aliphatic nitrogen-containing fungicides such as butylamine, Cymoxanil, Dodicin, Guazatine and Iminoctadine;

(b) Heterocyclic nitrogen-containing fungicides such as Ethoxyquin and Quinoxyphen (which contain quinoline), Bosalid, Pyroxychlor and Pyridinitril (which contain pyridine) and Buprimate, Ethirimol and Pyrimethanil (which contain pyrimidine);

(c) Azole-containing fungicides (which is by far the biggest class) including Carbendazim, Cypendazole, Debacarb, Fuberidazole, Mecorbinzid and Thiabendazole (containing benzimidazole), Prochloraz, Clotrimazole and triazoxide (containing imidazole) and azaconazole and hexaconazole (containing conazole); and

(d) others including morpholine-based fungicides such as Aldimorph, Carbamorph, Flumorph, and amine-based fungicides such as Anilazine, together with unclassified fungicides.

There are several modes of action of fungicides. The heterocyclic nitrogen-containing fungicides and the azole-based fungicides all contain sp^2-hydridized nitrogen and are thought to block the iron sites in cytochrome P450 enzymes (with exception of the benzimidazole-containing fungicides, which have a different mode of action) [26]. Blocking the action of the cytochrome P450-dependent 14-α-demethylase prevents the formation of ergosterol, which is required for the synthesis of fungal-cell membranes. High concentrations of azoles lead to the leakage of potassium from fungal cells and can also inhibit fungal respiration.

In addition to the azoles, benzimidazoles such as Benomyl and Carbendazim target the β-tubuline assembly. Morpholine-containing fungicides (Aldimorph and Tridemorph) and some simple amines such as diphenylamine (which is used as a post-harvest fungicide) and sec-butylamine (which can be used to protect stored oranges, potatoes and seeds from fungi and moulds) all inhibit sterol biosynthesis and target Δ14-reductase and Δ7 to Δ8 isomerase [27].

Imidazoline-based fungicides such as Fenamidone target the ubiquinol oxidase Q_o site and inhibit fungal respiration. Other amine-containing fungicides such as Fermizone (a pyrimidinone hydrazone) and diclomezine (a pyridazinone) have as yet undetermined methods of action.

The triazole ring is probably the single most important structural feature in the chemistry of fungicides, and it is thought that azole-containing and conazole-containing molecules are metabolized by first forming a serine conjugate, which then forms a glycine ester.

Fungicides are usually used to prevent fungal infections and are sprayed onto the leaves of plants to form a protective coating. This type of application is called non-systemic (or contact application) as the fungicide sits on the leaf and stem surfaces and does not enter the plant. Growing crops treated in this way require frequent retreatment as any new leaves and stem growth will not be protected by the previous fungicide treatment.

The structures of typical fungicides are given in Figure 8.6, showing the amine functions.

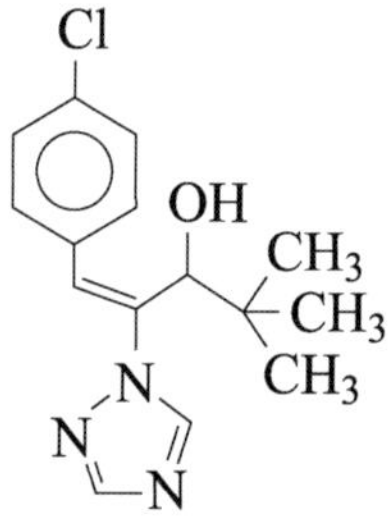

Unaconazole

(E)-(*R*,*S*)-1-(4-chlorophenyl)-4,4'-dimethyl-
2-(1H-1,2,4-triazol-1-yl)pent-1-en-3-ol

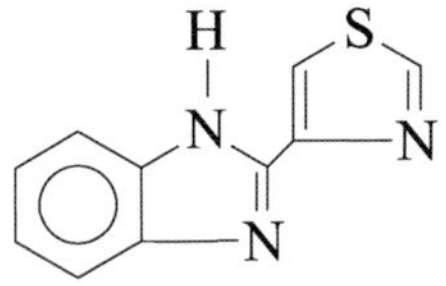

Thiabendazole

2-(thiazol-4-yl)-benzimidiazole

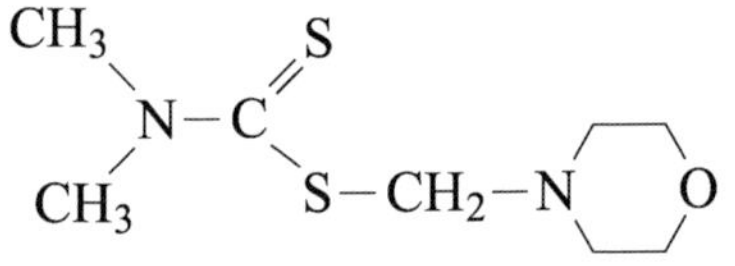

Carbamorph

4-morpholinomethyldimethyldithiocarbamate

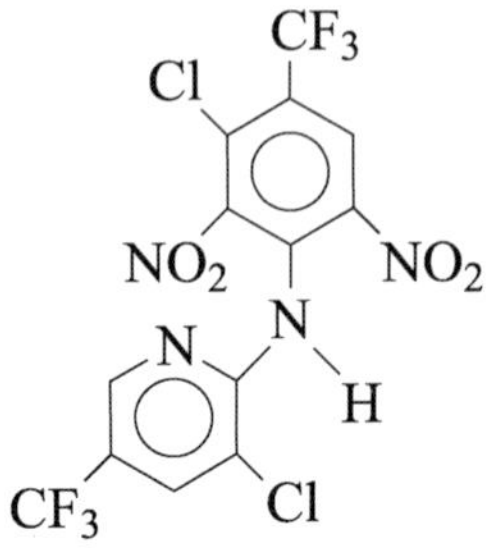

Fluazinim

3-chloro-*N*-[3-chloro-2,6-dinitro-
4-(trifluoromethyl)phenyl]-5-
(trifluoromethyl)-2-pyridinamine.

Figure 8.6. The structures of some amine-containing fungicides.

8.7.2 *Herbicides*

Unlike fungicides, which are mainly non-systemic in their method of action, herbicides penetrate the surface when sprayed onto the leaves or stems of a plant and are moved within the vascular system of the plant. Agrochemicals with this mode of action are called systemic, and most herbicides act in this way.

There are six main categories of amine-containing herbicides, which are:

(i) dinitroaniline herbicides,
(ii) triazine and chlorotriazine herbicides,
(iii) triazole herbicides,
(iv) quaternary ammonium herbicides,
(v) pyridine-based and pyridazine-based herbicides, and
(vi) imidazolinone herbicides.

aniline herbicides include Benefin, Butralin, Dinitramine, Isopropalin and Trifluralin. Their mode of action is the inhibition of the microtubule assembly [28].

Dinitroanilines such as Trifluralin are most probably metabolized in plants to form heterocyclic quinoxaline rings, as shown below.

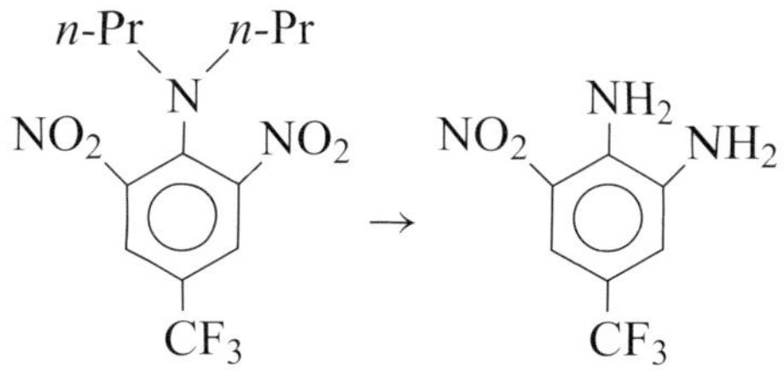

The first stage of the metabolic process is the dealkylation of the tertiary amine function, and then one of the nitro groups is also converted to an amine. At this stage two further steps are possible. Cyclization may take place to form a quinoxaline (route a), or the second nitro group may be converted to an amine (route b). The product from route b (3,4,5-triamino-1′,1′,1′-trifluorotoluene) has been identified as a soil-bound degradation product and is further dissipated by mineralization.

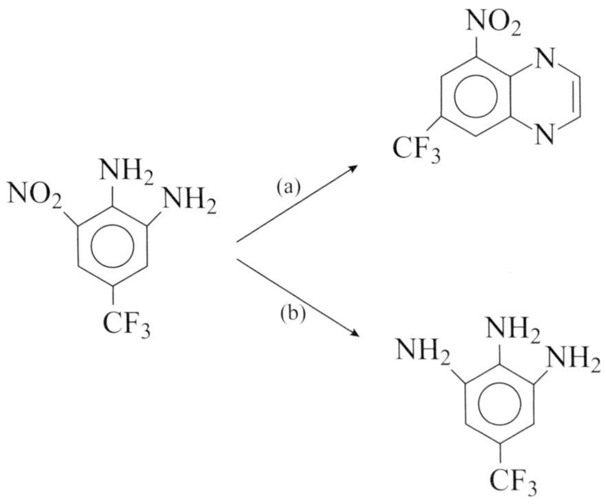

Two phenylenediamine-based herbicides, Dinitramine and Prodiamine, also include the 2,6-dinitroaniline molecular fragments.

Triazine herbicides and the commercially more-important chlorotriazine herbicides act by inhibiting photosynthesis at photosystem II. The two best-known chlorotriazine herbicides are probably Atrazine and Simazine, which differ only by a methylene unit attached to one amine unit. In the structures below, when R = ethyl the molecular structure is that of Simazine and when R = isopropyl the molecular

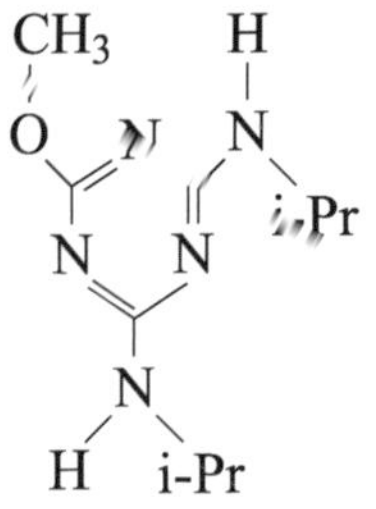

Prometon 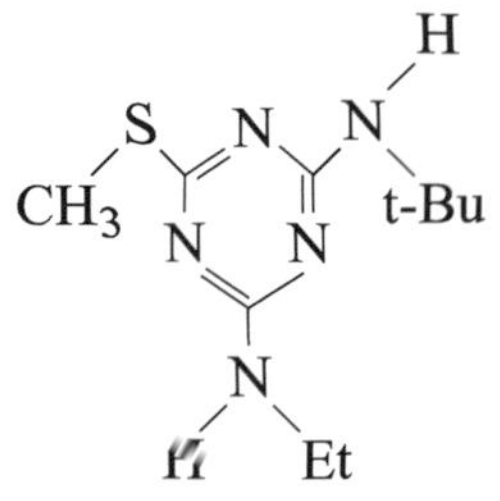   Terbutrin

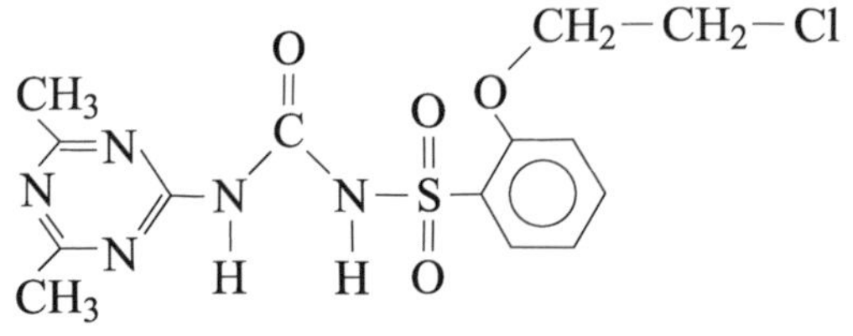

Triasulfuron

Figure 8.7. Triazole-based herbicides.

structure is that of Atrazine. Both Atrazine and Simazine are metabolized to the same 2-chloro-4,6-dihydroxy-1,3,5-triazine.

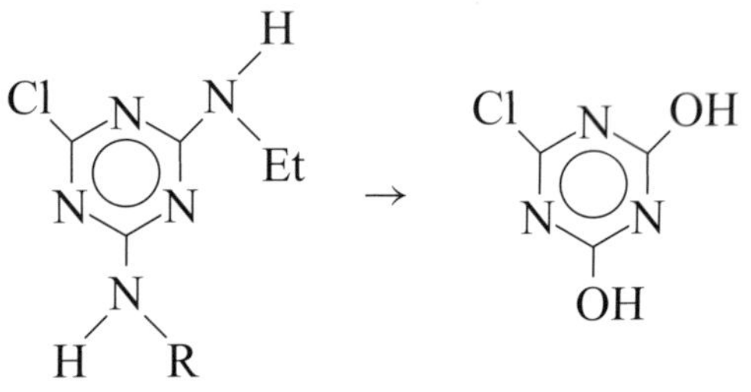

In general, the conversion of amino groups to hydroxy groups is a common feature of agrochemical metabolism and examples are known across the entire range of amine-based agrochemicals [29].

The triazole ring is also found in other herbicides such as the methoxytriazines, methoxythiotriazines and triazinylsulphonylureas, some examples of which are shown in Figure 8.7.

When triazinylsulphonylureas are metabolized, the aminotriazine fraction splits off from the rest of the molecule and is then hydrolysed to a triazine alcohol.

The precise biochemical target for triazole herbicides (which include Amitrole, Cafenstrazole, Flucarbazone, Propoxy carbazone and Sulfentrazole) is unknown, but is believed to involve the inhibition of carotenoid biosynthesis (*in vitro* this class of herbicides inhibits lycopene cyclase). Amitrole is metabolized by a deamination, followed by the formation of serine conjugate and then a glycine conjugate [30].

Herbicides based on quaternary amines include Chlormequat, Cyperquat, Diethamquat, Difenoquat, Diquat, Morfamquat and Paraquat. Generally this class of herbicides acts by inhibiting photosynthesis at photosystem I (Diquat and Paraquat), although other members of the series (such as Chlormequat) that are structurally similar to glycine can interfere with protein synthesis by forming conjugates with serine. These conjugates adsorb to de-esterified galacturonase sites on the cell wall and inhibit cation transport. Chlormequat chloride is also used as a growth retardant.

Chlormequat chloride

$$[(CH_3)N-CH_2-CH_2-Cl]^+Cl^-$$

Paraquat   Diquat

Diquat is metabolized to form 2-pyridinecarboxamide, which can then be further metabolized to 2-pyridinecarboxylic acid.

The best-known inhibitor of protein synthesis in plants is the herbicide Glyphosate, which inhibits EPSP (5-enolpyruvyl shikimate-3-phosphate) synthase and is widely used as a garden weedkiller [31]. The structure of Glyphosate is shown below and it can clearly be seen that it is an *N*-substituted glycine.

Glyphosate

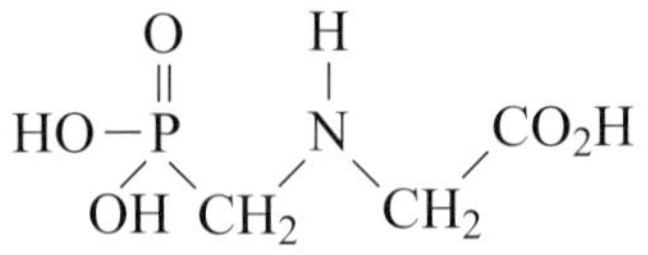

$$
\underset{\underset{OH}{\overset{\overset{O}{\parallel}}{HO-P}}}{}\underset{CH_2}{}\overset{\overset{H}{\mid}}{N}\underset{CH_2}{}CO_2H
$$

There are also many examples of pyridine and pyrimidine containing herbicides such as Cliodinate, Dithiopyr, Haloxydine, Picloram, Pyriclor and Tiazopyr for pyridine, and Chloridazon, Credazine, Dimidazon and Pyridate for pyridazine.

Pyridine-containing herbicides such as Dithiopyr and Thiazopyr act by inhibiting microtubule assembly [32], whereas pyridine carboxylic acids such as Clopyralid and Picloram act to inhibit auxins [33]. Pyridazines generally act in the same manner as pyridine carboxamides by inhibiting carotenoid biosynthesis at the phytoene desaturase step [34].

Picloram may be metabolized as a glucose ester conjugate.

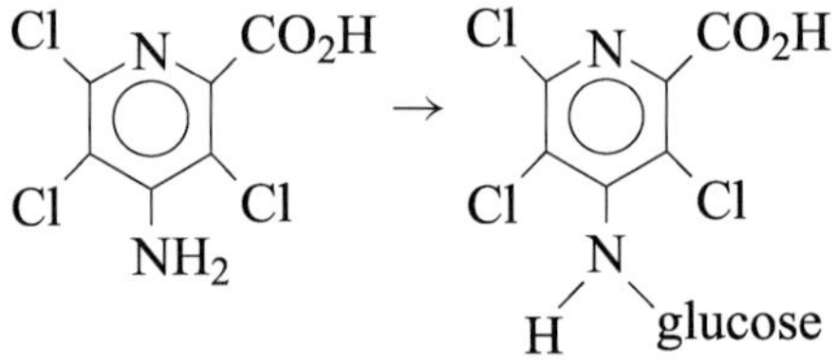

The application and synthesis of pyridazine-containing agrochemicals has been discussed in Section 8.6 on the applications of hydrazine.

The sixth class of amine-containing herbicides is based on imidazolinone, and includes Imazamox, Imazapyr and Imazaquin. These herbicides act by inhibiting acetolactase synthesis.

8.7.3 Pesticides

Next to fungicides, the second-largest market for agrochemicals is for pesticides (actually, more specifically, for insecticides). However, this class of agrochemicals does not contain many examples of amines as most pesticides are organophosphorus based.

Notable exceptions are the nicotinoid-based insecticides, which act as agonists of acetylcholine at the post-synaptic nicotinic ACh receptor (nAChR), leading to neuronal overstimulation and hyperactivity causing death [35]. The nicotinoid insecticides are further divided into three types: nitroguanines, such as Clothian-idin, Dinotefuran and Thiamethoxam; nitromethylenes, such as Nitenpyram and Nithiazine; and pyridylmethylamines, such as Acetamiprid, Imidacloprid, Niten-pyram and Thiacloprid.

Acetamiprid (N-((6-chloro-3-pyridinyl)methyl)-N'-cyano-N-methyl-ethanimi-damide) is the major ingredient in the insecticide formulation Assail and it has a neonicotinoid action. It is metabolized to form 2-chloro-5-hydroxymethylpyridine and N-cyanoamidine.

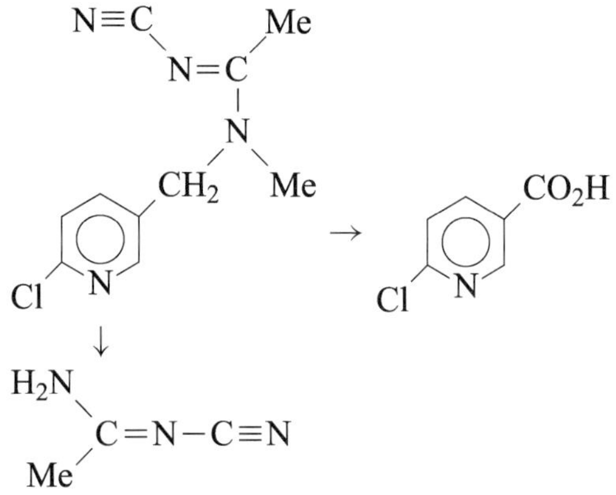

Another amine-containing pesticide is the rodenticide Crimidine, which has the following structure.

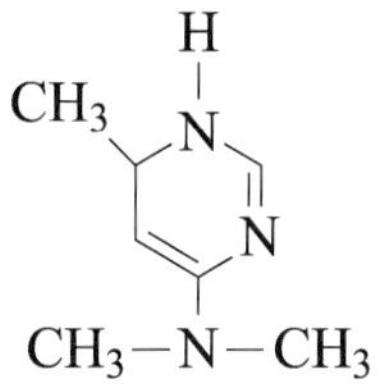

Apart from these examples and a few other organothiophosphate agrochemicals that contain pyridine (Chlorpyrifos) and pyrimidine (Butathiofos and Pirimiphos; and Chlorprazophos, which contains a pyrazolopyrimidine ring) there are few other examples of amine-based pesticides.

8.7.4 *Plant-growth regulators*

The naturally occurring phytohormone is 3-indoleacetic acid. Although it can be made from tryptophan and could be used as an agrochemical, the two most widely used growth promoters are 4-indoyl-3-yl-butyric acid (which is better known as the formulations Chryzoplus, Synergol or Seradix) and Ethychlozate (which is also sold as Figaron, shown below).

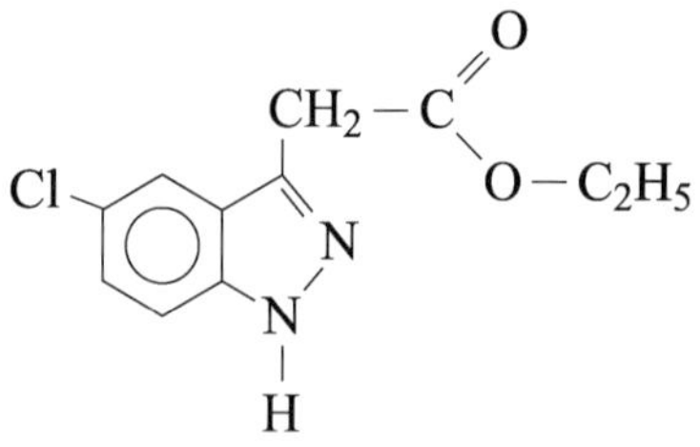

Figaron

Growth retardants are used to control crop growth to prevent cereals falling over before harvest by ensuring strong compact growth. Chlormequat chloride has been mentioned before as an amine-containing herbicide and it also acts to inhibit gibberellin biosynthesis [36]. Another gibberellin antagonist is Mepiquat (1,1-dimethylpiperidinum chloride), which is used in rape and cotton production. The presence of a quaternary nitrogen seems to give these agrochemicals an advantage over gibberellin when binding at the receptor. In fact, even non-quaternary nitrogen-based growth retardants contain at least one nitrogen atom, which may help to facilitate penetration of plant-cell membranes.

8.8 Azine and azo dyes

The total annual worldwide market for organic pigments is around 220 000 tonnes, which corresponds to an annual sales value of approximately USD4.2 billion. Of these figures, approximately 60% are from azo dyes, which despite some recent restrictions in their use within the European Union, still represent the largest single class of pigments today.

Unfortunately there is no systematic method for naming dyes, and many dyes have historical names given to them by their inventors or manufacturers. To avoid difficulties, each dye is assigned a colour number that is included in the *Colour Index*. However, this does not give the chemist any direct information about the molecular structure of the actual dyestuff.

Another method is to classify dyestuffs into ten groups according to their application. These groups are as follows.

   (i) Acid dyes – the sodium salts of carboxylic and sulphonic acids.
  (ii) Basic dyes – the acid salts of colour bases.
 (iii) Azoic dyes – insoluble azo dyes produced *in situ* for direct dyeing.
 (iv) Mordant dyes – dyes that require the presence of a substanting agent (a mordant) before they will dye fibres.
  (v) Metal-complex dyes – dyes prepared as metal complexes prior to dyeing.

 (vi) Vat dyes – dyes that are used as leuco compounds when dyeing and then are oxidized
      to form the correct colour after fibre impregnation.
 (vii) Sulphur dyes – macromolecules with di- and polysulphide cross linkages.
(viii) Disperse dyes – water-insoluble dyes that require a dispersent.
 (ix) Reactive dyes – dyes that contain a reactive group, which forms a chemical bridge
      with cellulosic fibres during dyeing.
  (x) Organic pigments – water-insoluble colouring additives that can, on reaction, be made
      water soluble and used as dyes.

Until the beginning of the nineteenth century the only organic dyes used for the colouration of textiles and food where those from plant or animal origin. Some well-known examples that were commonly used in the early Victorian era were madder root, which was used to produce a red dye, saffron stamen, which produced a yellow-orange dye, and woad or *Indigofera tinctoria*, which produced blue/purple dyes. The Mediterranian shellfish *Murex brandaris*, which had been known since the time of the Romans, was also used to produce a violet dye the use of which in the Middle Ages was reserved for the ruling classes.

The oldest synthetic organic dyes known are the azine family, which date back to 1834 when Runge discovered that when aniline was oxidized with dichromate dark green pigments could be obtained [37]. Runge's discovery remained a scientific curiosity until 1856 when William Henry Perkin, an 18-year-old student of Hofmann's, made an unexpected discovery.

Perkin was inspired by Hofmann to undertake a research project to try to discover a synthetic route to quinine. Perkin thought that it might be possible to do this starting from the oxidation of aniline. His first experiment using allylaniline failed, so he tried a second time using aniline and potassium dichromate. Perkin did not obtain quinine from this experiment and instead produced a black sludge. In the Victorian era the aim of the synthetic chemist was to try to produce crystalline products that could be characterized, so Perkin decided to dissolve the black sludge from his experiment in methylated spirit to see if he could remove the tar in the hope of obtaining a crystalline product. At this stage, Perkin produced a violet-coloured solution, which he found would dye silk mauve (one story about this experiment was that Perkin actually spilled some of the methylated-spirit solution onto his shirt). A 'blue plaque' to mark this historic event may be seen on the wall of 299 Oxford Street in central London. Perkin patented his process and invested his family's money to establish a chemical plant in Greenford, near London, to produce this mauve dye, which he called mauvine.

Perkin's success in producing mauvine was a consequence of a lucky experiment that did not go according to plan because the aniline source he used was heavily contaminated with toluidines. The structure of mauvine was not finally elucidated until 1994 when it was discovered to be a mixture of several chemicals, the most

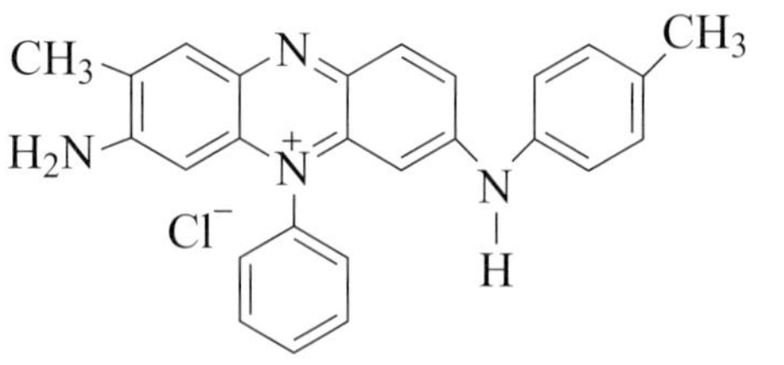

Figure 8.8. The structure of the most important constituent of mauvine.

Figure 8.9. The penny lilac postage stamp.

important of which is shown in Figure 8.8 [38]. In this figure the positive charge is shown on the central nitrogen atom for clarity but in reality it is dispersed over the whole molecule, leading to the electronic resonance that is the origin of mauvine's coloration.

Although initially mauvine could be used only to dye silk, a method of dyeing cotton with mauvine was soon discovered by using tannin derivatives as mordants. Mauvine rapidly became very popular in Victorian society, and Queen Victoria famously wore a dress dyed with mauvine to her daughter's wedding. One of the ribbons from this dress is kept in the Victoria and Albert Museum in London. Mauvine was also used as a dye for the penny lilac postage stamp, which is shown in Figure 8.9. The story of the development of mauvine and its impact on society has been reviewed by Garfield in his book on the subject [38]. Today it is relatively simple to reproduce Perkin's experiment in the laboratory, and a suitable method is given by Scaccia *et al.* [39] for the small-scale synthesis of mauvine.

A consequence of the discovery of mauvine was that several major industrial development programmes to create new dyes were launched across Europe

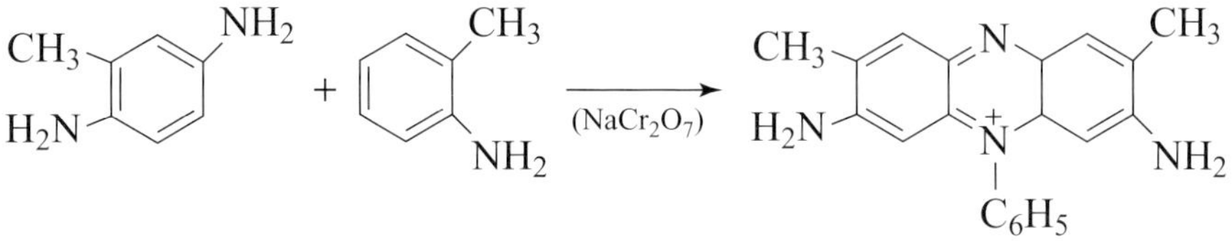

Figure 8.10. The synthesis of Safranine T.

(particularly at BASF in Ludwigshafen) and also in the USA. Although Perkin worked very quickly to commercialize mauvine, the method of its synthesis was not elucidated until 1858 by Greiss, who worked out the principles of diazotization [40]. Once the process of dye formation from aniline was understood, several new azine dyes, including the safranines, were developed. The safranines are a class of red basic dyes and can be used to dye wool, silk and cotton. They are more basic than mauvine but are prepared in a similar manner. The synthesis of Safranine T is shown in Figure 8.10.

All of the azine dyes are vattable, which means that they can be reduced to colourless leuco forms and then reoxidized back to the dye. All of the azine dyes are based on phenazine, phenoxazine or phenothiazine tri-cyclic ring structures, as shown below, but differ in their side chains, which control the extent of electronic resonance in the dyes and their colour.

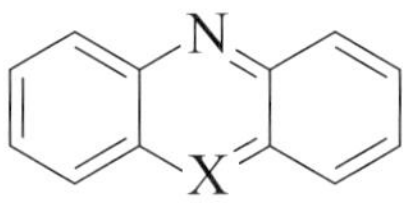

X = N = phenazine

X = O = phenoxazine

X = S = phenothiazine

The azine dyes were important in the nineteenth century but have since been replaced by the azo dyes, except for a few applications such as titration indicators. Azo dyes offer a better colour selection, are more stable to washing and sunlight, and are generally more economic to manufacture.

The azo dyes are characterized by the presence of the bridging $-N=N-$ group between an aromatic system and a heterocyclic, aromatic or aliphatic system. There are over 2000 azo dyes in the *Colour Index* and their annual sales value is over USD400 million. In the *Colour Index*, azo dyes are split into four categories: monoazo dyes (CI range 11 000 to 19 999), diazo dyes (CI 20 000 to 29 999), triazo dyes (CI 30 000 to 34 999) and polyazo dyes (CI 35 000 to 36 999). Azo dyes are also sub-classified according to their basic (cationic) or acidic (anionic properties).

Two simple acid dyes are Chrysoidine and Acid Red. Like all azo dyes, they are manufactured via diazotization and coupling of an aryl amine.

Chrysoidine

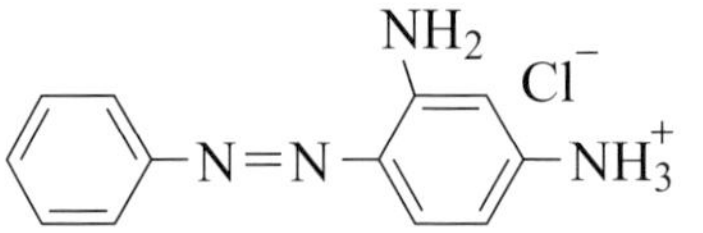

Acid Red

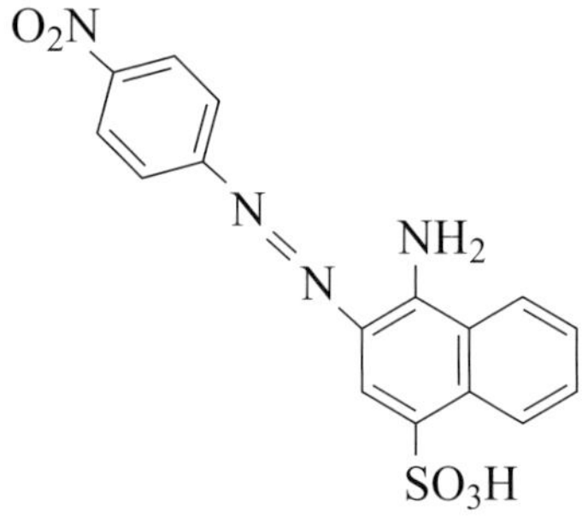

A typical basic dye is Methyl Orange, the molecular structure of which is given in Section 3.3.9. Azo dyes with hydroxyl groups *ortho* to the azo group are able to complex effectively with metals such as chromium. The complex is made by preboiling the fibre to be dyed in sodium dichromate solution in the presence of a reducing agent such as formic acid (this process is known as mordanting). Addition of the azo dye to the mordanted fibre produces a metal salt known as a 'lake', which dyes the fibre. Metal/azo complexes are also known for cobalt and for chromium; *ortho*-carboxylic groups and *ortho*-amino groups can also form this type of complex, as can *ortho*-hydroxy groups. Unfortunately some fibres are damaged by the mordanting process and the metal/azo dye complex has to be preformed before the dyeing process, which does not always work so well.

Another type of azo dye is the 'reactive azo dye', which contains a dichlorotri-azine side chain. These dyes are manufactured from the reaction of trichlorotriazine with the amino group of an azo dye. Once in aqueous alkaline solution, the two remaining chloro groups are hydrolysed to hydroxy groups, which can then react with cellulose fibres to form bridging bonds, as shown below.

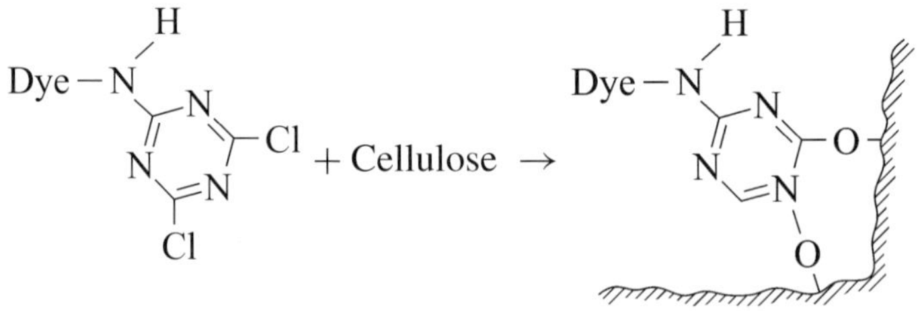

Components for the synthesis of azo dyes may be anilines, substituted anilines, naphthylamines, naphthylamine sulphonic acids (such as peri acid, tobias acid, laurent acid, kohl acid, etc.), aromatic diamines (which may be phenylene, naph-thylamine or biphenyl based), non-aromatic diamines in which two aromatic rings are bridged by a group such as −NH−, −CO−, or −NH−CO−, and heterocyclic amines such as triazole, thiazole, benzothiazole and benzimidazole derivatives.

The appropriate conditions required for the successful preparation of diazonium salts have been examined in Section 3.3, and the chemistry of the azo dyes has been the subject of several comprehensive reviews; see, for example, reference [41].

8.9 Indigo dyes

Although blue indigo dye is obtained from natural vegetable sources, it occurs as natural pro-dyes that have to be processed before the blue colour develops. Indigo is obtained from *Indigofera tinctoria* in India as the β-glucoside of indoxyl and in woad (*Isatis tinctoria*) as the indoxyl 5-ketogluconate. Both of these forms require hydrolysis to form indoxyl, followed by aerial oxidation to produce indigotin. Fractional crytallization from sulphuric acid produces pure indigotin. A related dye 'Tyrian Purple' was obtained from *Murex* and is actually the 6,6'-bromo derivative of indigo.

Today most indigo is produced synthetically and the total worldwide production is around 20 000 tonnes per year, which represents a sales value of around USD250 millon per year. The main application of indigo is in the production of blue denim for jeans, which evolved from work clothes to become a fashionable item of clothing over 50 years ago. Denim clothing has a perennial appeal and is likely to continue to be worn in the future the world over. Indigotin is a dark blue powder that is practically insoluble, but when mixed with alkaline sodium hyposulphite it forms a soluble leuco compound called indigotin white. Denim jeans are washed in the indigotin-white solution and then dried in air. This reoxidizes the indigotin white back to insoluble indigotin and produces the well-known blue colour of denim jeans.

There are three synthetic processes to indigotin starting from (i) anthranilic acid with chloroacetic acid, (ii) *N*-phenylglycine with sodium amide, and (iii) aniline with ethylene oxide. The second process is called the Heumann–Pfleger Process and is the most important commercial indigotin synthetic process used today [42]. The original process developed by Karl Heumann in 1890 did not use sodium amide, but instead used a combination of anhydrous sodium hydroxide and potassium hydroxide and gave a very poor yield of less than 10%. The addition of sodium amide in 1901 by Pfleger improved the yield to over 90%.

$$C_6H_5-NH_2 + Cl-CH_2-CO_2H \rightarrow$$

$$\downarrow \; +NaOH/KOH$$

$$+ NaNH_2 \rightarrow \qquad + \textit{xs } NaNH_2 \rightarrow \qquad + O_2 \rightarrow \text{ Indigotin}$$

Indigotin cannot be brominated directly to produce Tyrian Purple, but instead undergoes tetrasubstitution to yield 5,7,5′,7′-tetrabromo indigo, which is produced commercially although the total annual worldwide production is less than 1% of that of indigo. 5,7,5′,7′-Tetrabromo indigo is used for dyeing and also printing and has the trade names 4BD and Brilliant Indigo 4BN, respectively.

References

[1] Y. Kuznetsov, *Organic Inhibitors of Corrosion of Metals* (New York: Plenum Press, 1996).
[2] S. A. Lawrence and C. Kettemann, *Chimica Oggi*, **15** (1997), 33.
[3] D. T. Haynie, *Biological Thermodynamics* (Cambridge: Cambridge University Press, 2001), p. 101.
[4] O. Sten-Knudsen, *Biological Membranes* (Cambridge: Cambridge University Press, 2002), p. 73.
[5] H. G. Mautner, *Pharmac. Rev.*, **19** (1967), 107.
[6] J. D. Stolzenburg, *Fritz Haber* (Weinheim: Wiley VCH, 1994).
[7] E. Eckhardt, *Curr. Med. Chem. Anti-cancer Agents*, **2**:3 (2002), 419.
[8] W. A. Denny, *Curr. Med. Chem.*, **8**:5 (2001), 533.
[9] M. Hensel, A. Schneeweiss, H. P. Sinn, G. Egerer, M. Kornacker, E. Solomayer, R. Haas and G. Bastert, *Stem Cells*, **20** (2002), 32.
[10] T. Branna, *HAPPI*, December 1995, p. 80.
[11] H. D. Hueck, D. M. M. Adema and J. R. Weigmann, *Appl. Microbiol.*, **14** (1966), 308.

[12] M. Ash and I. Ash, *Handbook of Cosmetic and Personal Care Products* (Hampshire, UK: Gower, 1994).

[13] L. H. Tai, *Formulating Detergents and Personal Care Products* (Illinois: AOCS Press, 2000).

[14] M. Häring, *Helv. Chim. Acta*, **42** (1959), 1845.

[15] F. Pilati in *Comprehensive Polymer Science*, ed. G. C. Eastmond and A. Ledwith (Oxford: Pergamon, 1989).

[16] M. J. Rosen and M. Dahanayake, *Industrial Utilisation of Surfactants: Principles and Practice* (Illinois: AOCS Press, 2000).

[17] *Blue List Cosmetic Ingredients*, ed. F. H. Kemper, N.-P. Loepke and W. Umbach (Schussenried: ECV Press, 2000).

[18] R. P. Linstead, *J. Chem. Soc.* (1934), 1016.

[19] N. B. McKeown, *Phthalocyanine Materials* (Cambridge: Cambridge University Press, 1998).

[20] C. D. Brown, *Space Craft Propulsion* (Reston, VA: AIAA Education Series, 1996).

[21] M. Ozawa, *Process Metall.*, **7A** (1992), 603.

[22] US Patent 3637389 (1967, assigned to Agfa Gevaert).

[23] K. B. Sundaram and H. Chang, *J. Electrochem. Soc.*, **140** (1995), 1592.

[24] M. R. Andrew, *J. Appl. Electrochem*, **2** (1972), 327.

[25] V. K. Gouda and S. M. Sayed, *Corrosion Sci.*, **13** (1973), 647.

[26] Y. Uesgi in *Fungicidal Activity*, ed. D. H. Hutson and J. Miyamoto (Chichester: J. Wiley, 1998), p. 23.

[27] T. Roberts and D. H. Hutson, *Metabolic Pathways of Agrochemicals, Part 2*, (Cambridge: Royal Society of Chemistry, 1999).

[28] A. P. Appleby and B. E. Valverde, *Weed Technology*, **3** (1989), 198.

[29] F. Müller, *Agrochemicals: Composition, Production, Toxicology and Applications* (Weinheim: Wiley VCH, 2000).

[30] L. Franco and A. M. Municio., *Gen Pharm.*, **6** (1975), 163.

[31] J. E. Franz, M. K. Mao and J. A. Sikorski, *Glyphosate: A Unique Herbicide*, ACS Monograph 189.

[32] P. C. C. Feng, S. Raja-Rao and D. E. Schaefer, *Pestic. Sci.*, **45** (1995), 203.

[33] D. S. Frear, H. R. Swanson and E. R. Mansager, *J. Agric. Food Chem.* **37** (1989), 1408.

[34] S. D. West and S. J. Parka, *J. Agric. Food Chem.*, **40** (1992), 160.

[35] A. S. Perry, I. Yamamoto, I. Ishaaya and R. Y. Perry, *Insecticides in Agriculture and the Enviroment: Retrospects and Prospects* (Heidelberg: Springer Verlag, 1998).

[36] C. Intrieri and W. Ryugo, *J. Am. Soc. Hort. Sci.*, **99** (1974), 349.

[37] F. Runge, *Pogg. Ann.*, **31** (1834), 65.

[38] S. Garfield, *Mauve: How One Man Invented a Colour That Changed the World* (London: W. W. Norton, 2002). (See also O. Meth-Cohn and A. S. Traus, *Chem. in Britain*, **31** (1995), 547.)

[39] R. L. Scaccia, D. Coughlin and D. W. Ball, *J. Chem. Ed.*, **75** (1998), 769.

[40] P. Griess, *Ann*, **106** (1858) 123.

[41] P. F. Gordon and P. Gregory, *Organic Chemistry and Colour* (Heidelberg: Springer Verlag, 1983).

[42] F. Henesey, *J. Soc. Dyers Colour*, **54** (1938) 105.

Appendix 1
Molecular structures and isosurface electronic charges of selected amines

The determinations of molecular structures and electronic charges presented here were performed with HyperChem 6 and used Polak–Ribiere geometric optimization and AMBER for the calculation of the electrostatic potential at the molecular isosurface. Figure A1 shows the ammonia molecule NH_3 in its lowest energy conformation in the gas phase. The pyramidal shape of the molecule can be clearly seen, together with the localization of negative electrostatic charge at the molecular isosurface vertically above the nitrogen atom, corresponding to the non-bonded lone electron pair.

The maximum negative electrostatic charge at the isosurface corresponds to $-0.209 \, e/a_0^3$, where e is the charge on the electron and a_0 is a function of the radius from the molecular isosurface to the atomic surface.

Substitution of all three of the hydrogen atoms in ammonia for fluorine to produce nitrogen trifluoride gives a nearly planar molecule with only a slight pyramidal structure. However, the electronegative fluorine atoms act to draw the electron density associated with the lone electron pair completely away from the nitrogen atom. From Figure A2 it can be seen that there is not a single point of negative electrostatic potential on the molecule's isosurface and the lowest electrostatic potential is actually $0 \, e/a_0^3$.

The gas-phase basicity of nitrogen trifluoride is $\Delta G^{\ominus} = 538.6 \, \text{kJ mol}^{-1}$, which is far lower than the corresponding values measured for ammonia and other amines [1]. The gas-phase basicity refers to the general equation for protonation $\mathbf{B} + [\text{H}]^+ \rightarrow [\mathbf{BH}]^+$; values for some amines are given in Table A1.

Replacement of two hydrogen atoms on the ammonia molecule with two methylene groups to produce aziridine has the opposite effect to that in nitrogen trifluoride, and the methylene groups act to increase the electrostatic potential of the lone pair but not to the same extent that the three fluorine atoms act to decrease it. The highest point of negative electrostatic potential on the isosurface of the aziridine molecule is $-0.273 \, e/a_0^3$, as shown in Figure A3. The cyclic structure of the aziridine molecule

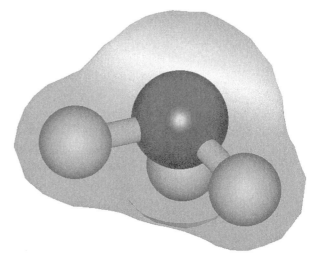

Figure A1. The ammonia molecule.

Table A1. *Gas-phase basicity values for some amines*

Amine	Gas-phase basicity (kJ mol^{-1})
Nitrogen trifluoride	538.6
Ammonia	819.0
Ethylamine	878.0
Methylamine	864.5
Pyrrole	843.8

also forces the nitrogen lone pair into a more exposed position that in ammonia and so it is more readily available for donation.

Aziridine, C_2H_5N, might be expected to be similar in properties to ethylamine but the two amines differ because the former is a secondary amine and the latter is a primary amine. The electron-donating effect of the ethyl group to the central nitrogen in ethylamine (see Figure A4) is less than that of two methylene groups in aziridine. The highest negative electrostatic potential on the surface of ethylamine is $-0.214 \, e/a_0^3$, which is between the values calculated for aziridine and ammonia.

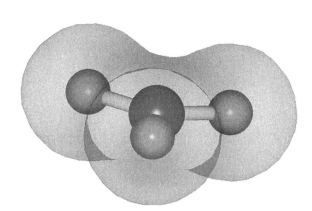

Figure A2. The nitrogen trifluoride molecule.

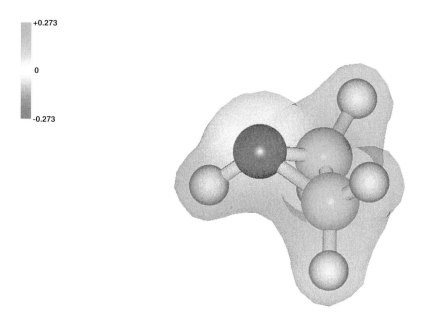

Figure A3. The aziridine molecule.

Therefore it can be seen that a β-methyl group has a less-pronounced effect on influencing the electronic charge on the nitrogen atom than two α-methylene groups. Long alkyl side chains do have other controlling effects on amine chemistry and are important in defining their hydrophilicity. By using the same calculation parameters as in the previous examples, the maximum negative isosurface electrostatic charges

Table A2. *Electrostatic maxima and pK$_a$ values for amines*

Molecule	Electrostatic charge	pK$_a$
Trifluoroamine	0.00	−5.8(c)
Diphenylamine	−0.101	0.78
Aniline	−0.0151	4.62
Ammonia	−0.209	4.75
Ethylamine	−0.214	10.63
Aziridine	−0.273	10.90
Diethylamine	−0.275	11.04

Note: (c) = calculated from thermodynamic data.

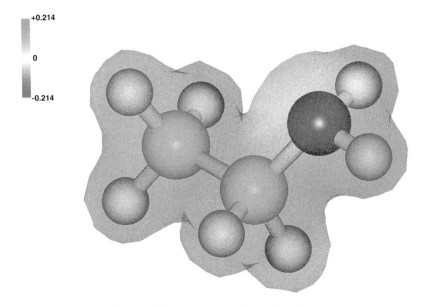

Figure A4. The ethylamine molecule.

were calculated for a range of amines and are presented in Table A2. It can be seen the there is a fair relationship between the level of negative electrostatic charge on amine molecules and the measured amine pK$_a$ values.

Electron-withdrawing phenyl and fluoro groups withdraw electron density from the central nitrogen atoms of amines and so decrease their basic properties. Conversely alkyl groups, which donate electronic charge, increase the electron density at the nitrogen atom and produce stronger bases.

The molecular structures of two heterocycles pyridine and imidazole were also calculated. The structure of pyridine is shown in Figure A5 and the electrostatic

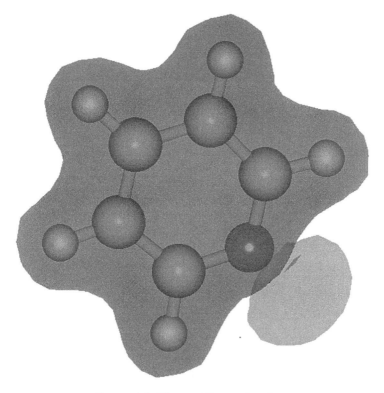

Figure A5. The pyridine molecule.

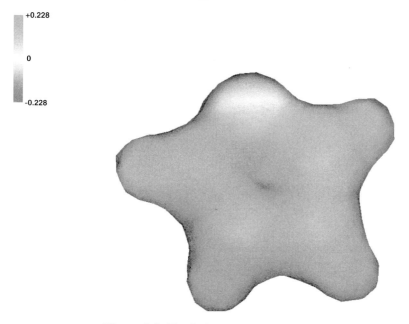

Figure A6. The imidazole molecule.

potential created by the nitrogen lone pair can be clearly seen sticking out perpendicular to the plane of the ring. A similar calculation for imidazole shows that the sp^2 nitrogen (the pyridine type) has a lobe of negative electrostatic charge, but that the other sp^3 nitrogen atom (the pyrrole type), which donates its lone pair to the aromatic π-electron cloud, does not and in fact cannot be distinguished from a C–H group from outside of the opaquely shaded isostructure in Figure A6.

The nitrogen atom is actually on the bottom left-hand side of the molecule and can be detected only by the molecular asymmetry as the N–H group has a slightly different profile to the three C–H groups.

Reference

[1] E. P. Hunter and S. G. Lias, *J. Phys. Chem. Ref. Data*, **27** (1998), 413.

Appendix 2
Physical properties of selected amines

Amine	CAS	Melting point (°C)	Boiling point (°C)	pK_a
Primary aliphatic				
Methylamine	[74-89-5]	−92	−6	10.64
Ethylamine	[7504-7]	−80	18	10.63
n-Amylamine	[110-58-7]	−50	105	10.63
Cyclohexylamine	[108-91-8]	−17	134	10.64
Benzylamine	[100-46-9]	10	184	9.34
Diphenylmethylamine	[91-00-9]	12	304	7.17
Ethylenediamine	[107-15-3]	8	116	9.98
Hydrazine	[302-01-2]	1.4	113.5	8.10
Secondary aliphatic				
Dimethylamine	[124-40-3]	−93	7	10.64
Diethylamine	[109-89-7]	−50	56	10.98
Piperidine	[110-89-4]	−9	105	11.22
Morpholine	[110-91-8]	−7	130	8.36
Tertiary aliphatic				
Trimethylamine	[75-50-3]	−117	3.5	9.76
Triethylamine	[121-44-8]	−115	89	10.65
Tri-n-butylamine	[102-82-9]	−70	216	10.89
Tribenzylamine	[620-40-6]	92	380	7.44
Primary aromatic				
Aniline	[62-53-3]	−6	184	4.62
α-Naphthylamine	[134-32-7]	50	301	3.92c
β-Naphthylamine	[91-59-8]	112	306	4.11(c)
2-Chloroaniline	[95-51-2]	−2	208	2.62
3-Chloroaniline	[108-42-9]	−10	236	3.32
4-Chloroaniline	[106-47-8]	70	230	3.81
2,4,6-Trichloroaniline	[634-93-5]	74	262	0.09
2-Aminophenol	[95-55-6]	174	153(s)	4.72

(cont.)

315

(cont.)

Amine	CAS	Melting point (°C)	Boiling point (°C)	pK_a
3-Aminophenol	[95-55-6]	123	164(d)	4.17
4-Aminophenol	[123-30-8]	186	284(d)	5.50
2-Aminobiphenyl	[90-41-5]	50	299	4.2
4-Aminobiphenyl	[92-67-1]	53	302	4.18
Secondary aromatic				
N-Methylaniline	[100-61-8]	−57	194	4.85
N-Ethylaniline	[103-69-5]	−63	206	5.11
Diphenylamine	[122-39-4]	54	302	0.78
Tertiary aromatic				
N,N-Dimethylaniline	[121-69-7]	193	228	5.06
N-Methyl, N-ethylaniline	[613-97-8]	201	125	5.98
Triphenylamine	[603-34-9]	127	365	−5.0(c)
Heterocyclic amines				
Pyridine	[110-86-1]	−42	115	5.14
2-Aminopyridine	[504-29-0]	59	204	7.2
3-Aminopyridine	[462-08-8]	57	248	6.7
4-Aminopyridine	[504-24-5]	161	273	9.2
Indole	[120-72-9]	53	254	−2.9
Quinoline	[91-22-55]	−19	239	4.85
Isoquinoline	[119-65-3]	23	240	5.14
Acridine	[260-94-6]	111(s)	345(c)	5.60
Hexamethylene tetramine	[100-97-0]	280	263(d)	6.3

Notes: (d) = decomposes, (s) = sublimes, (c) = calculated from thermodynamic data.

Appendix 3
Named reactions involving amines

This appendix contains further information about the named reactions and syntheses of amines that have been mentioned in this book. This list is not exhaustive as new reactions are continually being reported in the literature. In as many places as possible, the primary reference source is quoted; where this has not been possible, an important review has been cited. Trivial examples where the amine group is not directly involved in the reaction process have been omitted.

1. The Alper Reaction

Reagents $= CO/Pd(P(C_6H_5)_4$
H. Alper, *J. Chem. Soc. Chem. Commun.* (1983), 1270.

2. The Asinger Thiazole Synthesis

Reagents $= S_8/NH_3$
F. Asinger, *Liebigs Ann.*, **602** (1957), 37.

318   *Appendix 3*

3. The Baeyer–Drewson Indigo Synthesis

Reagents = (i) acetone (or pyruvic acid) and (ii) potassium hydroxide
A. Baeyer and V. Drewson, *Ber.*, **15** (1882), 2856.

4. The Baeyer Pyridine Synthesis

Reagents = (i) Me_2SO_4, (ii) 20% $HClO_4$ and (iii) $(NH_4)_2CO_3$
A. Baeyer, *Ber.*, **43** (1910), 2337.

5. The Bailey Cycloaddition

Reagents = KSCN, HOAc
J. Bailey, *J. Am. Chem. Soc.*, **39** (1917), 279.

6. The Bamberger Benzotriazine Synthesis

Reagents = (i) HNO_2, (ii) $H_2N-N=C(CH_3)-CO_2H/KOH$ and
(iii) $HOAc/H_2SO_4$
E. Bamberger, *Chem. Ber.*, **25** (1892), 3201.

7. The Bamberger Imidazole Cleavage

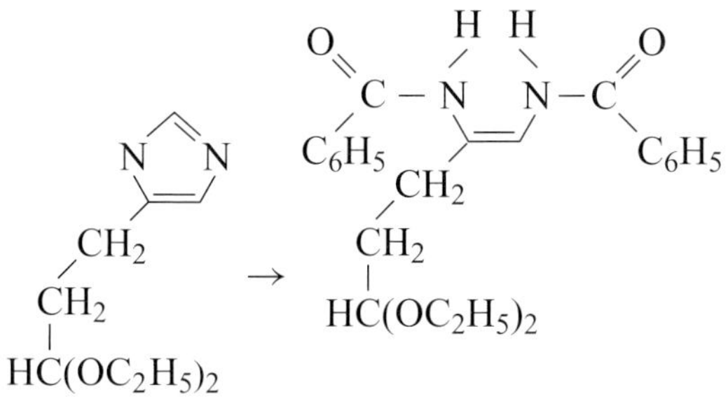

Reagents $= C_6H_5-CO-Cl/NaHCO_3$
E. Bamberger, *Liebigs Ann.*, **273** (1893), 342.

8. The Béchamp Reduction Reaction

$$Ar-NO_2 + 2Fe + 6HCl \rightarrow Ar-NH_2 + 2H_2O + 2FeCl_3$$

A. J. Béchamp, *Ann. Chim. Phys.*, **42** (1854), 186.

9. The Beckmann Rearrangement

$$R-C(R')=N-OH \rightarrow R-C(O)-NHR'$$

Reagent $=$ phosphorous pentachloride
E. Beckmann, *Ber.*, **19** (1886), 988.

10. The Bénary Reaction

$$RR'N-CH=CH-COR'' + R'''-Mg-X \rightarrow R'''-CH=CH-COR''$$
$$(+ \text{ other minor products})$$

E. Bénary, *Ber.*, **63** (1930), 1573.

11. The Benzidine Rearrangement

$$C_6H_5-NH-NH-C_6H_5 \rightarrow 4,4'-H_2N-C_6H_4-C_6H_4-NH_2$$

Reagent $=$ an acid source (acid catalysed)
A. W. Hofmann, *Proc. R. Soc. Lond.*, **12** (1863), 576.

12. The Bergmann Azalactone Synthesis

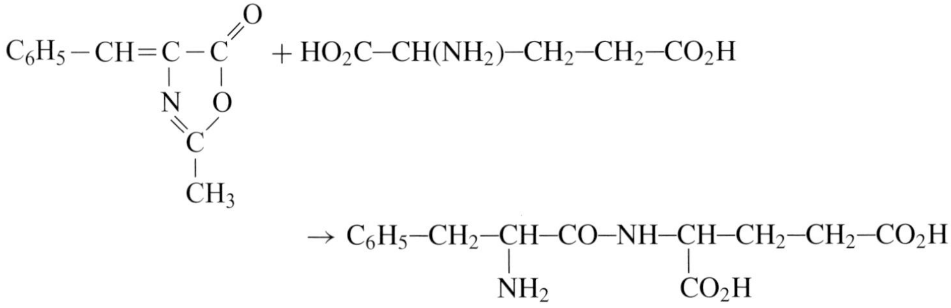

$$C_6H_5-CH=C-C \overset{O}{\diagup} \;\; + \; HO_2C-CH(NH_2)-CH_2-CH_2-CO_2H$$

with the ring:

N—O
‖
C
|
CH₃

$$\rightarrow C_6H_5-CH_2-\underset{\underset{NH_2}{|}}{CH}-CO-NH-\underset{\underset{CO_2H}{|}}{CH}-CH_2-CH_2-CO_2H$$

Reagents = (i) H_2/Pd and (ii) HCl
M. Bergmann, F. Stern and C. Witte, *Ann.*, **449** (1926), 277.

13. The Bergmann Degradation

$$R'CO-NH-CH(R)-C(O)N_3 \rightarrow R'CONH_2 + RCHO + CO_2$$
$$+ C_6H_5-CH_3 + NH_3$$

Reagents = (i) benzyl alcohol, (ii) H_2/Pd and H_2O
M. Bergmann, *Science*, **79** (1934), 439.

14. The Bernthsen Acridine Synthesis

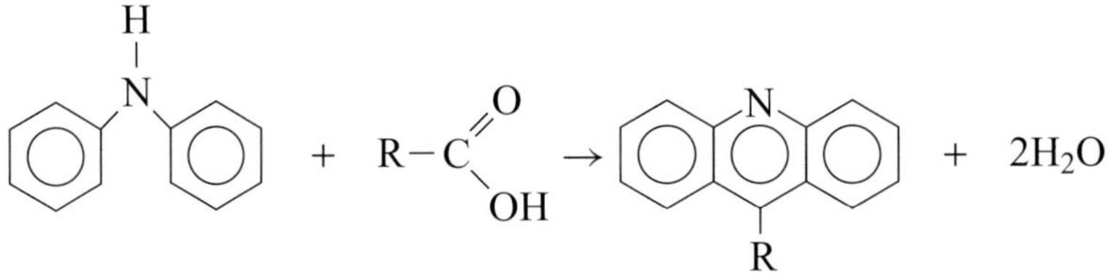

Reagents = zinc chloride
A. Bernthsen, *Ann.*, **192** (1884), 1.

15. The Betti Reaction

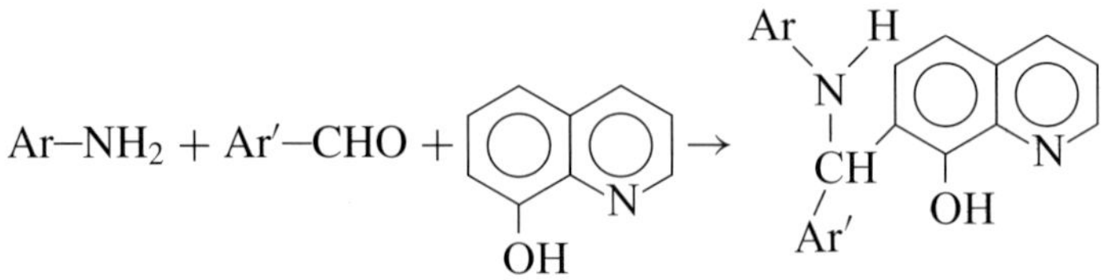

M. Betti, *Gazz. Chim. Ital.*, **30**: II (1900), 301.

16. The Biginelli Reaction

$$CH_3-CO-CH_2-C(O)O-C_2H_5 + R-CHO + H_2N-CO-NH_2$$

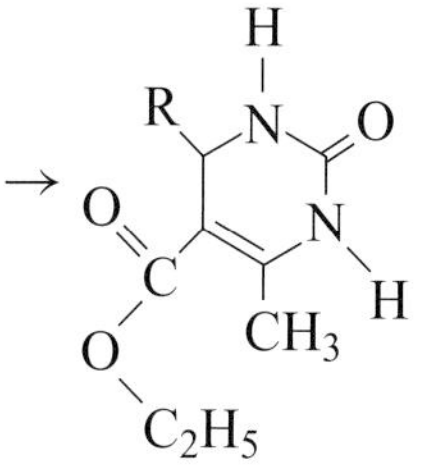

P. Biginelli, *Ber.*, **24** (1891), 1317.

17. The Bischler–Möhlau Indole Synthesis

$$C_6H_5-CO-CH_2Br + C_6H_5-NH_2 \rightarrow$$

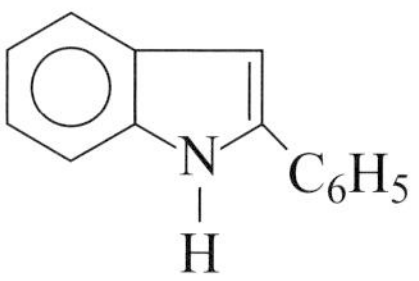

Reagent = *xs* $C_6H_5-NH_2$

A. Bischler and R. Möhlau, *Ber.*, **25** (1892), 2860.

18. The Bischler–Napieralski Reaction

$$C_6H_5-CH_2-CH_2-NH-CO-R \rightarrow \qquad + \quad H_2O$$

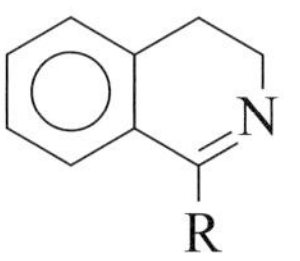

Reagents = zinc chloride or phosphorous pentoxide

A. Bischler and B. Napieralski, *Ber.*, **26** (1893), 1903.

19. The Blum Aziridine Synthesis

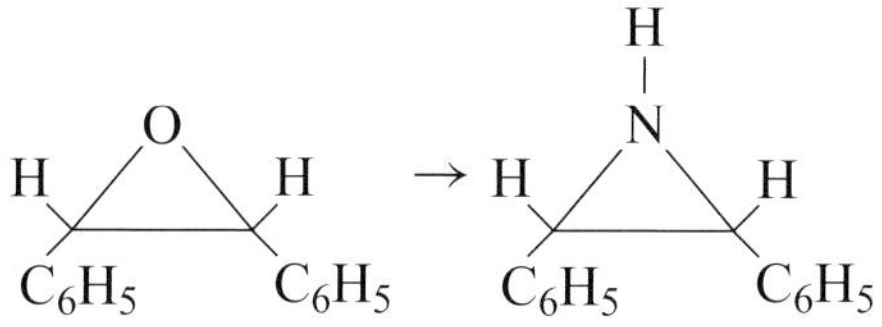

Reagents = (i) NaN_3 and (ii) $P(C_6H_5)_3$

J. Blum, *J. Org. Chem.*, **43** (1978), 397.

20. The Borsche–Dreschel Cyclization

$$C_6H_5-NH-N=C_6H_{10} \rightarrow$$

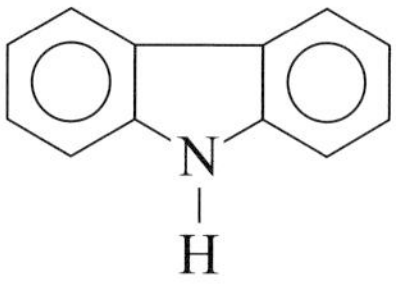

Reagents = (i) HCl and (ii) Pb_3O_4

E. Dreschel, *J. Prakt. Chem.*, **38** (1858), 69.

21. The Bosch–Meiser Urea Process

$$CO_2 + 2NH_3 \rightleftharpoons H_2N-CO_2-NH_2 \rightleftharpoons H_2N-CO-NH_2 + H_2O$$

C. Bosch and W. Meiser, US Patent 1429483 (1922).

22. The Boyland Synthesis

$$(CH_3)_2N-C_6H_5 \rightarrow$$

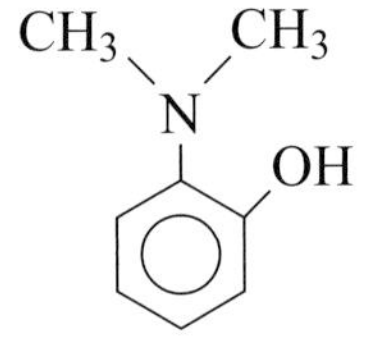

Reagents = (i) $KOH/K_2S_2O_8$ and (ii) HCl
E. Boyland and P. Sims, *J. Chem. Soc.* (1953), 3623.

23. The Brandi–Guarna Rearrangement

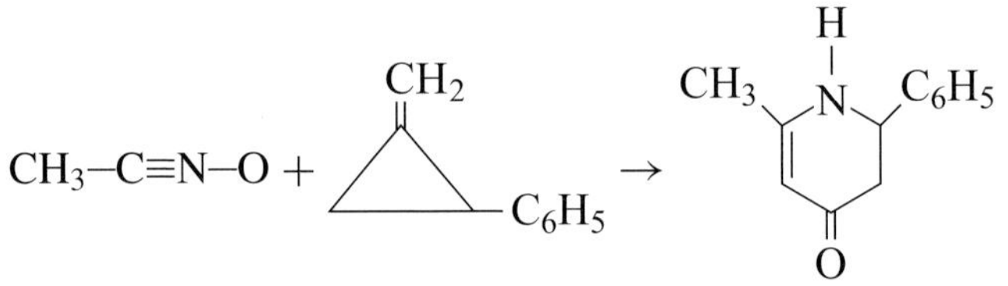

$$CH_3-C\equiv N-O + \rightarrow$$

A. Brandi and A. Guarna, *J. Chem. Soc. Chem. Commun.* (1985), 1518.

24. The von Braun Reaction

$$R_2NR + Br-CN \rightarrow R_2N-CN + R-Br$$

Reagent = phosphorous pentabromide
J. von Braun, *Ber.*, **40** (1907), 3914.

25. The Brederick Imidazole Synthesis

$$CH_3-CO-CO-CH_3 + H-CO-NH_2 \rightarrow$$

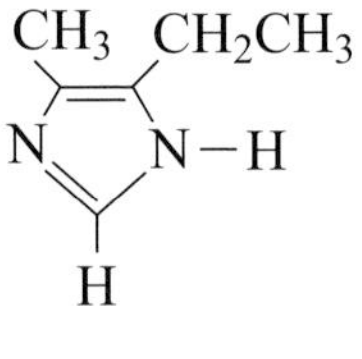

H. Brederick, *Ber.*, **86** (1953), 88.

26. The Brown Hydroboration

$$(CH_3)_2C=CH_2 \rightarrow (CH_3)_2C(H)-CH_2-NH_2$$

Reagents (i) BH_3/THF and (ii) H_2N-OSO_3H
H. C. Brown, *J. Am. Chem. Soc.*, **78** (1956), 2583.

27. The Bucherer Reaction

Reagent = $(NH_4)_2SO_3.NH_3$
H. T. Bucherer, *J. Prakt. Chem.*, **69** (1904), 49.

28. The Bucherer–Bergs Reaction

Reagents = (i) KCN and (ii) $(NH_4)_2CO_3$
H. T. Bucherer and W. Steiner, *J. Prakt. Chem.*, **140** (1934), 69.

29. The Bucherer Carbazole Synthesis

Reagent = $NaHSO_3$
H. T. Bucherer and F. Seyde, *J. Prakt. Chem.*, **77** (1908), 403.

30. The Buchner–Curtius–Schlotterbeck Reaction

$$R-CH_2-NH_2 + R'-CHO \rightarrow R-CH_2-COR' + N_2$$

E. Buchner and T. Curtius, *Ber.*, **18** (1885), 2371.

31. The Cadogan–Cameron-Wood Cyclization

Reagent = triethylphosphite
J. I. G. Cadogan, *J. Chem. Soc. Chem. Commun.* (1966), 491.

32. The Camps Quinoline Synthesis

Two products are possible, depending upon the R group present.
Reagent = NaOH
R. Camps, *Ber.*, **22** (1899), 3228.

33. The Chapmann Rearrangement

No reagents are required; the rearrangement is thermal.
A. W. Chapmann, *J. Chem. Soc.*, **127** (1925), 1992.

34. The Chichibabin Pyridine Synthesis

$$3R-CH_2-CHO + NH_3 \rightarrow$$

A. E. Chichibabin, *J. Russ. Phys. Chem. Soc.*, **37** (1906), 1229.

35. The Chichibabin Reaction

A. E. Chichibabin and O. E. Zeide, *J. Russ. Phys. Chem. Soc.*, **46** (1914), 216.

36. The Ciamician–Dennstedt Reaction

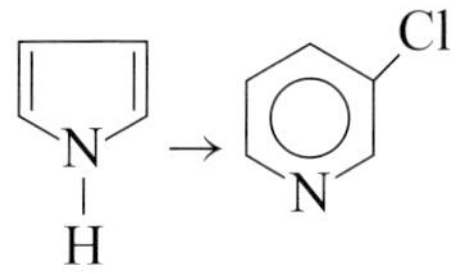

Reagents $= CHCl_3/NaOH$

G. L. Ciamician and M. Dennstedt, *Ber.*, **14** (1881), 1153.

37. The Combes Quinoline Synthesis

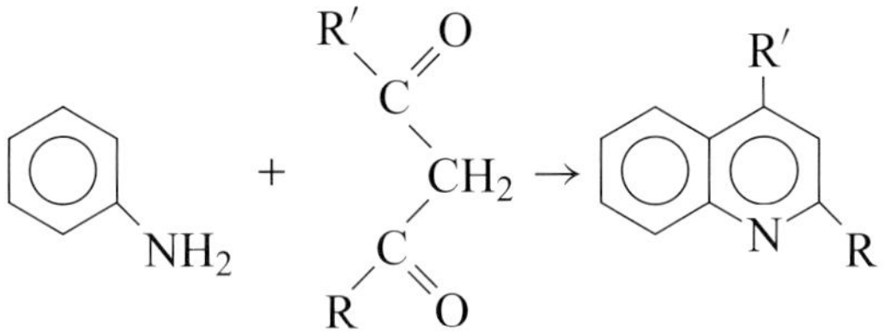

Reagent $= H_2SO_4$

A. Combes, *Bull. Soc. Chim. France*, **49** (1888), 89.

38. The Conrad–Limpach Reaction

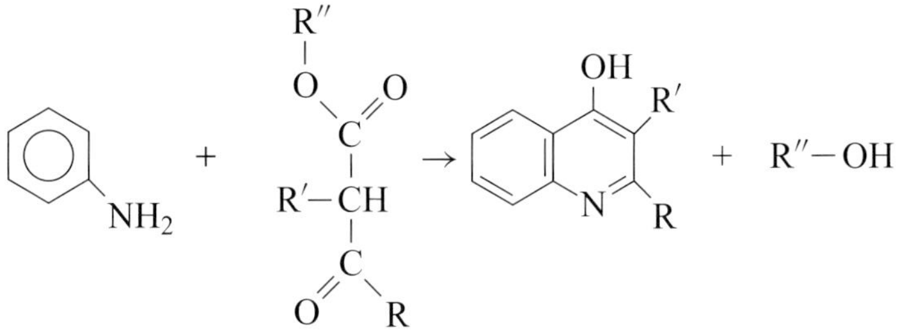

The reaction is a thermal condensation.

M. Conrad and L. Limpach, *Ber.*, **20** (1887), 944.

39. The Cope Elimination

$$R_2HC{-}CR_2{-}N(CH_3)_2{-}O \rightarrow R_2C{=}CR_2 + HO{-}N(CH_3)_2$$

The R groups may be different or the same.

A. C. Cope, T. T. Foster and P. H. Towle, *J. Am. Chem. Soc.*, **71** (1949), 3939.

40. The Cope–Mamloc–Wolfenstein Reaction

Reagents = (i) H_2O_2/CH_3OH and (ii) ΔH/vacuum
L. Mamloc and R. Wolfenstein, *Ber.*, **33** (1900), 159.

41. The Cornforth Rearrangement

This reaction is a thermal rearrangement in the presence of water.
J. W. Cornforth, *Chem. Penicillins*, **689** (1949), 705.

42. The Craig Reaction

Reagents = $HX/NaNO_2$, where X is a halogen
L. C. Craig, *J. Am. Chem. Soc.*, **56** (1934), 231.

43. The Curtius Rearrangement

$$R–CO–Cl + NaN_3 \rightarrow R–CO–N_3 \rightarrow R–N{=}C{=}O$$

This is a thermal rearrangement.
T. Curtius, *J. Prakt. Chem.*, **50** (1894), 275.

44. The Dakin–West Reaction

$$\underset{\underset{NH_2}{|}}{R–CH–CO_2H} \rightarrow \underset{\underset{NH–COCH_3}{|}}{R–CH–COCH_3}$$

Reagent = Ac_2O
H. D. Dakin and R. West, *J. Biol. Chem.*, **78** (1928), 745.

45. The Davidson Oxazole Synthesis

$$C_6H_5-CH(OH)-C(O)-C_6H_5 + C_6H_5-C(O)-Cl \rightarrow$$

Reagents = (i) pyridine and (ii) NH_4OAc
D. Davidson, *J. Org. Chem.*, **2** (1937), 328.

46. The Delépine Reaction

$$R-CH_2X + C_6H_{12}N_4 \rightarrow [R-CH_2-NH_3]^+[X]^-$$

Reagent = aqueous halogen acid, e.g. HX_{aq}
M. Delépine, *Compt. Rend.*, **124** (1897), 292.

47. The Demjanov Rearrangement

Reagents = (i) $NaNO_2$ and (ii) H_3PO_4
N. J. Demjanov, *J. Russ. Phys. Chem. Sci.*, **35** (1903), 26.

48. The Dimroth Rearrangement

Reagent = H_2O
O. Dimroth, *Ann.*, **459** (1927), 39.

49. The Doebner Reaction

Reagents = $C_6H_5-CHO/CH_3-CO-CO_2H$
O. Doebner, *Ann.*, **242** (1887), 265.

50. The Doebner–Miller Quinoline Synthesis

$$C_6H_5-NH_2 + RCH=CH-CHO \rightarrow$$

This is an acid-catalysed reaction.
O. Doebner and W. V. Miller, *Ber.*, **16** (1883), 2464.

51. The Duff Reaction

Reagents = $C_6H_{12}N_4$ + acid catalyst
J. C. Duff and E. J. Bills, *J. Chem. Soc.* (1932), 1987.

52. The Ehrlich–Sachs Reaction

$$C_6H_5-CH_2CN + C_6H_5-NO \rightarrow C_6H_5-C=N-C_6H_5 + H_2O$$
$$\overset{|}{C \equiv N}$$

This is a base-catalysed reaction.
P. Ehrlich and F. Sachs, *Ber.*, **32** (1899), 2341.

53. The Einhorn–Brunner Reaction

A. Einhorn, E. Bischkopff and B. Szelinski, *Ann.*, **343** (1905), 229.

54. The Emde Degradation

Reagents = Na/Hg and ethanol
H. Emde, *Ber.*, **42** (1909), 2590.

55. The Emmert Reaction

$$\text{[pyridine]} + RR'C{=}O \rightarrow \text{[pyridine]} \underset{R'}{\overset{R}{\underset{|}{C}}} OH$$

Reagents $=$ Al/Hg or Mg/Hg
B. Emmert and E. Asendorf, *Ber.*, **72** (1939), 1188.

56. The Eshweiler–Clarke Reaction

$$R{-}NH_2 + 2HCHO + 2HCOOH \rightarrow R{-}N(CH_3)_2 + 2CO_2 + 2H_2O$$

H. T. Clarke, H. B. Gillespsie and S. Z. Weisshaus, *J. Am. Chem. Soc.*, **55** (1933), 4571.

57. The Finegan Tetrazole Synthesis

$$F_3C{-}CN + C_8H_{17}{-}N_3 \rightarrow \text{[tetrazole ring]}$$

Reagent $=$ Cl–SO$_3$H
W. G. Finegan, *J. Am. Chem. Soc.*, **80** (1956), 3908.

58. The Fischer–Hepp Rearrangement

$$\text{[R-N-NO benzene]} \rightarrow \text{[R-N-H benzene, NO]}$$

Reagent $=$ HCl
O. Fischer and E. Hepp, *Ber.*, **19** (1886), 2991.

59. The Fischer Indole Synthesis

Reagent = zinc chloride
E. Fischer and F. Jourdan, *Ber.*, **16** (1883), 2241.

60. The Fischer Oxazole Synthesis

$$R-CH(OH)-CN + R'CHO \rightarrow$$

Reagents = diethylether/HCl
E. Fischer, *Ber.*, **29** (1896), 205.

61. The Fischer Phenylhydrazine Synthesis

$$[R-N_2]^+[NaSO_3]^- \rightarrow R-NH-NH_2$$

Reagents = (i) $NaHSO_3$ and (ii) HCl
E. Fischer, *Ber.*, **8** (1875), 589.

62. The Fischer Phenylhydrazone and Osazone Reaction

$$R-CH(OH)-CHO \rightarrow R-C(=NH-NH-C_6H_5)-CH=NHNHC_6H_5$$

Reagent = *xs* $C_6H_5-NH-NH_2$
E. Fischer, *Ber.*, **17** (1884), 579.

63. The Forster Reaction

$$R-NH_2 + C_6H_5-CHO \rightarrow R'-NH-R + C_6H_5-CHO$$

Reagents = (i) R'—Cl and (ii) H_2O
M. O. J. Forster, *J. Chem. Soc.*, **107** (1915), 260.

64. The Friedlander Quinoline Synthesis

$$+ R-CH_2-C(=O)R' \rightarrow$$

This reaction is base catalysed.
P. Friedlander, *Ber.*, **15** (1882), 2572.

65. The Gabriel–Colman Rearrangement

Reagents = (i) Cl–CH$_2$–CO$_2$R and (ii) NaOC$_2$H$_5$
S. Gabriel and J. Colman, *Ber.*, **33** (1900), 980.

66. The Gabriel–Heine Aziridine Isomerization

Reagents = (i) (C$_2$H$_5$)$_3$N and (ii) I$_2$
H. W. Heine, *J. Org. Chem.*, **23** (1958), 1554.

67. The Gabriel–Marckwald Reaction

$$H_2N–CH_2–CH_2–Br \rightarrow$$

Reagent = KOH
S. Gabriel and W. Marckwald, *Ber.*, **21** (1888), 1049.

68. The Gabriel Primary Amine Synthesis

$$\rightarrow R–NH_2 + C_6H_4(COOK)_2$$

Reagents = (i) R–X, where X = halogen, and (ii) water
S. Gabriel, *Ber.*, **20** (1887), 2224.

69. The Gastaldi Synthesis

$$2RCOCH(CN)NHSO_3K + 2HCl + [O] \rightarrow$$

G. Gastaldi, *Gazz. Chim. Ital.*, **51** (1921), 233.

70. The Gattermann Reaction

$$[C_6H_5-N_2]^+[X]^- + KCN \rightarrow C_6H_5-CN + N_2 + KCl$$

Reagent = copper powder
L. Gattermann, *Die Praxis des organischen Chemickers*, 40th edition (Berlin: de Gruyter and Co., 1961), p. 260. (This is referred to by chemists as 'Gattermann's Kochbuch'!)

71. The Gomberg–Bachmann Reaction

$$C_6H_5-NH_2 + C_6H_6 \rightarrow C_6H_5-C_6H_5$$

Reagents = (i) NaNO$_2$/HCl and (ii) NaOH
M. Gomberg and W. E. Bachmann, *J. Am. Chem. Soc.*, **46** (1924), 2339.

72. The Gould–Jacobs Quinoline Synthesis

Reagents = (i) (C$_2$H$_5$-O$_2$C)$_2$C=CH-O-C$_2$H$_5$ and (ii) NaOH/HCl
R. G. Gould and W. A. Jacobs, *J. Am. Chem. Soc.*, **61** (1939), 2890.

73. The Graebe–Ullmann Synthesis

Reagents = (i) HCl/NaNO$_2$ and (ii) thermal decomposition
C. Graebe and F. Ullmann, *Ann.*, **291** (1896), 16.

74. The Graham Diaziridine Synthesis

$$CH_3-C(NH_2)=NH.HCl \rightarrow$$

Reagents = (i) NaOCl and (ii) (CH$_3$)$_2$S=O
W. H. Graham, *J. Am. Chem. Soc.*, **87** (1965), 4396.

75. The Griess Deamination

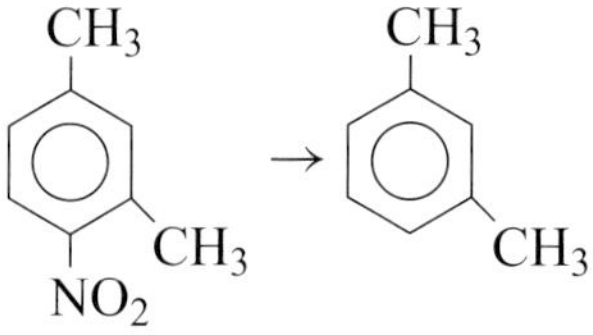

Reagents = (i) $NaNO_2/HCl$ and (ii) PO_2H_3

P. Griess, *Phil. Trans. R. Soc. Lond.*, **154** (1864), 683.

(The widely used Griess Reagent for the deamination of proteins is one part of 1% sulphanilamide with one part of 0.1% naphthylenediamine dihydrochloride in 5% phosphoric acid.)

76. The Groß–Eschenmoser Fragmentation

$$R_2N-CH_2-CH_2-CH_2-X \rightarrow [R_2N=CR'_2]^+[X]^- + R'_2C=CR'_2$$

Reagents = $R'-SO_2-NH-NH_2$, where $R' = CH_3$ or H

A. Eschenmoser and A. Frey, *Helv. Chim. Acta*, **35** (1952), 1660.

77. The Guarreschi–Thorpe Condensation

$$CH_3-C(=O)-CHR-CO_2R + RO_2C-CH_2-CN \rightarrow$$

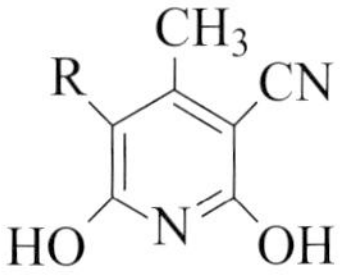

Reagent = alcoholic ammonia

H. Baron, F. G. P. Renfry and J. F. Thorpe, *J. Chem. Soc.*, **85** (1904), 1726.

78. The Gutknecht Pyrazine Synthesis

$$2R-C(=O)-CH_2-R' \rightarrow$$

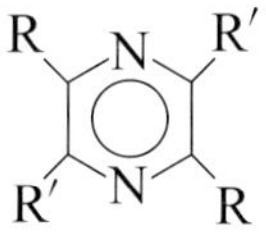

Reagents = (i) HNO_2, (ii) H_2/catalyst and (iii) [O]

H. Gutknecht, *Ber.*, **12** (1879), 2290.

79. The Hantzsch Pyridine Synthesis

$$CH_3C(=O)-C=CH-CH_3 + CH_3C(=O)-CH_2-CO_2C_2H_5$$
$$| \atop CO_2C_2H_5$$

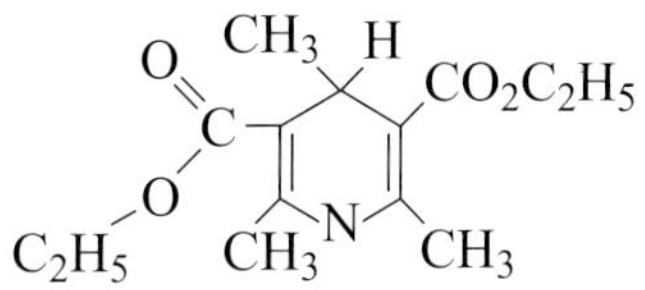

Reagent = aqueous ammonia

A. Hantzsch, *Ann.*, **215** (1882), 1.

80. The Hantzsch Pyrrole Synthesis

$$CH_3-C(=O)-CH_2Cl + CH_3-C(=O)-CH_2-CO_2C_2H_5$$

$$\rightarrow$$ [pyrrole ring with $CO_2-C_2H_5$ at position 3, CH_3 at positions 2 and 5, N—H]

Reagent = ammonia
A. Hantzsch, *Ber.*, **23** (1890), 1474.

81. The Hassner Azirine (and Aziridine) Synthesis

$$C_6H_5-CH=CH-CH_3 \xrightarrow{\text{Route A}}$$ [aziridine ring with C_6H_5 and CH_3 substituents, N—H]

$$\downarrow \text{Route B}$$

[azirine ring with C_6H_5 and CH_3 substituents, N]

Reagents = Route A (i) ICl/NaN$_3$ and (ii) LiAlH$_4$
        Route B (i) ICl/NaN$_3$ and (ii) t-BuOK
A. Hassner, *Accts Chem. Rev.*, **4** (1971), 9.

82. The Herz Reaction

$$C_6H_5-NH_2 \rightarrow$$ [benzene ring with NH_2, Cl, and S^-Na^+ substituents]

Reagents = (i) S$_2$Cl$_2$ and (ii) NaOH
R. Herz, German Patent DE360690 (1914).

83. The Herzig–Meyer Determination

$$RRN-CH_3 + HI \rightarrow CH_3I + RRNH$$

J. Herzig and H. Meyer, *Ber.*, **27** (1894), 319.

84. The Heumann–Pfleger Indigo Synthesis

$$C_6H_5-NH-CH_2-CO_2H \rightarrow$$

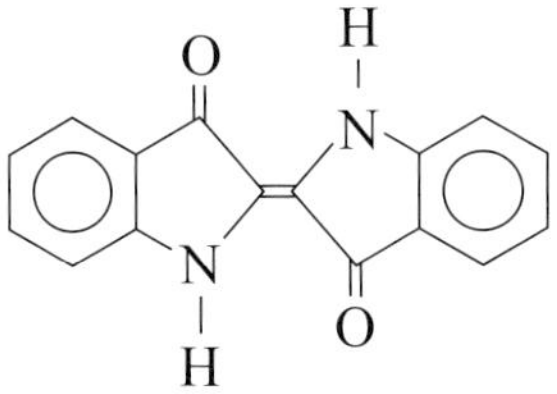

Reagent = $NaNH_2$
K. Heumann, *Ber.*, **23** (1890), 3043.

85. The Hilbert–Johnson Reaction

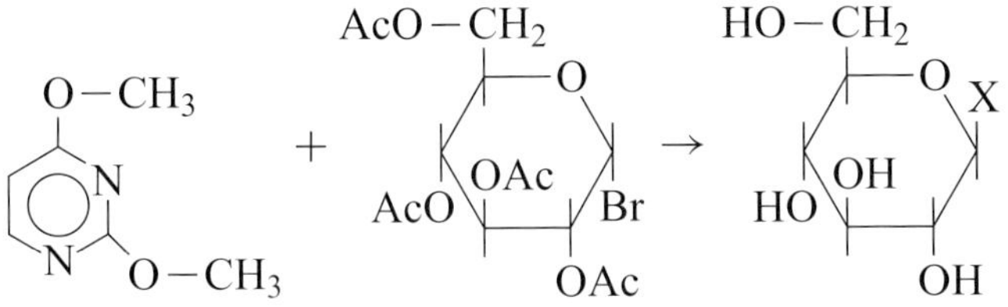

where X = 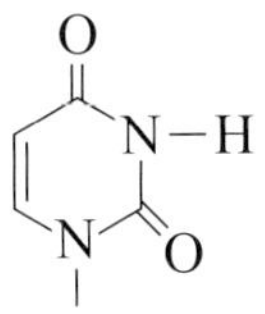   or

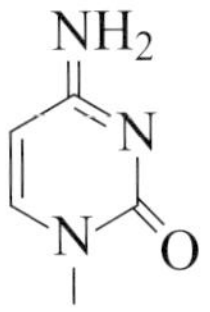

Reagents = HCl/C_2H_5OH or NH_3/C_2H_5OH
T. B. Johnson and G. E. Hilbert, *Science*, **69** (1929), 579.

86. The Hinsberg Oxindole Synthesis

$$C_6H_5-NH-R + (NaO_2SO-CH(OH)-)_2 \rightarrow$$

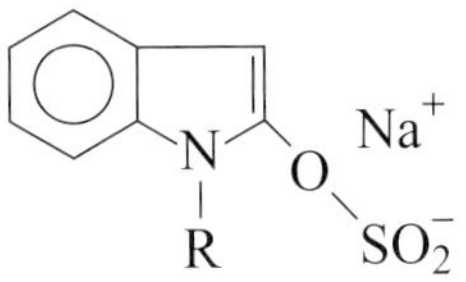

Reagents = sodium bisulphite and glyoxal
O. Hinsberg, *Ber.*, **21** (1888), 110

87. The Hinsberg Test

$$C_6H_5SO_2Cl + RR'NH \rightarrow C_6H_5-SO_2-NRR'$$

When R = H the product is soluble in NaOH solution.
When R = an alkyl or aryl group the product is insoluble in NaOH solution.
O. Hinsberg, *Ber.*, **23** (1890), 2962.

88. The Hoch–Campbell Aziridine Synthesis

Reagent = LiAlH$_4$

J. Hoch, *Comptes Rend.*, **198** (1934), 1865.

89. The Hofmann Degradation

$$R-CONH_2 \rightarrow R-NH_2$$

Reagents = bromine and sodium hydroxide
A. W. Hofmann, *Ber.*, **14** (1881), 2725.

90. The Hofmann Isocyanide Synthesis

$$R-NH_2 + CHCl_3 + 3NaOH \rightarrow R-NC + 3NaCl + 3H_2O$$

A. W. Hofmann, *Ann.*, **146** (1868), 107.

91. The Hofmann–Löffler–Freytag Reaction

This reaction is acid catalysed, or with *hν*.
A. W. Hofmann, *Ber.*, **16** (1883), 558.

92. The Hofmann–Martius Rearrangement

This reaction proceeds by thermal conversion.
A. W. Hofmann and P. Martius, *Ber.*, **4** (1871), 742.

93. The Hofmann Reaction

$$NH_3 + R-X \rightarrow R-NH_2 + HX$$

where X = halogen
A. W. Hofmann, *Ann.*, **78** (1851), 253.

94. The Huisgen Tetrazole Rearrangement

$$Ar-\underset{N=N}{\overset{N-N\overset{H}{\diagup}}{\diagdown}}\Big| \;+\; Ar'-C\overset{O}{\underset{Cl}{\diagup}} \;\rightarrow\; Ar-\underset{O}{\overset{N-N}{\diagdown}}-Ar'$$

R. Huisgen, *Chem. Ber.*, **93** (1960), 2106.

95. The Japp Oxazole Synthesis

$$C_6H_5-CH(OH)-CO-C_6H_5 + KCN \rightarrow \quad \underset{C_6H_5}{\overset{C_6H_5}{\diagup}}\underset{O}{\overset{N}{\diagdown}}$$

F. R. Japp, *J. Chem. Soc.*, **63** (1893), 469.

96. The Japp–Klingemann Reaction

$$CH_3-C(=O)-CH(CH_3)-CO_2H + [ArN_2]^+[X]^-$$
$$\rightarrow CH_3-C(=O)-C(CH_3)=NH-NHAr$$

This reaction is base catalysed.
F. R. Japp and F. Klingemann, *Ann.*, **247** (1888), 190.

97. The Jourdan Diphenylamine Synthesis

F. Jourdan, *Ber.*, **18** (1885), 1444.

98. The Kaiser–Johnson–Middleton Dinitrile Cyclization

Reagents = HBr/HOAc
F. Johnson, *J. Org. Chem.*, **29** (1964), 153.

99. The Kaluza Synthesis

$$C_6H_5-NH_2 \rightarrow C_6H_5-N=C=S$$

Reagents = (i) $(C_2H_5)_3N/CS_2$, (ii) $Cl-CO_2C_2H_5/CHCl_3$ and (iii) $(C_2H_5)_3N$
H. Kaluza, *Monatschr.*, **33** (1912), 363.

100. The Kametani Oxidation

Reagents = Cu_2Cl_2/O_2 and (ii) pyridine
T. Kametani, *Synth.* (1977), 245.

101. The Kishner Cyclopropane Synthesis

This reaction proceeds by thermal decomposition.
N. M. Kishner and A. Zavadovskii, *J. Russ. Phys. Chem.*, **43** (1911), 1132.

102. The Knoop–Osterlin Amino Acid Synthesis

$$R-CO-CO_2H + NH_3(aq) \rightarrow R-CH-CO_2H$$
$$|$$
$$NH_2$$

Reagents = Pt or Raney nickel catalyst + H_2
F. Knoop and H. Oesterlin, *Z. Physiol. Chem.*, **148** (1925), 294.

103. The Knorr Pyrazole Synthesis

$$O=CR-CHR'-C(=O)R'' + NH_2-NHAr \rightarrow$$

L. Knorr, *Ber.*, **16** (1883), 2587.

104. The Knorr Pyrrole Synthesis

$$H_2N-CHR-C(=O)R' + R''-CO-CH_2-CO_2C_2H_5 \rightarrow$$

L. Knorr, *Ber.*, **17** (1884), 1635.

105. The Knorr Quinoline Synthesis

$$C_6H_5-NH_2 + O=CR-CHR'-CO_2C_2H_5 \rightarrow$$

Reagents = sulphuric acid (concentrated)
L. Knorr, *Ann.*, **236** (1886), 69.

106. The König Benzoxazine Synthesis

Reagent = $CH_3-NH-CH_2-CH_2-Br.HBr$
K. H. König, *Ber.*, **92** (1959), 257.

107. The Kresze Amination Reaction

$$R_2-C=CH-CH_2-C_6H_5 \rightarrow R_2C=CH-CH-C_6H_5$$
$$H-N-CO_2CH_3$$

Reagent = SCl_2 and $CH_3O_2C-NCl_2$
G. Kresze, *Liebigs Ann.* (1975), 1725.

108. The Kröhnke Aldehyde Synthesis

$$R-CH_2-X + pyridine \rightarrow$$

$$R-CH_2-\overset{+}{N} \rightarrow R-CH=N- \rightarrow R-CHO$$

Reagents = (i) *p*-nitrosodimethylaniline and (ii) acid hydrolysis to aldehyde
F. Kröhnke *et al., Ber.*, **69** (1936), 2006.

109. The Ladenburg Rearrangement.

This reaction proceeds by thermal rearrangement.
A. Landenburg, *Ber.*, **16** (1883), 410.

110. The Leimgruber–Batcho Indole Synthesis

Reagents = (i) $(CH_3O)_2CH-N(CH_3)_2$ and (ii) Al/Hg
A. D. Batcho and W. Leimgruber, US Patent 3976639 (1975).

111. The Lehmstedt–Tãnãsescu Reaction

Reagents = H_2SO_4/HNO_2
K. Lehmstedt and I. Tãnãsescu, *Bull. Soc. Chim. France*, **41** (1927), 528.

112. The Leuckart–Wallach Reaction

$$RR'NH + R''R'''C=O \rightarrow RR'N-CHR''R'''$$

Reagent = formic acid
R. Leuckart, *Ber.*, **18** (1885), 2341.

113. The Lossen Rearrangement

$$R-CO-NH-OH \rightarrow R-NH_2$$

Reagent = water
W. Lossen, *Ann.*, **161** (1872), 347.

114. The MacDonald Porphyrin Synthesis

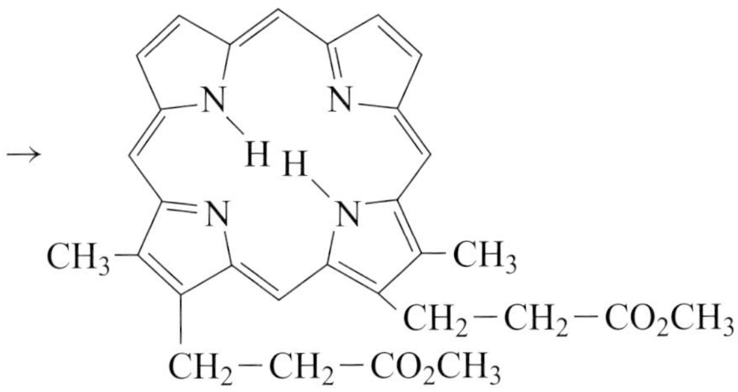

Reagents = HI/HOAc
S. P. MacDonald, *J. Am. Chem. Soc.*, **82** (1960), 4384.

115. The MacFadyen–Stevens Reaction

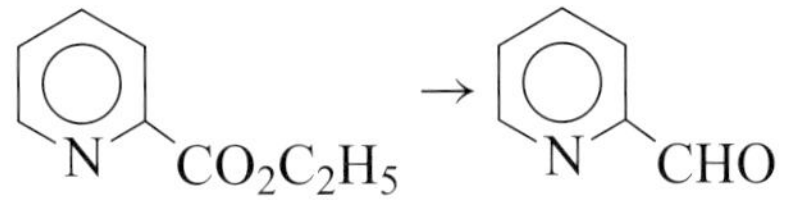

Reagents = (i) NH_2–NH_2, (ii) C_6H_5–SO_2Cl and (iii) Na_2CO_3
J. S. McFadyen and T. S. Stevens, *J. Chem. Soc.* (1936), 584.

116. The Madelung Indole Synthesis

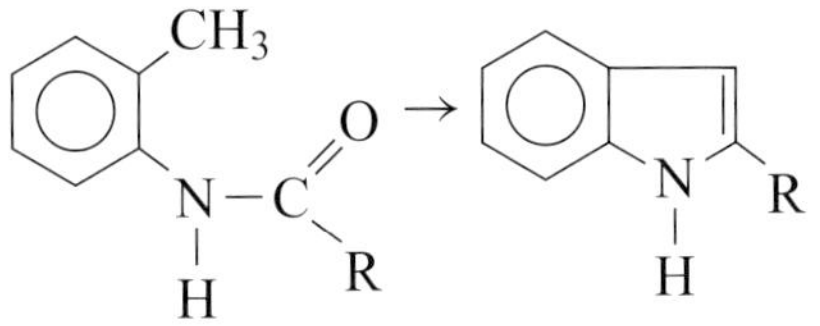

Reagents = $NaOC_2H_5$ + high temperature and pressure
W. Madelung, *Ber.*, **45** (1912), 1128.

117. The Makosza Substitution

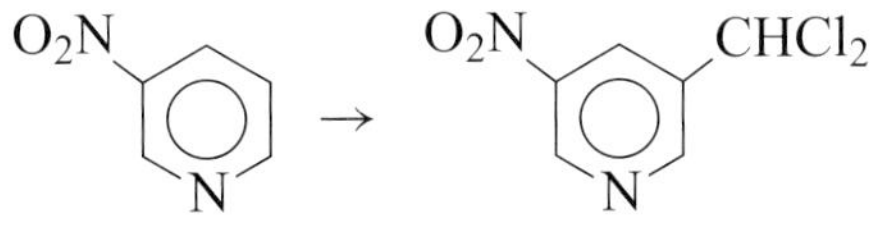

Reagents = $CHCl_3$/$NaOCH_3$/NH_3
M. Makosza, *J. Org. Chem*, **48** (1983), 3860.

118. The Mannich Reaction

$$(CH_3)_2NH + HCHO + CH_3COCH_3 \rightarrow (CH_3)_2N\text{–}CH_2\text{–}CH_2\text{–}COCH_3 + H_2O$$

C. Mannich and W. Krosche, *Arch. Pharm.*, **250** (1912), 647.

119. The Martinet Dioxindole Synthesis

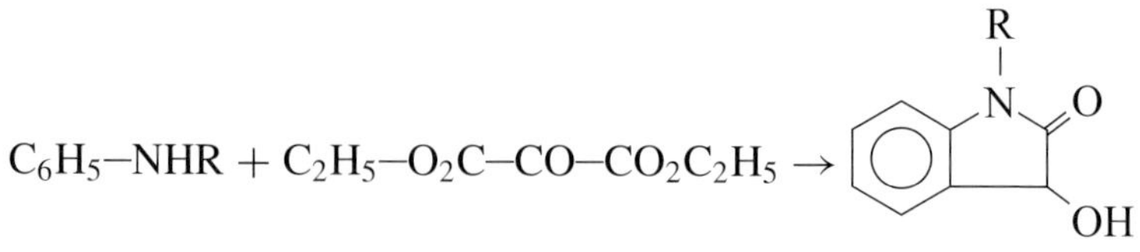

$$C_6H_5-NHR + C_2H_5-O_2C-CO-CO_2C_2H_5 \rightarrow$$

Reagent = water
A. Guyot and J. Martinet, *Comptes Rend.*, **156** (1913), 1625.

120. The Meisenheimer Rearrangement

$$(CH_3)_2-\underset{R}{N}\rightarrow O \rightarrow (CH_3)_2N-OR$$

This reaction proceeds by thermal rearrangement.
J. Meisenheimer, *Ber.*, **52** (1919), 1667.

121. The Menschutkin Reaction

$$RR'R''N + R'''X \rightarrow [RR'R''R'''N]^+[X]^-$$

N. Menschutkin, *Z. Physik. Chem.*, **5** (1890), 589.

122. The Michaelis Phosphonylation Reaction

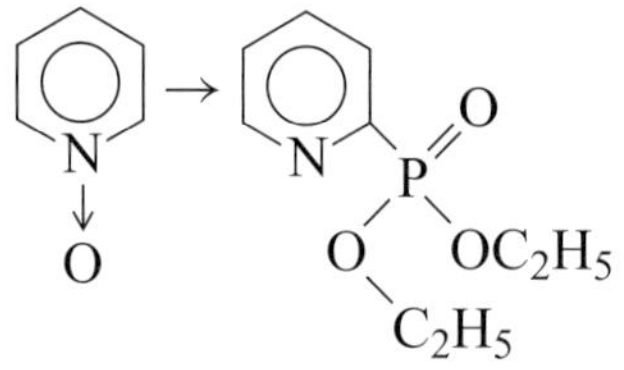

Reagents = (i) $(CH_3)_2SO_4$ and (ii) $Li-PO(OC_2H_5)_2$
A. Michaelis, *Ber.*, **31** (1898), 1048.

123. The Mignonac Reaction

$$R-COR' + NH_3 + H_2 \rightarrow RR'CH-NH_2 + H_2O$$

Reagents = Ni catalyst and absolute ethanol
G. Mignonac, *Comptes Rend.*, **172** (1921), 223.

124. The Minischi Substitution

$$(CH_3)_2N-Cl + C_6H_5-H \rightarrow (CH_3)_2N-C_6H_5$$

Reagents = $FeSO_4$/HOAc
F. Minischi, *Tetrahedron Lett.* (1965), 433.

125. The Neber Rearrangement

$$R-CH_2-C=NOSO_2C_6H_5 \quad \rightarrow \quad R-CH-COR' \quad + C_6H_5-SO_3H$$

with R' below the left structure and NH$_2$ below the right structure.

Reagents = (i) KOC$_2$H$_5$ and (ii) H$_2$O
P. W. Neber, and A. V. Friedolsheim, *Ann.*, **449** (1926), 109.

126. The Nenitzeschu Indole Synthesis

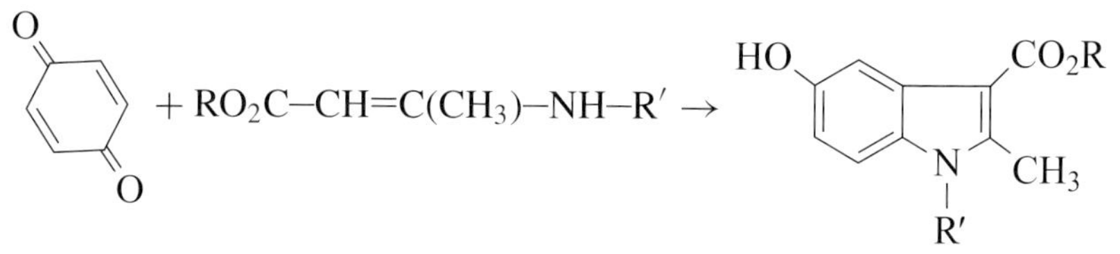

$+ RO_2C-CH=C(CH_3)-NH-R' \rightarrow$

C. D. Nenitzeschu, *Bull. Soc. Chim. Romania*, **11** (1929), 37.

127. The Niementowski Quinazoline Synthesis

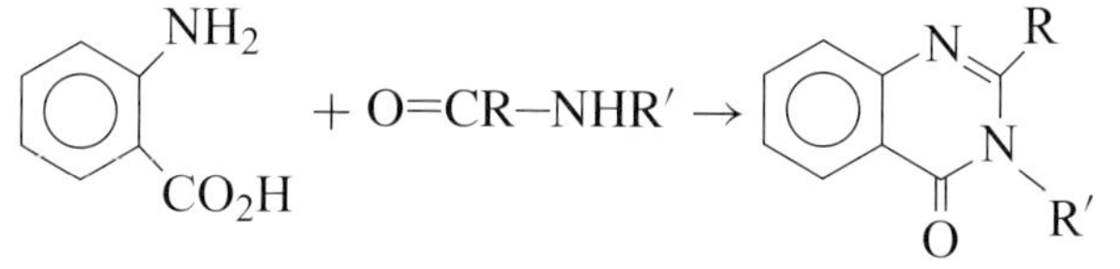

$+ O=CR-NHR' \rightarrow$

S. V. Niementowski, *J. Prakt. Chem.*, **51** (1895), 564.

128. The Niementowski Quinoline Synthesis

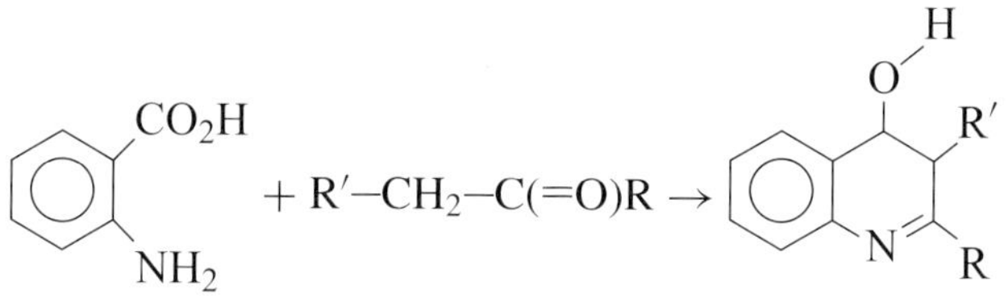

$+ R'-CH_2-C(=O)R \rightarrow$

S. V. Niementowski, *Ber.*, **27** (1894), 1394.

129. The Orton Rearrangement

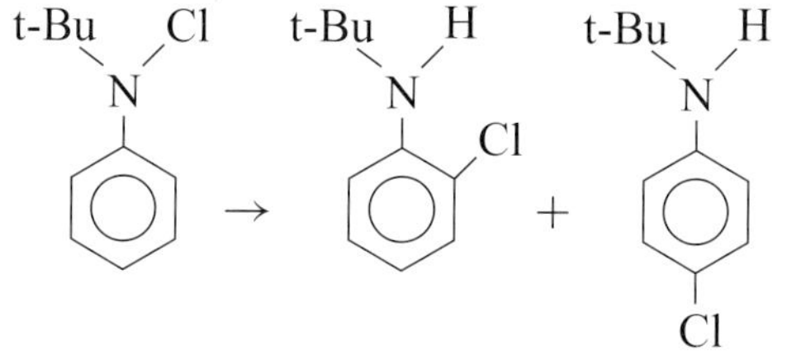

Reagent = NaOH
K. J. Orton, *J. Chem. Soc.*, **95** (1909), 1465.

130. The Paal–Knorr Pyrrole Synthesis

$$RC(=O)-CH_2-CH_2-C(=O)R' + NH_3 \rightarrow$$

Ammonia may be replaced with a primary amine.
L. Knorr and C. Paal, *Ber.*, **18** (1885), 299 and 367.

131. The Padwa Pyrroline Synthesis

Reagents = (i) $C_6H_5-CH_2-NH_2$ and (ii) $NaOCH_3$
A. Padwa, *Tetrahedron Lett.* (1988), 2417.

132. The Passerini Condensation

Reagents = (i) F_3C-CO_2H and (ii) NaOH
M. Passerini, *Gazz. Chim. Ital.*, **51** (1921), 126.

133. The Pechmann Pyrazole Synthesis

$$H-C\equiv C-H + H_2C-N_2 \rightarrow$$

H. V. Pechmann, *Ber.*, **31** (1898), 2950.

134. The Pellizzari Reaction

$$R-C(OH)=NH + HO-C(R)=N-NH-C_6H_5 \rightarrow$$

G. Pellizzari, *Gazz. Chim. Ital.*, **41** (1911), 20.

135. The Petrenko-Kritschenko Piperidone Synthesis

$R-CO_2C-CH_2-CO-CH_2-CO_2R + 2C_6H_5-CHO + R'-NH_2$

$\rightarrow$

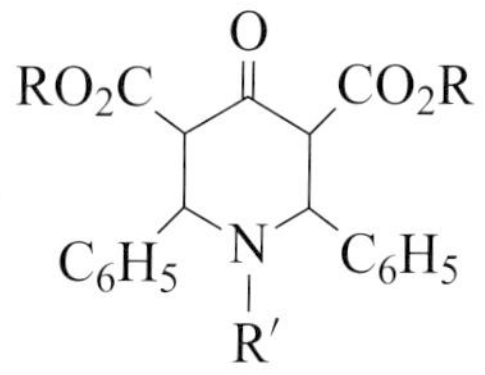

P. Petrenko-Kritschenko and S. Schöttle, *Ber.*, **39** (1906), 1358.

136. The Pfitzinger Reaction

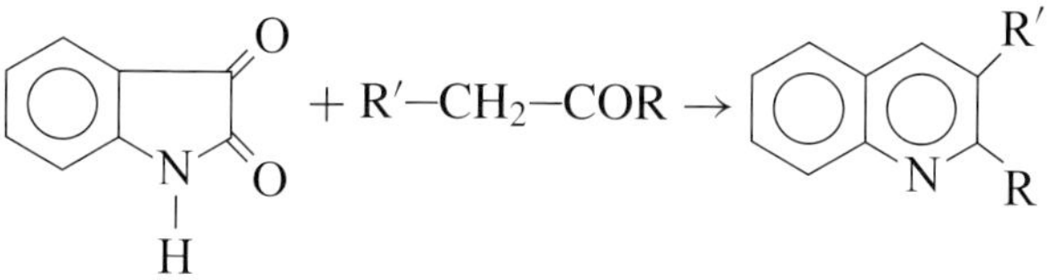

$+ R'-CH_2-COR \rightarrow$

This reaction proceeds by a thermal decarboxylation.
W. Pfitzinger, *J. Prakt. Chem.*, **33** (1886), 100.

137. The Pictet–Gams Isoquinoline Synthesis

$C_6H_5-CH(OH)-CH_2-NH-COR \rightarrow$

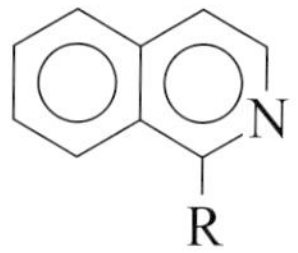

Reagent $= P_2O_5$
A. Pictet and A. Gams., *Ber.*, **43** (1910), 2384.

138. The Pictet–Hubert Reaction

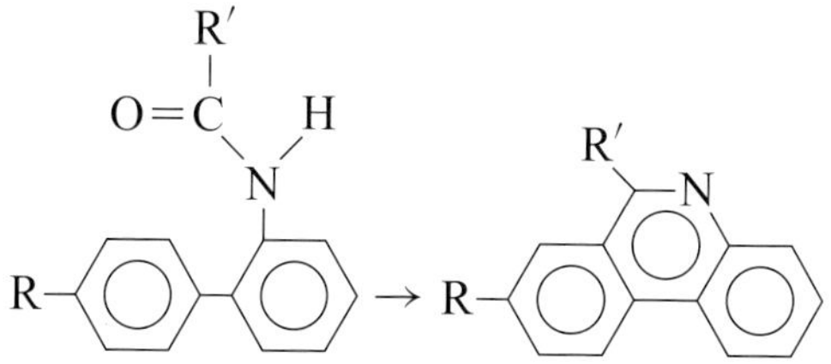

$\rightarrow$

Reagents $=$ ZnCl$_2$ or POCl$_3$ in nitrobenzene
A. Pictet and A. Hubert, *Ber.*, **29** (1896), 1182.

139. The Pictet–Spengler Reaction

RO, RO — CH$_2$CH$_2$NH$_2$ + R′—CHO → RO, RO — (tetrahydroisoquinoline with N—H, R′ H)

Reagent = HCl
A. Pictet and T. Spengler, *Ber.*, **44** (1911), 2030.

140. The Piloty–Robinson Synthesis

$$CH_3-CH_2-C(C_2H_5)=N-N=C(C_2H_5)-CH_2-CH_3 \rightarrow C_2H_5-\underset{\underset{H}{N}}{pyrrole}-C_2H_5$$

with CH$_3$, CH$_3$ substituents

Reagent = HCl
O. Piloty, *Ber.*, **43** (1910), 489.

141. The Pinner Triazine Synthesis

$$2Ar-C(NH_2)=NH + COCl_2 \rightarrow$$

OH triazine with Ar, N, Ar substituents

Ar = aryl group only (see also the Schroeder–Grundmann Reaction)
A. Pinner, *Ber.*, **23** (1890), 2919.

142. The Piria Reaction

NO$_2$—C$_6$H$_5$ + NaHSO$_3$ → NH$_2$—C$_6$H$_4$—SO$_3$Na + NH$_2$—C$_6$H$_5$

Reagent = HCl/H$_2$O
R. Piria, *Ann.*, **78** (1851), 31.

143. The Polonovski Reaction

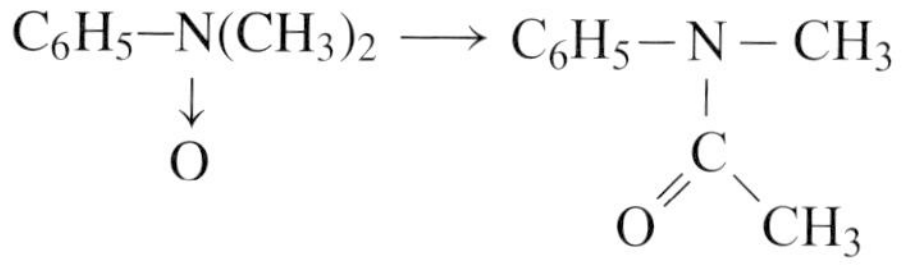

$$C_6H_5-N(CH_3)_2 \longrightarrow C_6H_5-N-CH_3$$

Reagent = Ac_2O
M. Polonovski and M. Polonovski, *Bull. Soc. Chem. Fr.*, **41** (1927), 1190.

144. The Pomeranz–Fritsch Reaction

$$C_6H_5-CO_2H + H_2N-CH_2-CH(OC_2H_5)_2 \rightarrow$$

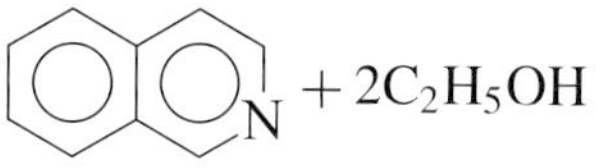

$+ 2C_2H_5OH$

C. Pomerantz, *Monatschr.*, **14** (1893), 116; P. Fritsch, *Ber.*, **26** (1893), 419.

145. The Pudovik Reaction

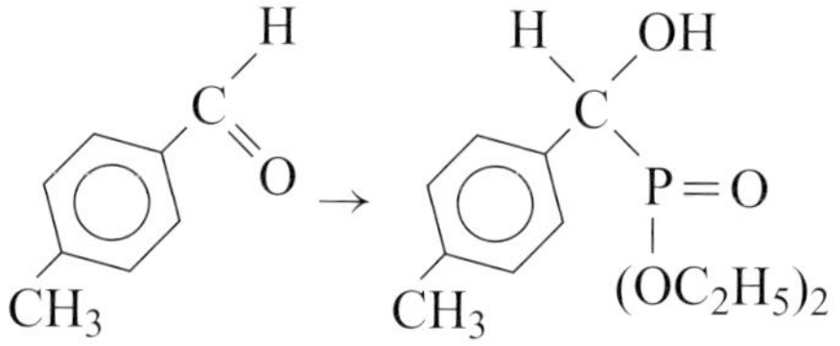

Reagents = (i) $H-PO-(OC_2H_5)_2$ and (ii) LLB/THF
A. N. Pudovik, *Synth.* (1979), 79.

146. The Reissert Indole Synthesis

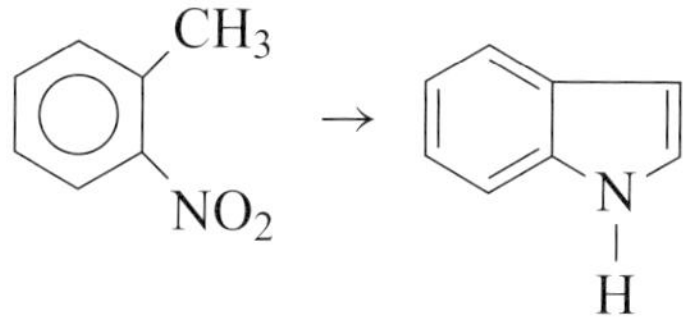

Reagents = (i) $(CO_2C_2H_5)_2/NaOC_2H_5$ and (ii) Zn/AcOH
A. Reissert, *Ber.*, **30** (1897), 1030.

147. The Reissert Reaction

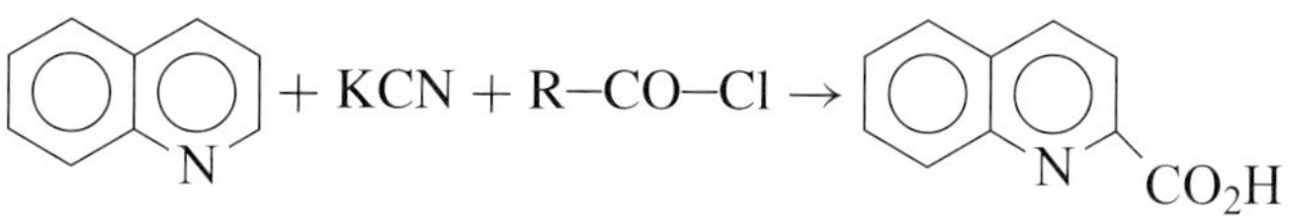

$+ KCN + R-CO-Cl \rightarrow$

Reagent = HCl
A. Reissert, *Ber.*, **38** (1905), 1603.

148. The von Richter Cinnoline Synthesis

Reagents = (i) $NaNO_2$/HCl and (ii) H_2O
V. von Richter, *Ber.*, **16** (1883), 677.

149. The Riehm Quinoline Synthesis

$$C_6H_5-NH_2 + 2CH_3CO-CH_3 \rightarrow$$

P. Riehm *et al.*, *Ber.*, **18** (1885), 2245.

150. The Ritter Reaction

$$R_3C-OH + HCN \rightarrow H-CO-NH-CR_3$$

Reagents = acidic media
J. J. Ritter and P. P. Minieri, *J. Am. Chem. Soc.*, **70** (1948), 4045.

151. The Robinson–Schopf Reaction

$$OHC-CH_2-CH_2-CHO + CH_3NH_2 + HO_2C-CH_2-CO-CH_2-CO_2H$$

R. Robinson, *J. Chem. Soc.*, **111** (1917), 762.

152. The Rosenmund–Braun Reaction

Reagents = CuCN/DMF
K. W. Rosenmund, *Ber.*, **52** (1916), 1749.

153. The Rothenmund Reaction

P. Rothenmund, *J. Am. Chem. Soc.*, **57** (1935), 2010.

154. The Rowe Rearrangement

Reagent = dilute aqueous acid
F. M. Rowe *et al.*, *J. Chem. Soc.* (1926), 690.

155. The Sandemeyer Diphenylurea Isatin Synthesis

$(C_6H_5-NH)_2C=S \rightarrow$

Reagents = (i) KCN/PbCO$_3$, (ii) (NH$_4$)$_2$S, (iii) AlCl$_3$ and (iv) HCl
T. Sandemeyer, *Z. Farb. Textile Chem.*, **2** (1903), 129.

156. The Sandemeyer Isonitrosoacetanilide Isatin Synthesis

$C_6H_5-NH-C-CH=N-OH \rightarrow$
$\quad\quad\quad\quad ||$
$\quad\quad\quad N-C_6H_5$

Reagents = (i) H$_2$SO$_4$ and (ii) H$_2$O
T. Sandemeyer, *Helv. Chim. Acta.*, **2** (1919), 234.

157. The Sandemeyer Reaction

$$[C_6H_5{-}N_2]^+[X]^- + KCN \rightarrow C_6H_5{-}CN + N_2 + KCl$$

Reagent = cuprous salts
T. Sandemeyer, *Ber.*, **17** (1884), 1633.

(If this reaction is conducted with cupric salts as catalysts then it is called the Körner–Contardi Reaction. See G. Körner and A. Contardi, *Atti. Accad. Nazl. Lincei*, **23** (1914), 464.)

158. The Schiemann Reaction

$$Ar{-}NH_2 + HNO_2 + HBF_4 \rightarrow Ar{-}F + N_2 + BF_3$$

This reaction proceeds by a thermal decomposition.
G. Balz and G. Schiemann, *Ber.*, **60** (1927), 1186.

159. The Scheiner Aziridine Synthesis

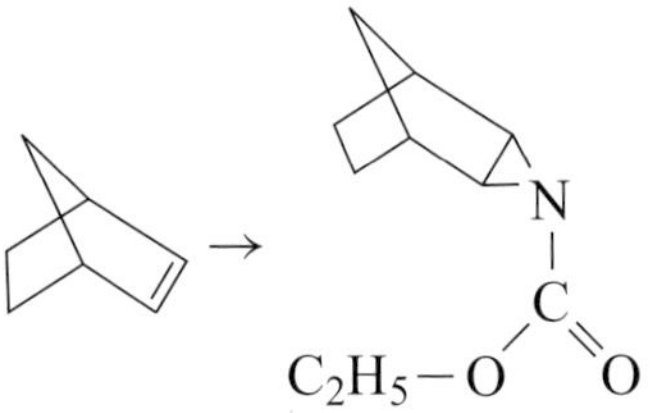
N
C
C2H5–O
O

Reagents = $N_3CO_2C_2H_5$/pentane
P. Scheiner, *J. Org. Chem.*, **26** (1961), 1923.

160. The Schmidt Reaction

$$R{-}CO_2H + HN_3 \rightarrow R{-}NH_2 + CO_2 + N_2$$

Reagent = acid catalysed (usually H_2SO_4)
R. F. Schmidt, *Ber.*, **57** (1924), 704.

161. The Schmitz Diaziridine Synthesis

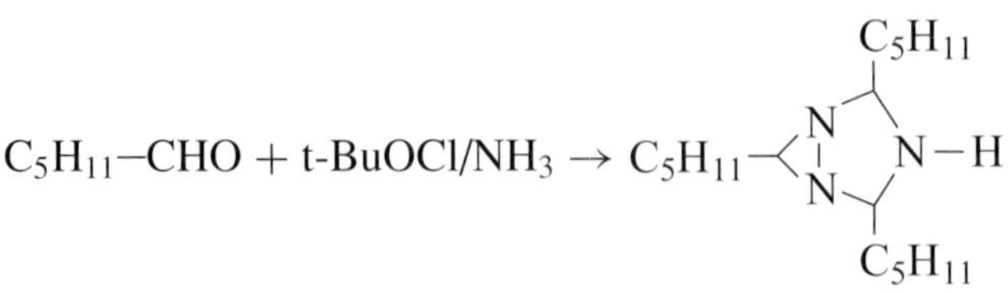
C5H11–CHO + t-BuOCl/NH3 → C5H11
C5H11
N
N
N–H
C5H11

E. Schmitz, *Angew. Chemie*, **71** (1959), 127.

162. The Schroeder–Grundmann Reaction

$$Cl_2CH-C(NH_2)=NH + COCl_2 \rightarrow$$

(triazine structure with OH, CHCl₂, CHCl₂ substituents)

H. Schroeder and C. Grundmann, *J. Am. Chem. Soc.*, **78** (1956), 2447.

163. The Schweizer Synthesis

(phthalimide structure) $\rightarrow C_6H_5-CH=CH-CH_2-NH_2$

Reagents = (i) C_6H_5-CHO/NaH and (ii) H_2N-NH_2
E. W. E. Schweizer, *J. Org. Chem.*, **31** (1966), 467.

164. The Schweizer Rearrangement

$$C_6H_5-C=N-NH_2 \rightarrow$$
$$\quad\quad |$$
$$\quad CO_2C_2H_5$$

(pyrazole structure with CH_3, O, C_6H_5, Ar substituents)

Reagents = (i) $[H-C\equiv C-CH_2]^+[P(C_6H_5)_3]^-$ and (ii) $Ar-N=C=O$
E. E. Schweizer, *J. Org. Chem.*, **43** (1978), 4328.

165. The Semmler–Wolff Reaction

(cyclohexenone oxime with two CH₃ groups) $\rightarrow$ (aniline with NH₂ and two CH₃ groups)

Reagents = acidic conditions
L. Wolff, *Ann.*, **322** (1902), 351.

166. The Sheverdina–Kocheshkov Reaction

(pyridine with Br) $\rightarrow$ (pyridine with NH₂)

Reagents = methyl lithium/methoxamine
N. I. Sheverdina and Z. Kocheshkov, *J. Gen. Chem. USSR*, **8** (1938), 1825.

167. The Shibasaki Cyclization

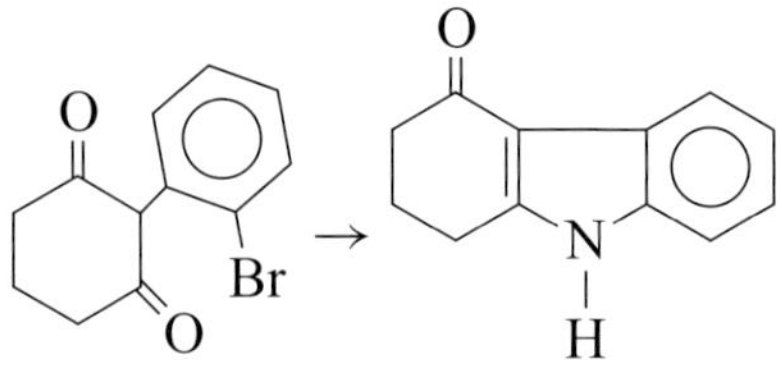

Reagent = iodine
M. Shibasaki and M. Mori, *Tetrahedron Lett.*, **28** (1987), 6187.

168. The Skraup Reaction

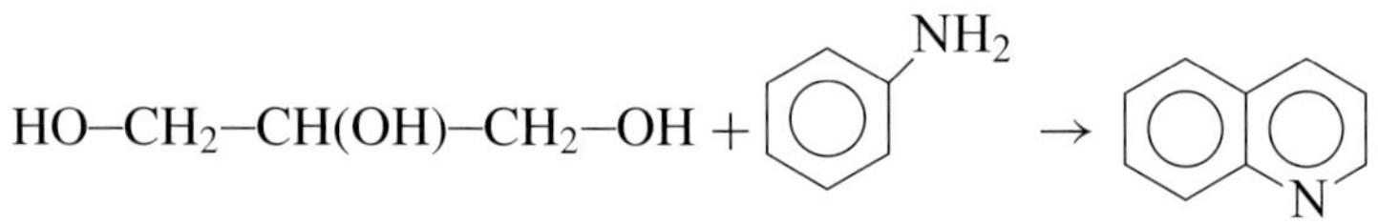

$$HO-CH_2-CH(OH)-CH_2-OH +$$

Reagents = H_2SO_4 and $C_6H_5-NO_2$
Z. H. Skraup, *Ber.*, **13** (1880), 2086.

169. The Smiles Rearrangement

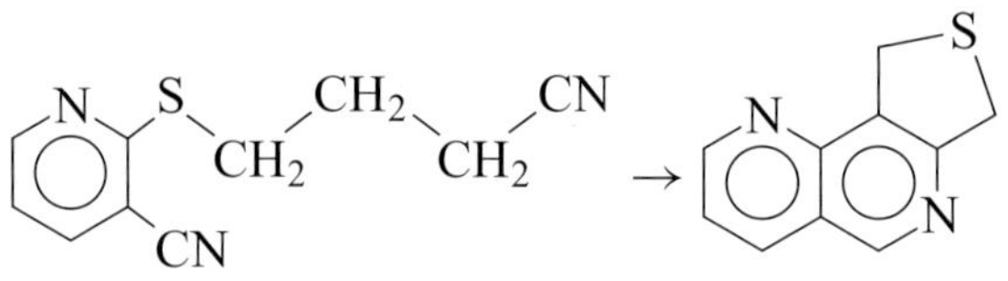

Reagents = t-BuOK
S. Smiles, *J. Chem. Soc.* (1931), 2364.

170. The Sommelet–Hauser Reaction

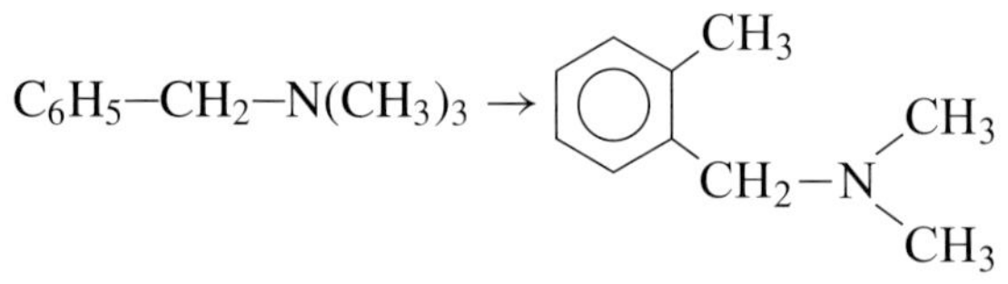

$$C_6H_5-CH_2-N(CH_3)_3 \rightarrow$$

Reagent = $NaNH_2$
M. Sommelet, *Comptes Rend.*, **205** (1937), 56.

171. The Staedel–Rügheimer Pyrazine Synthesis

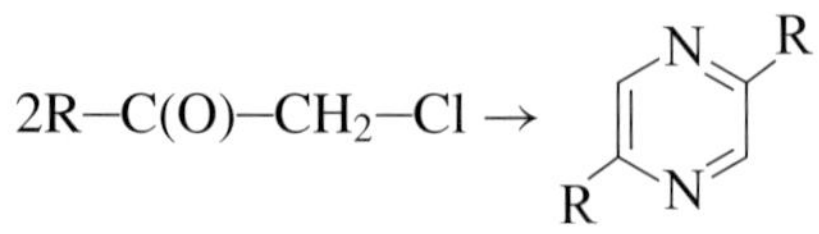

$$2R-C(O)-CH_2-Cl \rightarrow$$

Reagents = (i) NH_3 and (ii) $KMnO_4$
W. Staedel and L. Rügheimer, *Ber.*, **9** (1876), 563,

172. The Staudinger Reaction

$$R_3P + N_3X \rightarrow R_3P{=}N{-}X$$

H. Staudinger and J. Meyer, *Helv. Chim. Acta*, **2** (1919), 635.

173. The Stollé Synthesis

$$C_6H_5{-}NHR + Cl{-}CRR'{-}COCl \rightarrow$$

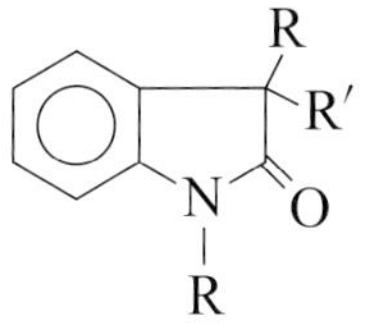

Reagents = (i) HCl and (ii) AlCl$_3$
R. Stollé, *Ber.*, **46** (1913), 3915.

174. The Stork Enamine Reaction

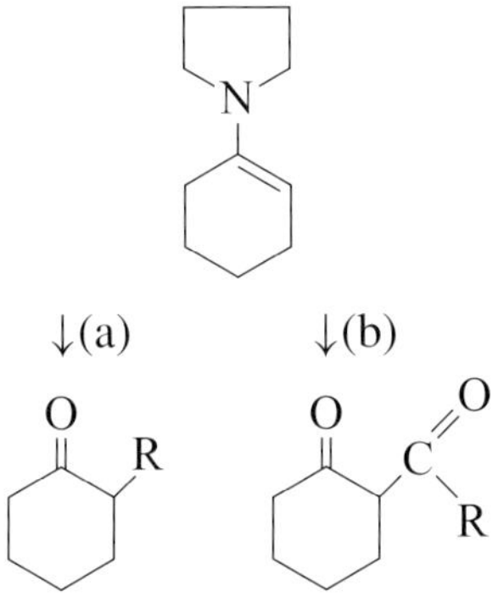

↓(a)   ↓(b)

Reagents = (a) (i) $[R]^+$ and (ii) H$_2$O
(b) (i) RCOCl and (ii) H$_2$O
G. Stork, R. Terrell and J. Szmuszkovicz, *J. Am. Chem. Soc.*, **76** (1954), 2029.

175. The Strecker Amino Acid Synthesis

$$R{-}CHO + HCN/NH_3 \rightarrow R{-}\underset{\underset{NH_2}{|}}{CH}{-}CO_2H$$

Reagents = water (for hydrolysis)
A. Strecker, *Ann.*, **75** (1850), 27.

(There have been many more-recent modifications of this process by Erlen-meyer, Tiemann, Zelinsky–Stadnikoff and Knoevenagel–Bucherer. See, for example, Y. Izumi, J. Chibata and T. Itoh, *Angew. Chemie Int. Edn.*, **90** (1978), 176.)

176. The Strecker Degradation

$$CH_2-CH(NH_2)-CO_2H \rightarrow CH_3-CHO + NH_3 + CO_2$$

Reagents = carbonyls (formaldehyde) or inorganic oxidants
A. Strecker, *Ann.*, **123** (1862), 363.

177. The Suzuki Reduction

$$C_6H_5-CN \rightarrow C_6H_5-CH_2-NH_2$$

Reagents = $NaBH_4/COCl_2$
S. Suzuki and Y. Suzuki, *Tetrahedron Lett.* (1969), 4555.

178. The Thorpe Reaction

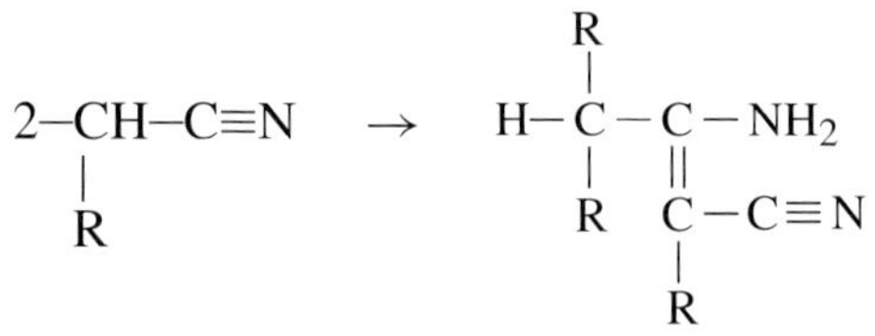

Reagents = (i) $NaOC_2H_5$ and (ii) HCl
H. Baron, F. G. P. Remfry and Y. F. Thorpe, *J. Chem. Soc.*, **85** (1904), 1726.

179. The Tiemann Rearrangement

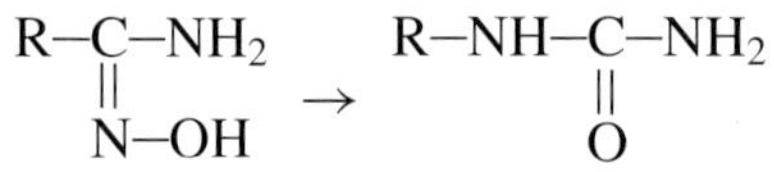

Reagents = (i) $C_6H_5SO_2Cl$ and (ii) H_2O
F. Tiemann, *Ber.*, **24** (1891), 4162.

180. The Timmis Pteridine Synthesis

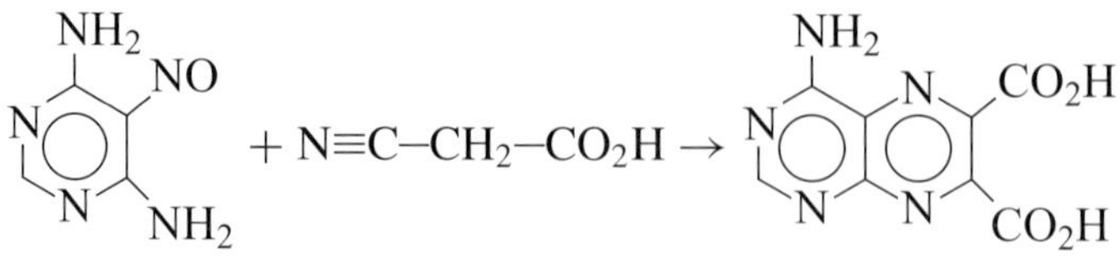

Reagents = $Na^+[C_2H_5-(O)CH_2-CH_2-OH]^-$
G. M. Timmis, *J. Chem. Soc.* (1954), 2881.

181. The Traube Purine Synthesis

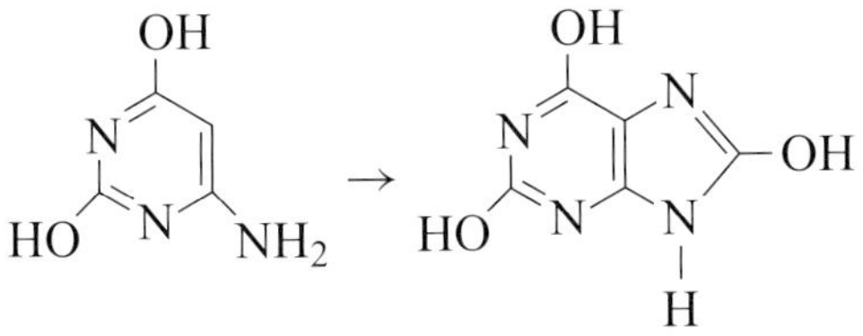

Reagents = (i) $HCl/NaNO_2$, (ii) $(NH_4)_2S$ and (iii) $Cl-CO_2R$
W. Traube, *Ber.*, **33** (1900), 1371.

182. The Ugi Reaction

$$R^1-CO_2H + R^2-NH_2 + R^3-CHO + R^4-N{\equiv}C \rightarrow$$

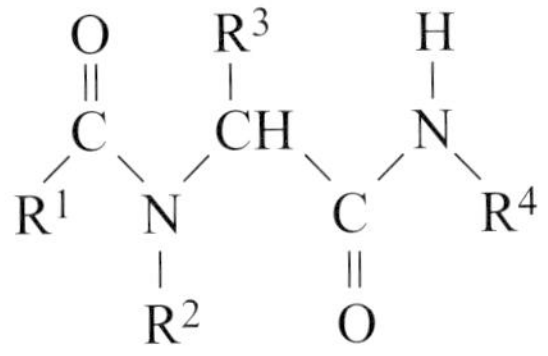

I. Ugi, *Angew. Chem. Int. Ed.*, **74** (1962), 9.

183. The Ullmann Reaction

$$2\,C_6H_5-I + Cu \rightarrow C_6H_5-C_6H_5 + CuI$$

F. Ullmann, *Ann.*, **332** (1904), 38.

184. The Ullmann–Fedvadjan Reaction

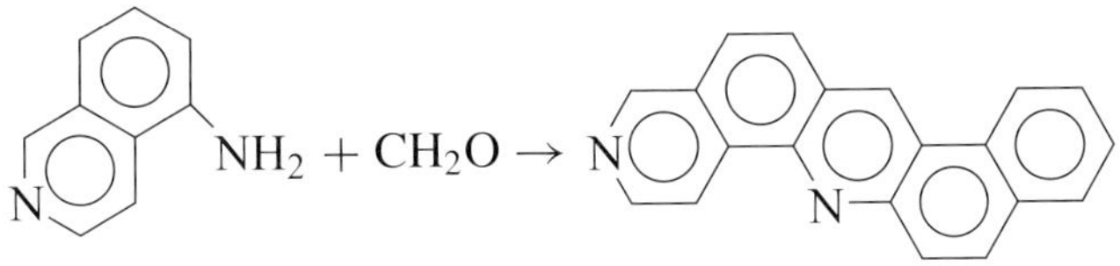

This reaction proceeds at 250 °C and high applied pressure.
F. Ullmann and A. Fedvadjan, *Ber.*, **36** (1903), 1027.

185. The Umpulong Reaction
Technically speaking, there is no single reaction called the Umpulong Reaction but instead there are several types of reaction where the expected polarities are reversed. The term Umpulong was first used by Seebach (D. Seebach, *Angew. Chem. Int. Ed.*, **18** (1979), 239) to describe reactions that worked backwards, or gave unexpected results. Electrophilic aminations could technically

be described as examples of Umpulong Reactions, especially when mediated by transition-metal complexes, for example as follows.

$$[R_3C]^- + Y-NR_2 \rightarrow R_3C-NR_2 + [Y]^-$$

Of course many important discoveries in chemistry in the past were made serendipitously when a reaction did not go according to plan. In the present author's experience, the results of such experiments were usually either a sticky black tar in the bottom of a flask, or (if I was very unlucky) a stain on the lab ceiling instead! However, the sometimes unpredictable nature of chemical reactions makes life more interesting for the research chemist and can sometimes lead to Nobel Prizes, fame and fortune.

186. The Urech Hydantoin Synthesis

$$HO_2C-CHR-NH_2 + KOCN \rightarrow$$

Reagent = HCl
F. Urech, *Ann.*, **165** (1873), 99.

187. The Van Slyke Determination

$$R-NH_2 + NaNO_2 + CH_3CO_2H \rightarrow R-OH + CH_3CO_2Na + H_2O + N_2$$

This determination also works for amino acids. Measurement of the evolved nitrogen allows determination of the amine content present.
D. D. Van Slyke, *Ber.*, **43** (1910), 3170.

188. The Vilsmeier–Haack Reaction

$$(CH_3)_2N-CHO + POCl_3 + C_6H_5-N(CH_3)_2 \rightarrow$$

Reagent = water
A. Vilsmeier and A. Haack, *Ber.*, **60** (1927), 119.

189. The Voight Amination

$$R-C(=O)-C(OH)HR' + R''-NH_2 \rightarrow R-CH-C(=O)R'$$
$$\underset{\displaystyle H-N-R''}{|}$$

Reagents $= P_2O_5/200\,^{\circ}C$
K. Voight, *J. Prakt. Chem.*, **34** (1886), 1.

190. The Wallach Rearrangement

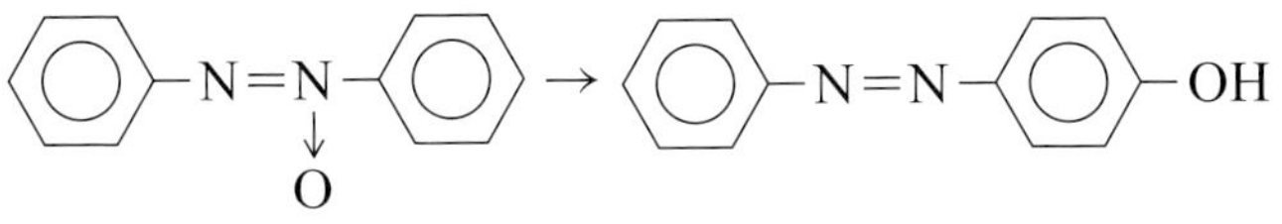

Reagent = acid source
O. Wallach and L. Belli, *Ber.*, **13** (1880), 525.

191. The Watanabe Cyclization

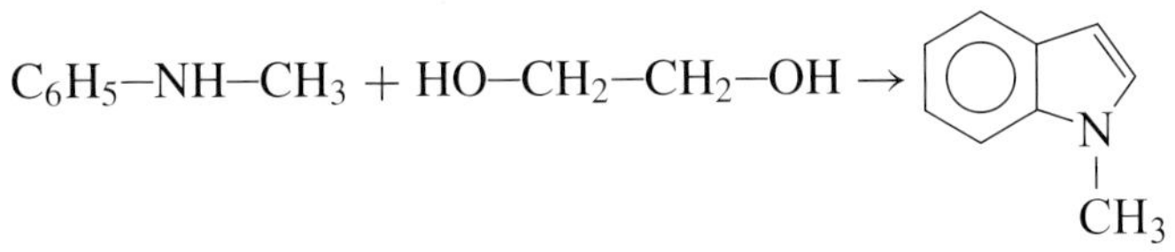

Reagents $= RuCl_2P(C_2H_5)_3/dioxane$
Y. Watanabe, *J. Org. Chem.*, **49** (1984), 3359.

192. The Wencker Aziridine Synthesis

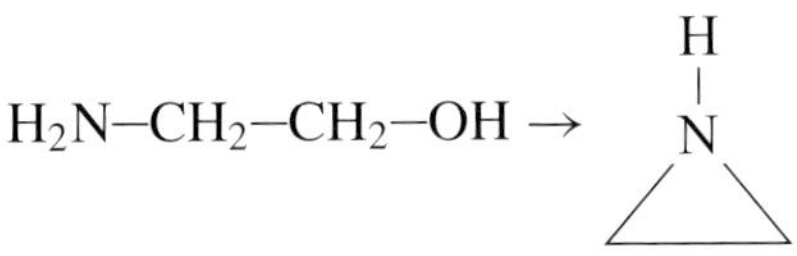

Reagents = (i) H_2SO_4 and (ii) NaOH
H. Wencker, *J. Am. Chem. Soc.*, **57** (1935), 2338.

193. The Westphal Condensation

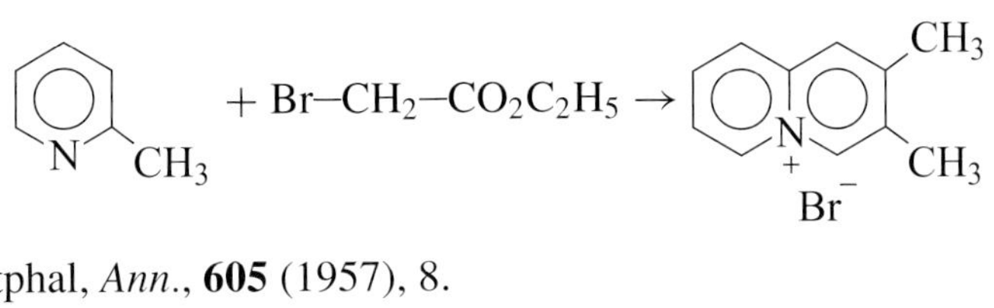

O. Westphal, *Ann.*, **605** (1957), 8.

194. The Widman–Stoermer Synthesis

O. Widman, *Ber.*, **17** (1884), 722.

195. The Willgerodt–Kindler Reaction

Reagents = $TsOH/S_8$ or $i\text{-}Pr\text{–}OH/S_8$
C. Willgerodt, *Ber.*, **20** (1887), 2467; K. Kindler, *Ann.*, **193** (1923), 431.

196. The Wöhler Urea Synthesis

$$NH_4CHO \rightarrow (NH_2)_2C{=}O$$

Reagent = water
F. Wohler, *Ann.*, **12** (1828), 253.

197. The Wolff Rearrangement

$$R\text{–}CO\text{–}CH\text{–}N_2 \rightarrow R\text{–}CH{=}C{=}O \text{ (a)} \rightarrow R\text{–}CH_2\text{–}CO_2R'$$
$$\text{(b)} \rightarrow R\text{–}CH_2\text{–}CO\text{–}NHR'$$

Reagents = (a) $R'\text{–}OH$/metal catalyst
            (b) $R'\text{–}NH_2$/metal catalyst
L. Wolff, *Ann.*, **394** (1912), 25.

198. The Wolff–Kishner Reduction

$$R\text{–}CO\text{–}R' + H_2N\text{–}NH_2 + KOH \rightarrow R\text{–}CH_2\text{–}R' + N_2$$

L. Wolff, *Ann.*, **394** (1912), 86.

199. The Zinin Rearrangement

Reagent = HCl
N. Zinin, *J. Prakt. Chem.*, **36** (1845), 93.

200. The Zinner Hydroxylamine Synthesis

Reagents = $(C_2H_5)_3N/HClO_4$
G. Zinner, *Ber.*, **91** (1958), 305.

Index